SMARTBOOK

enable
protect
inform
influence
attack

information advantage

ACTIVITIES, TASKS & CAPABILITIES

The Lightning Press
Norman M Wade

The Lightning Press

2227 Arrowhead Blvd.
Lakeland, FL 33813
24-hour Order/Voicemail: 1-800-997-8827
E-mail: SMARTbooks@TheLightningPress.com
www.TheLightningPress.com

INFO2 SMARTbook: Information Advantage (Activities, Tasks & Capabilities)

We no longer regard information as a separate consideration or the sole purview of technical specialists. As a dynamic of combat power, Army forces fight for, defend, and fight with information to create and exploit information advantages—the use, protection, and exploitation of information to achieve objectives more effectively than enemies and adversaries do. INFO2 chapters and topics include information advantage (enable, protect, inform, influence, attack), information in joint operations (OIE: operations in the information environment), information capabilities (PA, CA, MILDEC, MISO, OPSEC, CO, EW, Space, STO), information planning (information environment analysis, IPB, MDMP, JPP), information preparation, information execution, fires & targeting, and information assessment.

Copyright © 2024 The Lightning Press
ISBN: 978-1-935886-97-6

All Rights Reserved
No part of this book may be reproduced or utilized in any form or other means, electronic or mechanical, including photocopying, recording or by any information storage and retrieval systems, without permission in writing by the publisher. Inquiries should be addressed to The Lightning Press.

SMARTbook is a trademark of The Lightning Press.

Notice of Liability
The information in this SMARTbook and quick reference guide is distributed on an "As Is" basis, without warranty. While every precaution has been taken to ensure the reliability and accuracy of all data and contents, neither the author nor The Lightning Press shall have any liability to any person or entity with respect to liability, loss, or damage caused directly or indirectly by the contents of this book. If there is a discrepancy, refer to the source document. This SMARTbook does not contain classified or sensitive information restricted from public release. "The views presented in this publication are those of the author and do not necessarily represent the views of the Department of Defense or its components."

Credits: Cover image - Soldiers from A Co., 1st Bn, 111th Inf, 56th Stryker BCT conduct a night live-fire during Exercise Decisive Strike 2019 (U.S. Army photo by Staff Sgt. Frances Ariele Tejada). All other images courtesy Dept. of the Army and/or Dept. of Defense.

Printed and bound in the United States of America.

View, download FREE samples and purchase online:
www.TheLightningPress.com

(INFO2) Notes to Reader

Information is central to everything we do—it is the basis of intelligence, a fundamental element of command and control, and the foundation for communicating thoughts, opinions, and ideas. Information is the building block for intelligence and is the basis for situational understanding, decision making, and actions across all warfighting functions. As a **critical resource**, Army forces fight for, defend, and fight with information while attacking a threat's (adversary or enemy) ability to do the same.

We **no longer regard information as a separate consideration** or the sole purview of technical specialists. Instead, we view information as **a resource that is integrated into operations** with all available capabilities in a **combined arms approach** to **enable** command and control; **protect** data, information, and networks; **inform** audiences; **influence** threats and foreign relevant actors; and **attack** the threat's ability to exercise command and control.

Army forces create and exploit **informational power** similarly to the joint force through five **information activities** (enable, protect, inform, influence, and attack). Army forces also consider information as a dynamic of combat power employed with mobility, firepower, survivability, and leadership to achieve objectives during armed conflict. As a **dynamic of combat power**, Army forces fight for, defend, and fight with information to create and exploit **information advantages**—the use, protection, and exploitation of information to achieve objectives more effectively than enemies and adversaries do.

The **joint force** uses information to perform many simultaneous and integrated activities. The joint force employment of information is of central importance because it may provide an operational advantage.

The elevation of information as a **joint function** impacts all operations and signals a fundamental appreciation for the military role of information at the strategic, operational, and tactical levels within today's complex operational environment (OE).

Operations in the information environment (OIE) are military actions involving the integrated employment of multiple information forces to affect drivers of behavior by **informing** audiences; **influencing** foreign relevant actors; **attacking and exploiting** relevant actor information, information networks, and information systems; and by **protecting** friendly information, information networks, and information systems.

SMARTbooks - Intellectual Fuel for the Military!

SMARTbooks: Reference Essentials for the Instruments of National Power (D-I-M-E: Diplomatic, Informational, Military, Economic)! Recognized as a "whole of government" doctrinal reference standard by military, national security and government professionals around the world, SMARTbooks comprise a comprehensive professional library.

SMARTbooks can be used as quick reference guides during actual operations, as study guides at education and professional development courses, and as lesson plans and checklists in support of training. Visit **www.TheLightningPress.com**!

INFO2: Information Advantage SMARTbook (Activities, Tasks, & Capabilities), 2nd Ed.

ADP 3-13   *JP 3-04*   *JP 3-0*   *FM 3-13*

Plus more than a dozen primary references on information capabilities & more!

Intro: Nature of Information & the OE

Information is central to all activity Army forces undertake. It is fundamental to command and control (C2) and is the basis for situational understanding, decision making, and actions across all warfighting functions. Information is the building block for intelligence—the product resulting from the collection, processing, integration, evaluation, analysis, and interpretation of available information concerning foreign nations, hostile or potentially hostile forces or elements, or areas of actual or potential operations. As a **critical resource**, Army forces fight for, defend, and fight with information while attacking a threat's (adversary or enemy) ability to do the same.

Chap 1: Information Advantage (ADP 3-13)

An **information advantage** is a condition when a force holds the initiative in terms of situational understanding, decision making, and relevant actor behavior. There are several forms of information advantage. The **information advantage framework** presents a framework for creating and exploiting information advantages. Within this framework, Army forces integrate all relevant military capabilities through the execution of five information activities (enable, protect, inform, influence, and attack).

We no longer regard information as a separate consideration or the sole purview of technical specialists. Instead, we **view information as a resource** that is integrated into operations with all available capabilities in a **combined arms approach** to **enable** command and control; **protect** data, information, and networks; **inform** audiences; **influence** threats and foreign relevant actors; and **attack** the threat's ability to exercise command and control.

Army forces create and exploit **informational power** similarly to the joint force through five information activities (enable, protect, inform, influence, and attack). Army forces also consider information as a **dynamic of combat power** employed with mobility, firepower, survivability, and leadership to achieve objectives during armed conflict.

Chap 2: Information in Joint Operations: OIE (JP 3-04)

Information is a resource of the informational instrument of national power at the strategic level. Information is also a critical military resource. The joint force uses information to perform many simultaneous and integrated activities. The joint force employment of information is of central importance because it may provide an operational advantage.

The elevation of information as a **joint function** impacts all operations and signals a fundamental appreciation for the military role of information at the strategic, operational, and tactical levels within today's complex operational environment (OE).

Operations in the information environment (OIE) are military actions involving the integrated employment of multiple information forces to affect drivers of behavior by **informing** audiences; **influencing** foreign relevant actors; **attacking and exploiting** relevant actor information, information networks, and information systems; and by **protecting** friendly information, information networks, and information systems.

Chap 3: Information Capabilities (PA, CA, MILDEC, MISO, OPSEC, CO, EW, Space, STO)

In addition to planning all operations to benefit from the inherent informational aspects of physical power and influence relevant actors, the JFC also has additional means with which to leverage information in support of objectives. Leveraging information involves the generation and use of information through tasks to inform relevant actors; influence relevant actors; and/or attack information, information systems, and information networks.

Chap 4: Information Planning

Planning is the art and science of understanding a situation, envisioning a desired future, and laying out effective ways of bringing that future about. Commanders, supported by their staffs, ensure information activities are fully integrated into plans and orders through the military decision-making process. This includes integrating information activities into the concept of operations and supporting schemes, to include schemes of intelligence, information collection, maneuver, fires, and protection.

Chap 5: Information Preparation

Preparation consists of those activities performed by units and Soldiers to improve their ability to execute an operation (ADP 5-0). Preparation creates conditions that improve friendly force opportunities for success. It requires commander, staff, and Soldier actions to ensure the force is ready to execute operations.

Chap 6: Information Execution

Execution is the act of putting a plan into action by applying combat power to accomplish the mission and adjusting operations based on changes in the situation (ADP 5-0). Commanders, staffs, and subordinate commanders focus their efforts on translating decisions into action. They direct action to apply combat power, of which information is a dynamic, to achieve objectives and accomplish missions.

Chap 7: Fires & Targeting

The **fires warfighting function** is the related tasks and systems that **create and converge effects in all domains** against the threat to enable actions across the range of military operations. These tasks and systems create **lethal and nonlethal effects** delivered from both Army and Joint forces, as well as other unified action partners.

Targeting is the process of selecting and prioritizing targets and matching the appropriate response to them, considering operational requirements and capabilities (JP 3-0). Information is integrated into the targeting cycle to produce effects in and through the information environment that support objectives.

Chap 8: Information Assessment

Information activities and tasks must be continually assessed to judge whether they achieve the desired outcome. Assessment is not a discrete step of the operations process. Assessing information activities and tasks is continuous and informs the other activities of the operations process. Staffs assess information activities and tasks while working in functional and integrating cells, and while participating in cross-functional meetings such as working groups and boards. The purpose of assessing information activities and tasks is to equip the commander with the analysis necessary to make better decisions.

(INFO2) References

The following references were used in part to compile *INFO2: The Information Operations & Capabilities SMARTbook*. All military references used to compile SMARTbooks are in the public domain and are available to the general public through official public websites and designated as approved for public release with unlimited distribution. The SMARTbooks do not contain ITAR-controlled technical data, classified, or other sensitive material restricted from public release. SMARTbooks are reference books that address general military principles, fundamentals and concepts rather than technical data or equipment operating procedures.

* See Editor's Note on Changes in Information Terminology on p. 1-2. Based on changes to joint information doctrine, <u>Army forces will no longer use the terms information operations, information-related capabilities, or information superiority</u>. The <u>Army is currently revising all its doctrine</u>, to include FM 3-13, to account for these changes and the Army's new information advantage framework. As such, the INFO2 SMARTbook retains the original terminology as referenced from the original source, while recognizing this terminology is changing (references <u>marked with an asterisk</u>, where possible.)

Joint Publications

JP 3-0	Jun 2022	Joint Campaigns and Operations
JP 3-04	Sept 2022	Information in Joint Operations
JP 3-12	Dec 2022	Joint Cyberspace Operations
JP 3-85	May 2020	Joint Electromagnetic Spectrum Operations
JP 3-13.2	Dec 2011	Military Information Support Operations (w/Change 2)
JP 3-13.3	Jan 2016	Operations Security
JP 3-14	Oct 2020	Space Operations
JP 3-57	Jul 2018	Civil-Military Operations
JP 3-60	Sept 2018	Joint Targeting
JP 3-61	Aug 2016	Public Affairs (w/Change 1)

Army Doctrine Publications (ADPs)

ADP 3-13	Nov 2023	Information

Army Techniques Publications (ATPs)

ATP 3-13.1*	Oct 2018	The Conduct of Information Operations
ATP 3-13.5	Dec 2021	Soldier and Leader Engagement

Field Manuals (FMs)

FM 3-0	Oct 2022	Operations
FM 3-12	Aug 2021	Cyberspace and Electromagnetic Warfare
FM 3-13*	Dec 2016	Information Operations
FM 3-13.4	Feb 2019	Army Support to Military Deception
FM 3-14	Oct 2019	Army Space Operations
FM 3-57	Jul 2021	Civil Affairs Operations
FM 3-61	Feb 2022	Communication Strategy and Public Affairs Operations

(INFO2) Table of Contents

(INTRO) Nature of Information & the OE

I. Information Explained 0-1
   A. Data and Information 0-1
   B. Assignment of Meaning 0-1
      How Humans Assign Meaning to Data 0-2
      - Inherent Informational Aspects 0-2
      - Drivers of Behavior 0-3
      Automated Systems 0-4

II. Information in the Security Environment 0-5

III. Information Within an Operational Environment (OE) 0-6
   Domains and Dimensions of an Operational Environment (OE) 0-6
   Informational Considerations (of the OE) 0-7
   A. Human Dimension 0-8
      - Informational Considerations 0-8
   B. Information Dimension 0-8
      - Informational Considerations 0-8
   C. Physical Dimension 0-10
      - Informational Considerations 0-10

Chap 1 — Information Advantage

Information Advantage (Overview) 1-1
   I. Information Advantage (Overview) 1-1
      ADP 3-13, Information (Nov '23): Introduction & Overview 1-2
      - Editor's Note on Changes in Information Terminology* 1-2
      - Information Advantage Framework 1-3
   II. Information in Multidomain Operations 1-4
      Relative Advantages 1-4
      - Human Advantage 1-4
      - Information Advantage 1-5
      - Physical Advantage 1-5
   III. Informational Power 1-6
      - Information and the Instruments of National Power 1-6
      - Joint Informational Power 1-6
      - Information as a Dynamic of Combat Power 1-7

IV. Information Activities (Defined/Overview)..1-8
V. Information Activities and the Tenets of Operations...................................1-10
VI. Principles of Information Advantage...1-12
- Offensively Oriented..1-12
- Combined Arms...1-12
- Commander Driven...1-13
- Soldier Enabled...1-13
VII. Warfighting Function Contributions 1-14
VIII. Information Advantages (Across Strategic Contexts)...............................1-16
- Information Advantage (Examples)..1-19

I. Enable..1-21
I. Establish, Operate, and Maintain Command and Control Systems...............1-21
II. Conduct the Operations Process and Coordinate Across Echelons............1-23
III. Conduct the Integrating Processes..1-24
- Intelligence Preparation of the Operational Environment (IPOE)...............1-24
- Information Collection 1-24
- Targeting 1-25
- Risk Management 1-25
- Knowledge Management 1-25
IV. Enhance Understanding of an Operational Environment 1-26
A. Analyze the Operational and Mission Variables......................................1-26
- Operational Variables - PMESII-PT 1-26
- Mission Variables - METT-TC (I)...1-26
B. Identify and Describe Relevant Actors 1-27
- Relevant Human Actors 1-27
- Relevant Automated Systems...1-27
C. Identify Behaviors of Relevant Actors 1-27
Considerations for Enhancing C2...1-28
- Apply Principles of Mission Command...1-28
- Ensuring Digital Readiness 1-28
- Developing and Maintaining Digital Literacy 1-28

II. Protect 1-29
I. Secure and Obscure Friendly Information...1-29
II. Conduct Security Activities 1-30
A. Conduct Security Operations...1-30
B. Implement Physical Security..1-30
C. Implement Personnel Security Program..1-31
D. Conduct Counterintelligence (CI)...1-31
III. Defend the Network, Data, and Systems..1-34
A. Conduct Cyberspace Security 1-34
B. Conduct Defensive Cyberspace Operations..1-34
C. Conduct Communications Security..1-34
D. Conduct Electromagnetic Protection 1-34
Protect Considerations 1-32

III. Inform 1-35
INFORM Considerations...1-36
- Tell the Truth...1-36
- Timely Release of Information and Operations Security 1-36
- Department of Defense Principles of Information 1-37
- Compliance with Law and Policy..1-37
Commander's Communication Synchronization (CCS)...................................1-39
I. Inform and Educate Army Audiences...1-38
A. Inform Internal Audiences 1-38
B. Educate Soldiers...1-38

II. Inform United States Domestic Audiences 1-40
　　　　A. Conduct Public Communication 1-40
　　　　B. Conduct Community Engagement 1-40
　　　　C. Correct Misinformation and Counter Disinformation Related to 1-41
　　　　Army Forces or Operations
　　III. Inform International Audiences 1-42
　　　　A. Conduct Community Engagement Outside the Continental U.S. 1-42
　　　　B. Conduct Soldier and Leader Engagement (SLE) 1-43
　　　　C. Conduct Civil Affairs Operations 1-43
　　　　D. Correct Misinformation and Counter Disinformation Within 1-43
　　　　International Audiences

IV. Influence 1-45
　　I. Influence Threat Perception and Behaviors 1-45
　　　　A. Conduct Deception Activities 1-46
　　　　B. Conduct Military Information Support Operations (MISO) 1-46
　　II. Influence Other Foreign Audiences 1-47
　　　　A. Conduct Soldier and Leader Engagement (SLE) 1-47
　　　　B. Conduct Military Information Support Operations (MISO) 1-47
　　　　C. Conduct Civil Affairs Operations 1-47
　　Influence Considerations 1-48
　　　　- Deliberate Versus Incidental Influence 1-48
　　　　- Language, Regional, And Cultural Expertise 1-48
　　　　- Authorities 1-48

V. Attack 1-49
　　Information Attack Methods 1-49
　　　　- Physical Destruction 1-50
　　　　- Electromagnetic Attack (EA) 1-50
　　　　- Cyberspace Attacks 1-51
　　　　- Space Operations 1-51
　　I. Degrade Threat Command and Control 1-52
　　　　A. Affect Threat Understanding of an Operational Environment 1-52
　　　　B. Affect Threat Networks and Systems 1-52
　　II. Affect Threat Information Warfare Capabilities 1-53
　　　　- Threat Information Warfare 1-54
　　Attack Considerations 1-56
　　　　- Timelines for Preparatory Activities 1-58
　　　　- Precision and Scalability 1-58

Integration 1-57
　　I. Joint and Multinational Information Advantage 1-57
　　　　A. Information Joint Function 1-57
　　　　　　- Joint Information Advantage 1-57
　　　　　　- Leveraging the Inherent Informational Aspects of Operations 1-57
　　　　　　- Operations in the Information Environment 1-58
　　　　　　- Army Forces and the Information Joint Function 1-59
　　　　B. Multinational Considerations 1-58
　　II. Army Information Activities During Operations 1-60
　　Information Activities and the Operations Process 1-60
　　Integration of the Information Activities 1-61
　　　　A. Planning 1-60
　　　　B. Preparing 1-62
　　　　C. Executing 1-63
　　　　D. Assessing 1-63
　　Information Training and Education 1-64
　　　　- Common Training and Education 1-64
　　　　- Technical Training and Education 1-64

Chap 2: Information in Joint Operations (OIE)

I. Joint Force Use of Information 2-1
JP 3-04, Information in Joint Operations, Sept '22 (Summary of Changes) 2-2
- I. Military Operations and Information 2-1
- II. The Operational Environment (OE) and the Information Environment (IE) 2-1
- III. Information Advantage 2-1
- IV. Informational Power 2-4
- V. Relevant Actors 2-6
- VI. Joint Force Use of Narrative 2-6

II. Information (as a Joint Function) 2-7
- I. Information (as a Joint Function) 2-7
- II. Information Use Across the Competition Continuum 2-8
- III. Joint Force Capabilities, Operations, Activities for Leveraging Information 2-10
- IV. Information Joint Function Tasks 2-12
    - Tasks and Outcomes of the Information Joint Function 2-13
    - A. Understand How Information Impacts the OE 2-12
        - Analysis of Informational, Physical, & Human Aspects of the OE 2-15
        - Identify and Describe Relevant Actors 2-15
        - Identify Likely Behavior of Relevant Actors 2-15
    - B. Support Human and Automated Decision Making 2-16
        - Facilitate Shared Understanding 2-16
        - Collaboration 2-17
        - Knowledge Management (KM) & Information Management (IM) 2-17
        - Information and Intelligence Sharing 2-17
        - Protect Friendly Information, Information Networks & Systems 2-16
        - Build, Protect, and Sustain Joint Force Morale and Will 2-16
    - C. Leverage Information 2-19
        - Inform Domestic, International, and Internal Audiences 2-19
        - Influence Relevant Actors 2-19
        - Attack And Exploit 2-19
- V. Key Considerations (Information Function) 2-20

III. Unity of Effort (Information Forces) 2-21
- Authorities 2-21
- Responsibilities 2-22
- I. Service Organizations 2-23
- II. Department of State (DOS) Organizations 2-24
- III. Combatant Commanders (CCDRs) 2-26
- IV. The Joint Force 2-28
    - Joint Organizations 2-29
    - A. JFC's Staff 2-28
    - B. Information Planning Cell & CFT 2-30
- V. Information Forces 2-32
- VI. Interorganizational Collaboration 2-34
    - A. Coordination and Synchronization 2-34
    - B. USG Organizations 2-34
    - C. Civil-Military Operations Centers (CMOCs) 2-35
    - D. Joint Interagency Coordination Group (JIACG) 2-35
    - E. Joint Interagency Task Force (JIATF) 2-35
- VII. Multinational Partner Considerations 2-36

4-Table of Contents

IV. Operations in the Information Environment (OIE) 2-37
I. Operations in the Information Environment (OIE) 2-37
II. Organizing for Operations in the Information Environment............................2-38
   A. Organizations and Personnel...2-38
   B. Information Forces 2-38
   C. Facilitate the Integration of Information into Joint Force Operations 2-38
   D. OIE Unit Core Activities 2-40
II. The Joint Functions and OIE 2-40
   A. Command & Control (C2) Joint Function 2-42
   B. Information Joint Function..2-42
   C. Intelligence Joint Function 2-42
   D. Fires Joint Function 2-43
   E. Movement and Maneuver Joint Function...2-44
   F. Protection Joint Function..2-44
   G. Sustainment Joint Function 2-44

V. OIE: Planning, Coordination, Execution, & Assessment........2-45
Guide for the Integration of Information in Joint Operations..............................2-46
I. Planning..2-45
   - Operational Design 2-50
   - Joint Planning Process (JPP)..2-50
   - Information Planners 2-50
   - Information Cross-Functional Team (CFT)...2-50
II. Execution 2-52
   - The Synchronization Matrix...2-52
   - Commander's Critical Information Requirements (CCIRs) 2-52
   - The Narrative 2-54
   - Information and KM...2-54
   - The Information Staff Estimate..2-54
III. Assessment 2-54

VI. Information & the Joint Planning Process (JPP) 2-55
Step 1—Planning Initiation 2-56
Step 2—Mission Analysis 2-56
   A. Analyze Higher HQs' Planning Directives and Strategic Guidance 2-56
      - The Operational Mission Narrative 2-56
   B. Review Commander's Initial Planning Guidance 2-58
   C. Determine Known Facts and Develop Planning Assumptions..................2-58
   D. Determine and Analyze Operational Limitations.......................................2-58
   E. Determine Specified, Implied, and Essential Tasks and 2-59
      Develop the Mission Statement
   F. Conduct Initial Force and Resource Analysis 2-59
   G. Develop Military Objectives 2-59
   H. Develop COA Evaluation Criteria...2-60
   I. Develop Risk Assessment...2-60
   J. Determine Initial CCIRs (PIRs/FFIRs) 2-61
   K. Prepare Staff Estimates 2-62
   L. Prepare and Deliver Mission Analysis Brief...2-62
Step 3—COA Development 2-63
   A. Review Objectives and Tasks and Develop Ways to Accomplish Tasks.2-63
   B. Select and Prioritize 2-64
   C. Identify the Sequencing (Simultaneous, Sequential, or 2-65
      Combination) of Actions for each COA
Step 4—COA Analysis and Wargaming...2-65
Step 5—COA Comparison 2-66
Step 6—COA Approval 2-66
Step 7—Plan or Order Development 2-66

Chap 3 (Information) CAPABILITIES

Information Capabilities..3-1
   I. Joint Force Capabilities, Operations, & Activities for Leveraging Information 3-1
   II. Army Doctrine 3-1
      - Information Operations & the IRCs (* as previously defined).....................3-2
   III. INFO Capabilities (INFO2 SMARTbook Overview) 3-4

I. Public Affairs (PA) 3-5
   Public Affairs Guidance (PAG)..3-5
   I. Public Affairs and the Operational Environment (OE)..3-6
      - Public Perception 3-7
   II. (Army) Public Affairs Core Tasks 3-9
   III. Public Affairs Fundamentals 3-10
      - Principles of Information..3-10
      - Tenets of Public Affairs...3-10
      - PA and Commander's Communication Synchronization (CCS) 3-11
   IV. Audiences, Stakeholders, and Publics 3-14
   V. Narrative, Themes, and Messages..3-15
   VI. PA Actions in the Joint Planning Process 3-16

II. Civil Affairs and Civil-Military Operations (CMO) 3-17
   I. Civil Affairs and Civil-Military Operations..3-17
   II. Civil-Military Operations Range of Activities 3-18
      - Strategic Aspects of Civil-Military Operations 3-19
      A. Military Government..3-18
      B. Support to Civil Administration (SCA) 3-26
      C. Populace and Resources Control (PRC) 3-26
      D. Foreign Humanitarian Assistance (FHA) 3-26
      E. Foreign Assistance..3-26
      F. Overseas Humanitarian, Disaster, and Civic Aid (OHDACA) 3-26
   III. Civil-Military Operations and the Levels of War 3-20
   IV. CMO in Joint Operations 3-22
   V. Civil-Military Operations Center (CMOC) 3-24

III. Military Deception (MILDEC) 3-27
   I. Functions of Military Deception...3-27
   II. Categories of Deception 3-27
      A. Military Deception (MILDEC) 3-28
      B. Tactical Deception (TAC-D)...3-28
      C. Deception in Support of Operations Security (DISO) 3-28
   III. Deception Means 3-29
   IV. Types of Military Deception...3-30
   V. Army Tactical Deception Planning 3-32

IV. Military Information Support Operations (MISO)...................3-33
   I. MISO Purpose 3-34
   II. MISO Missions...3-34
   III. Information Roles & Relationships..3-35
   IV. Example Joint MISO Activities 3-38

V. Operations Security (OPSEC) 3-39
   I. Purpose of Operations Security..3-39
   II. Characteristics of Operations Security 3-40

6-Table of Contents

III. OPSEC and Intelligence (JIPOE) 3-40
IV. Implement Operations Security (OPSEC) 3-41
IV. The Operations Security Process 3-42
    A. Identify Critical Information 3-42
    B. Threat Analysis 3-42
    C. Vulnerability Analysis 3-43
    D. Risk Assessment 3-43
V. Operations Security Indicators 3-44
    - Association 3-44
    - Profile 3-44
    - Contrast 3-44
    - Exposure 3-44

VI. Cyberspace and the Electromagentic Spectrum 3-45
Cyberspace Operations 3-45
Electromagentic Warfare (EW) 3-45
Electromagentic Spectrum (EMS) 3-45
Cyberspace Electromagnetic Activities (CEMA) & Information Advantage 3-46

VI(a). Cyberspace Operations (CO) 3-47
The Cyberspace Domain 3-48
Cyberspace Missions & Actions 3-51
I. Cyberspace Operations (CO) 3-47
    A. Department of Defense Information Network Operations (DODIN) 3-50
    B. Defensive Cyberspace Operations (DCO) 3-50
    C. Offensive Cyberspace Operations (OCO) 3-52
II. Cyberspace Actions 3-53
    A. Cyberspace Security 3-53
    B. Cyberspace Defense 3-53
    C. Cyberspace Exploitation 3-53
    D. Cyberspace Attack 3-54

VI(b). Electromagnetic Warfare (EW) 3-55
I. Electromagnetic Warfare (EW) 3-55
    A. Electromagnetic Attack (EA) 3-56
    B. Electromagnetic Protection (EP) 3-58
    C. Electromagnetic Support (ES) 3-60
    * Electronic Warfare Reprogramming 3-60
II. Spectrum Management 3-57

VII. Space Operations 3-61
I. Space Operations 3-61
II. Space Capabilities 3-62
III. Unity of Effort 3-64
    - Space Control 3-64
    - Space Superiority 3-64
IV. Army Space Capabilities 3-66
V. Combined Space Tasking Order (CSTO) 3-68

VIII. Additional Capabilities 3-69
A. Integrated Joint Special Technical Operations (IJSTO) 3-70
B. Special Access Programs (SAP) 3-70
C. Personnel Recovery (PR) 3-70
D. Physical Attack 3-70
E. Physical Security 3-71
F. Presence, Profile, and Posture (PPP) 3-71
G. Soldier and Leader Engagement (SLE) 3-72
H. Police Engagement 3-72
I. Social Media 3-72

Chap 4 — (Information) PLANNING

PLANNING (Overview) 4-1
   A. IO & Army Design Methodology (ADM)..4-2
   B. IO & the Military Decision-making Process (MDMP).............................4-2

I. Synchronization of Information-Related Capabilities 4-3
   I. Commanders' Responsibilities 4-3
      Commander's Responsibilities (Overview) 4-4
         - Commander's Narrative..4-4
         - Commander's Intent 4-5
         - Guidance 4-5
         - Concept of Operations..4-5
         - Risk Assessment 4-5
   II. Staff Responsibilities 4-3
      Key IO Planning Tools and Outputs 4-3
      A. IO Running Estimate..4-6
      B. Logic of the Effort..4-8
      C. CCIRs and EEFIs 4-9
         - Commander's Critical Information Requirements (CCIRs)....................4-9
         - Priority Intelligence Requirements (PIRs)..4-9
         - Friendly Force Information Requirements (FFIRs) 4-9
         - Essential Elements of Friendly Information (EEFIs) 4-9
   IV. IO Input to Operation Orders and Plans 4-10
      A. Mission Statement 4-11
      B. Scheme of Information Operations 4-12
      C. IO Objectives & IRC Tasks 4-14
      D. IO Synchronization Matrix..4-16

II. Information Environment Analysis 4-17
   IO and Intelligence Preparation of the Battlefield (IPB) 4-17
   Analyze and Depict the Information Environment..4-17
   Information Environment Analysis (Overview)..4-18
   IPB Considerations for the Information Environment.......................................4-20
   Step 1: Define the Information Environment...4-22
   Step 2: Describe the Information Environment Effects 4-23
   Step 3: Evaluate the Threat's Information Situation 4-28
   Step 4: Determine Threat Courses of Action 4-34

III. IO & the MDMP...4-35
   Step I. Receipt of Mission 4-36
   Step II. Mission Analysis 4-39
   Step III. Course of Action Development...4-50
   Step IV. Course of Action Analysis & War-Gaming 4-56
   Step V. Course of Action Comparison 4-58
   Step VI. Course of Action Approval..4-60
   Step VII. Orders Production, Dissemination, and Transition.............................4-60

IV. Appendix 15 (IO) to Annex C (Operations)............................4-61

Chap 5 — (Information) PREPARATION

PREPARATION (Overview) 5-1
IO Preparation Activities 5-1
    A. Improve Situational Understanding 5-2
    B. Revise and Refine Plans and Orders 5-2
    C. Conduct Coordination and Liaison 5-3
        - Examples of Internal/External Coordination 5-4
    D. Initiate Information Collection 5-3
    E. Initiate Security Operations 5-6
    F. Initiate Troop Movements 5-6
    G. Initiate Network Preparation 5-6
    H. Manage and Prepare Terrain 5-6
    I. Conduct Confirmation Briefings 5-6
    J. Conduct Rehearsals 5-6

Chap 6 — (Information) EXECUTION

EXECUTION (Overview) 6-1
I. Information Operations Working Group 6-1
II. IO Responsibilities Within the Various Command Posts 6-4
III. Assessing During Execution 6-5
    A. Monitoring IO 6-5
    B. Evaluating IO 6-10
IV. Intelligence Support 6-6
    Intelligence "Push" and "Pull" 6-6
    Information Operations and the Intelligence Process 6-7
    Intelligence Preparation of the Operating Environment (IPOE) 6-7
V. Decision Making During Execution 6-8
    A. Executing IO as Planned 6-8
    B. Adjusting IO to a Changing Friendly Situation 6-8
    C. Adjusting IO to an Unexpected Threat Reaction 6-9

Chap 7: Fires & Targeting

I. Fires (INFO Considerations) 7-1
- I. The Fires Warfighting Function 7-1
- II. Fires Overview 7-2
- III. Execute Fires Across the Domains 7-4
- IV. Joint Fires (OIE Considerations) 7-6
  - Joint Force Capabilities, Operations, and Activities for 7-6
    LEVERAGING INFORMATION
  - Nonlethal Effects 7-7
- V. Foundations of Fire Support (FS) 7-8
  - Attack & Delivery Capabilities 7-8
- VI. Scheme of Information Operations 7-10

II. Targeting (INFO Integration) 7-11
- Targeting Methodology 7-11
- I. Decide, Detect, Deliver, Assess (D3A) 7-12
  - D - Decide 7-12
  - D - Detect 7-13
  - D - Deliver 7-13
  - A - Assess 7-13
- II. Targeting Tasks during the MDMP 7-16
- III. Dynamic Targeting (F2T2EA) 7-20

Chap 8: (Information) ASSESSMENT

ASSESSMENT (Overview) 8-1
- I. Assessment Framework 8-1
- II. IO Assessment Considerations 8-2
  - Measures of Effectiveness (MOE) 8-2
  - Measures of Performance (MOP) 8-3
  - Indicators 8-3
- III. Assessment Rationale 8-4
- IV. Principles That Enhance the Effectiveness of IO Assessment 8-4
- V. Assessment Focus 8-5
- VI. Assessment Methods 8-6
- VII. Assessment Process 8-6
  - A. Monitoring Information Operations 8-6
  - B. Evaluating Information Operations 8-7
    - MOEs, MOPs, and Indicators 8-7
- VIII. Assessment Products 8-8

(INTRO) Nature of Information & the OE

Ref: ADP 3-13, Information (Nov '23), chap. 1.

I. Information Explained

Information is central to all activity Army forces undertake. It is fundamental to command and control (C2) and is the basis for situational understanding, decision making, and actions across all warfighting functions. Information is the building block for intelligence—the product resulting from the collection, processing, integration, evaluation, analysis, and interpretation of available information concerning foreign nations, hostile or potentially hostile forces or elements, or areas of actual or potential operations (JP 2-0). As a critical resource, Army forces fight for, defend, and fight with information while attacking a threat's (adversary or enemy) ability to do the same.

A. Data and Information

The effective use of information to create and exploit information advantages begins with a common understanding of the terms data and information. Data is any signal or observation from the environment. An observation of an enemy force or a radar sounding are examples of data. A series of facts used for statistical analysis is also referred to as data. It can include facts such as lists of daily fuel and ammunition expenditures of subordinate units. In the context of computer science, data is electromagnetic encoded information for repeatability, meaning, and procedural use by automated means. Data can be collected, quantified, stored, and transmitted in electronic or other tangible forms; however, data is most useful when processed and assigned meaning by humans or human-designed algorithms (programs).

> Information is data in context to which a receiver (human or automated system) assigns meaning.

Information is data in context to which a receiver processes and assigns meaning. Receivers include humans and automated systems that acquire data in a variety of ways—observations, spoken or written words, database retrieval, or other sensing mechanisms. Humans assign meaning to contextual data and use that information to understand, make decisions, communicate, and act. Automated systems—a combination of hardware and software—process and assign meaning to contextual data to support decision making, control their own functions, or control the functions of other systems.

B. Assignment of Meaning

The assignment of meaning to data is receiver centric. For example, a company commander may interpret an enemy platoon moving into an assault position as the lead element of the enemy's main attack. The battalion commander may interpret the same observation differently, discerning the enemy platoon is a feint based on other reporting from the area of operations. A multitude of factors influences how a receiver interprets data to make sense of a situation or activity. For humans, factors range from education and experience to culture and beliefs. Automated systems assign meaning to data based on human programming, and in some cases, artificial intelligence and machine learning.

How Humans Assign Meaning to Data

Ref: ADP 3-13, Information (Nov '23), pp. 1-2 to 1-4.

How humans progressively assign meaning to data into understanding can be visualized as a hierarchy as shown in Figure 1-1. At the lowest level of the hierarchy is data. At the highest level is understanding. Processing transforms data into information. Analysis then refines information into knowledge. Humans then apply judgement to transform knowledge into understanding. It is this understanding that informs decision making and ultimately behavior.

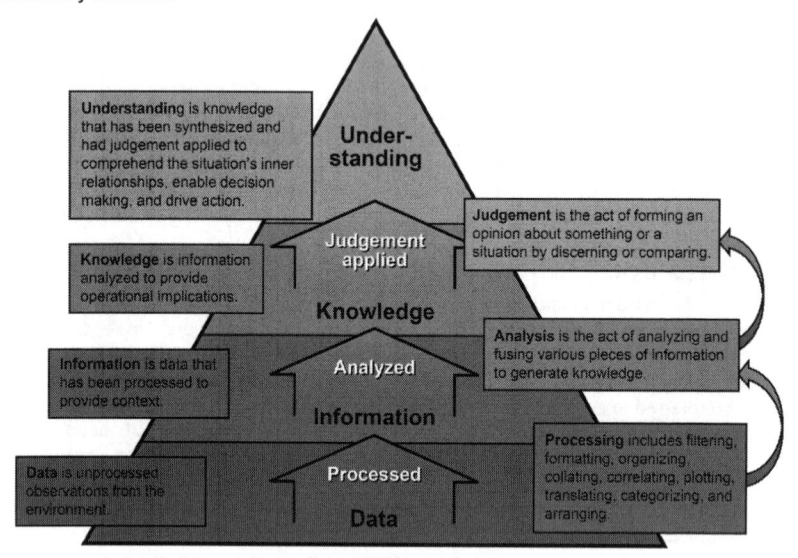

Understanding is knowledge that has been synthesized and had judgement applied to comprehend the situation's inner relationships, enable decision making, and drive action.

Knowledge is information analyzed to provide operational implications.

Information is data that has been processed to provide context.

Data is unprocessed observations from the environment.

Judgement is the act of forming an opinion about something or a situation by discerning or comparing.

Analysis is the act of analyzing and fusing various pieces of information to generate knowledge.

Processing includes filtering, formatting, organizing, collating, correlating, plotting, translating, categorizing, and arranging.

Understanding → Judgement applied → Knowledge → Analyzed → Information → Processed → Data

Ref: ADP 3-13, (Nov '23), fig. 1-1. Cognitive hierarchy.

The meaning of information that leads to understanding and decision making relies on both the information itself (data and its context) and factors that influence how a receiver interprets that information. The premise of receiver-centric meaning is that individuals interpret symbols, messages, actions, and events differently. To increase the likelihood of a receiver interpreting the information in the way it was intended, the sender considers the factors that influence how a receiver assigns meaning. Two models help describe factors that affect how humans assign meaning to data:

Inherent Informational Aspects

All operations and activities have inherent informational aspects—features and details of a situation or an activity that can be observed. Humans use these inherent informational aspects to derive meaning from that situation or activity. When not directly observed, these aspects can be communicated to, inform, or influence an audience.

Inherent informational aspects include, but are not limited to, physical attributes of the capabilities and forces involved; the duration, location, and timing of the situation or activity; and any other characteristics that convey information to an observer. Inherent informational aspects, along with the context within which the activity occurs (for example, the background, setting, or surroundings), are processed through an individual's worldview to make sense of what is happening. Commanders purposefully design operations to optimize their inherent informational aspects, to include revealing or concealing signatures to influence relevant actor's perceptions and behavior.

- **Duration**: The period during which an activity or situation lasts. For example, an exercise occurring for one day or three weeks.
- **Location**: A position or site in which the activity or situation takes place usually marked by a distinguishing feature. For example, the situation or activity takes place on key terrain or a culturally significant site.
- **Timing**: The precise moment or the range of times in which the activity or situation takes place. For example, a cordon and search conducted during the night.
- **Platform**: The equipment or capability used during an activity or situation. For example, a force patrolling on foot or in armored vehicles.
- **Size**: The physical magnitude, extent, or bulk; the relative or proportionate dimensions of the force being presented. For example, an infantry company or an armored brigade in an assembly area.
- **Posture**: The state or condition at a given time in a particular circumstance; the position or bearing of the force.

Drivers of Behavior

In addition to inherent informational aspects, a combination of many other factors influences how humans interpret data to make sense of an idea or a situation. These factors drive behavior because they ultimately affect how humans decide and act on information. Understanding these factors is essential to leaders effectively using information to inform or influence audiences. Examples of drivers of human behavior include:

- **Attitude**—a positive or negative evaluation of a thing based on thoughts, behavior, and social content.
- **Bias**—a tendency to simplify information through a filter of personal experience and preferences that can cause errors in thinking.
- **Cognition**—the process by which knowledge and understanding are developed in the mind, to include retrieving stored information and processing that information.
- **Culture**—the customs, arts, social institutions, religious traditions, and achievement of a particular nation, people, or other social group.
- **Desire**—a strong feeling of wanting to have something or wishing for something to happen derived from factors such as affiliation, self-esteem, safety, security, freedom, & power.
- **Emotion**—an internal, unconscious mental reaction subjectively experienced and often manifested in physiological reactions and behavior. Emotional appeals can be highly effective because they bypass logic and critical thinking. However, forecasting the response (in order to measure it) is challenging.
- **Expertise**—in-depth knowledge and skill developed from experience, training, and education.
- **Instinct**—an innate, typically fixed pattern of behavior derived from desires such as a will to live, procreation, and pleasure.
- **Language**—shared communication that enables a population or group to interpret or make sense of data and information. Awareness of the attributes of a culture's language can provide insight to a culture's norms, attitudes, and beliefs.
- **Memory**—the mental storage of things learned and retained from activities and experiences. Memories are subject to deterioration and inaccurate recall. These inaccuracies can affect behavior just as much as accurate memories can.
- **Narrative**—a way of presenting or understanding a situation or series of events that reflects and promotes a particular point of view or set of values.
- **Perception**—the organization, identification, and interpretation of sensory information influenced by factors such as experiences, education, faith, and values.

Humans

How humans progressively assign meaning to data into understanding can be visualized as a hierarchy as shown in Figure 1-1. At the lowest level of the hierarchy is data. At the highest level is understanding. Processing transforms data into information. Analysis then refines information into knowledge. Humans then apply judgement to transform knowledge into understanding. It is this understanding that informs decision making and ultimately behavior.

See fig. 1-1. Cognitive hierarchy (previous page).

The meaning of information that leads to understanding and decision making relies on both the information itself (data and its context) and factors that influence how a receiver interprets that information. The premise of receiver-centric meaning is that individuals interpret symbols, messages, actions, and events differently. To increase the likelihood of a receiver interpreting the information in the way it was intended, the sender considers the factors that influence how a receiver assigns meaning. Two models help describe factors that affect how humans assign meaning to data:

- Inherent informational aspects.
- Drivers of behavior.

See previous pages (pp. 0-2 to 0-3) for an overview and further discussion of how humans assign meaning to data.

Automated Systems

Automated systems are a combination of hardware and software that allow computer systems, network devices, or machines to function with limited human intervention. Modern militaries depend on automated systems to perform basic functions such as communications, administration, data analytics, battle tracking, navigation, and detection. Examples of automated systems include integrated air defense, fire control, and supply management systems. These systems have varying degrees of autonomy depending on their programming. The functionality of these systems ranges from basic automation of simple, repetitive tasks to sophisticated artificial intelligence and machine learning.

Typically, automated systems assign meaning to data based on programming—a set of instructions that determine the actions of an automated system based on environmental conditions and inputs to the system. In some instances, automated systems enabled by artificial intelligence assign meaning and make decisions through environmental experience—apart and beyond base programming. For example, information-focused automated systems can rapidly search, sort, and collate publicly available information; identify events, issues, and trends in public sediment from around the world; and issue alerts to users.

Automated systems can quickly sort through volumes of data that would overload human decision makers and provide a concise analysis or take an action. When connected by a network, automated systems can exchange data across the globe at nearly light speed. Automated systems, however, are vulnerable to attack. Cyberspace attacks such as data poisoning, electromagnetic attacks, and physical signals such as sounds, visuals, and vibrations can impair an automated system.

II. Information in the Security Environment

Ref: ADP 3-13, Information (Nov '23), pp. 1-6 to 1-7.

The strategic security environment consists of national, international, and global factors that affect the decisions of senior civilian and military leaders with respect to employing U.S. instruments of national power. Continued advances in information technologies constantly impact the strategic security environment. Governments, institutions, militaries, commercial organizations, and individuals rely on communications networks to perform basic functions. Informational technologies enable and accelerate global human-to-human, human-to-computer, and computer-to-computer interactions resulting in an exponential growth in the amount of information created, processed, and shared.

Smart phones, the internet, and social media accelerate and expand collective awareness of events, issues, and concerns within and outside an operational area. These developments have dramatically increased the speed at which information can affect an OE. Billions of people are connected in an instantly responsive network, through which information and ideas are shared worldwide. These ideas ignite passions, spark new perspectives, and crystallize deeply held beliefs that influence how governments, militaries, organizations, and people act. The exponential growth in computer capability and global connectivity continues to shape the way people interact and how forces fight. These devices and the internet provide threats with an enormous amount of digital information concerning friendly forces, to include location, intentions, timings, and tactics.

Strategic Competition *(See pp. 1-54 to 1-55.)*

A central challenge to U.S. security is the reemergence of long-term, strategic competition. Adversaries, to include China, Russia, Iran, and North Korea, use informational power to try to gain regional influence and control the global narrative well ahead of potential armed conflict in what are otherwise considered by most societies as times of peace. Competition for information and ideas is continuous and persistent. Adversaries rely on enduring campaigns of influence to achieve their objectives. Media manipulation and censorship, disinformation campaigns transmitted through social media, physical presence and activities, and public diplomacy compose some of a threat's perception management activities used to influence both internal and external audiences.

Nation states remain the principal actors on the global stage, but nonstate actors also threaten the security environment with increasingly sophisticated information capabilities. Terrorists, transnational criminal organizations, hackers, and other malicious nonstate actors seek to transform global affairs with increased abilities. Terrorism remains a persistent threat driven by ideology and unstable political and economic structures. Some of these groups act within the United States. Transnational criminal organizations affect the U.S. economy and negatively affect the nation's security. Hackers are not bound by geography; they can hold worldwide government and commercial entities hostage and steal data and information to sell to threats.

Today's operational environment presents threats to the Army and joint force that are significantly more dangerous in terms of capability and magnitude than those we faced in Iraq and Afghanistan. Major regional powers like Russia, China, Iran, and North Korea are actively seeking to gain strategic positional advantage.

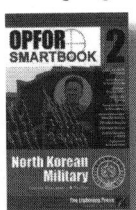

III. Information Within an Operational Environment (OE)

Within the broader strategic security environment, Army forces conduct operations in specific OEs. An operational environment is the aggregate of the conditions, circumstances, and influences that affect the employment of capabilities and bear on the decisions of the commander (JP 3-0).

For Army forces, an OE includes portions of the **land, maritime, air, space, and cyberspace domains** understood through **three dimensions (human, information, and physical).** The land, maritime, air, and space domains are defined by their physical areas. The cyberspace domain, a man-made network of networks, transits and connects the other domains through the electromagnetic spectrum as represented by the dots shown in Figure 1-3.

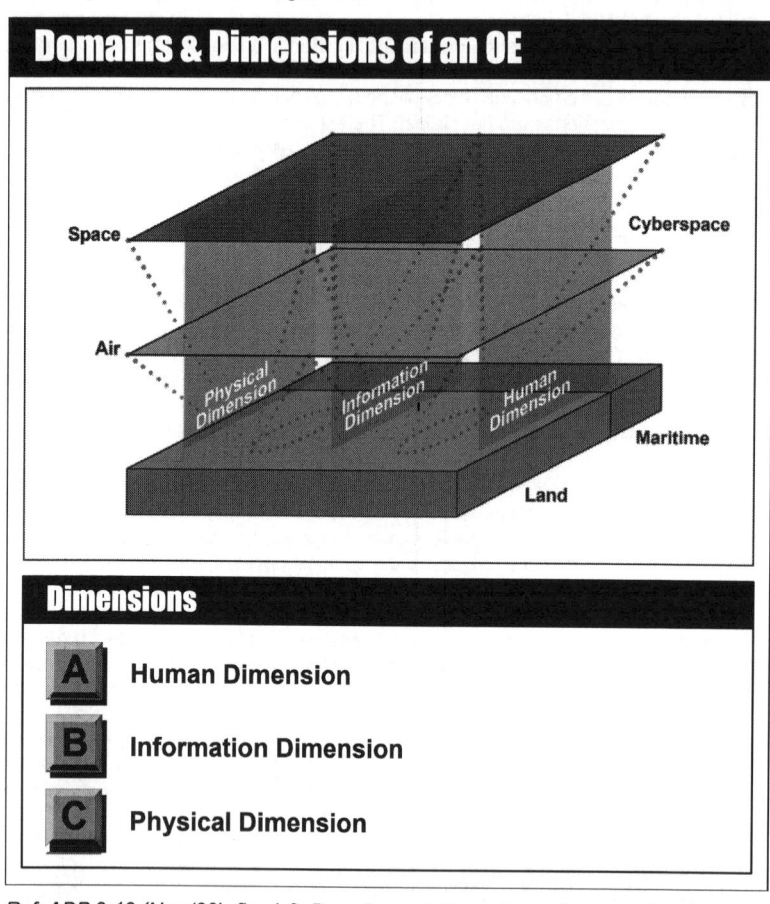

Domains & Dimensions of an OE

Space, Cyberspace, Air, Physical Dimension, Information Dimension, Human Dimension, Maritime, Land

Dimensions

- **A** Human Dimension
- **B** Information Dimension
- **C** Physical Dimension

Ref: ADP 3-13 (Nov '23), fig. 1-3. Domains and dimensions of an operational environment.

An OE consists of the totality of factors, specific circumstances, and conditions that impact the conduct of operations. Understanding an OE enables leaders to better identify problems; anticipate potential outcomes; and understand the results of various friendly, enemy, adversary, and neutral party actions and the effects these actions have on achieving objectives.

Informational Considerations (of the OE)

Ref: ADP 3-13, Information (Nov '23), pp. 1-8 to 1-9.

The interrelationship among the land, maritime, air, space, and cyberspace domains requires cross-domain understanding. As such, Army leaders seek to understand an OE through the human, information, and physical dimensions inherent to each domain. While used to understand all aspects of an OE, analysis of the human, information, and physical dimensions also helps leaders identify and understand informational considerations.

See pp. 2-12 to 2-16 for related discussion of joint doctrine's "analysis of informational, physical, and human aspects of the environment" from JP 3-04.

> **Informational considerations** are those aspects of the human, information, and physical dimensions that affect how humans and automated systems derive meaning from, use, act upon, and are impacted by information (FM 3-0).

Informational Considerations

Human Dimension	Information Dimension	Physical Dimension
Relevant Actors *(p. 1-27.)* - Military leaders - Civilian leaders - Key influencers - Groups - Organizations - Populations **Drivers of Behavior** *(p. 0-3.)* - Attitude - Expertise - Bias - Instinct - Cognition - Language - Culture - Memory - Desire - Narrative - Emotion - Perception	**Ideas** *(See p. 3-15.)* - Narratives - Messages - Themes **Data** **Software** **Information** - Friendly - Neutral - Threat **Malign and Benign Information** *(See p. 0-9.)* - Misinformation - Propaganda - Disinformation - Information for effect	**Inherent Informational Aspects of Opns** *(See p. 0-2.)* - Duration - Platform - Location - Size - Timing - Posture **Terrain** **Weather** **Electromagentic Radiation** **Communications** - Computer networks - Internet - Cellular networks - Print - Television - Radio - Satellite constellations **Bandwidth** **Storage**

Ref: ADP 3-13 (Nov '23), fig. 1-4. Example informational considerations.

Leaders analyze informational considerations from friendly, threat, and neutral perspectives to aid them in developing ways to use, protect, and attack data, information, and capabilities. This analysis enhances several aspects of planning, to include the selection of objectives and targets; approaches to influence foreign relevant actors; and identification of force protection measures. Figure 1-4 depicts potential informational considerations by dimension.

> **NOTE:** The Army's model of an OE established in FM 3-0 no longer includes an information environment. The term <u>informational considerations</u> is similar to the joint term and definition of <u>information environment</u>. The information environment is the aggregate of social, cultural, linguistic, psychological, technical, and physical factors that affect how humans and automated systems derive meaning from, act upon, and are impacted by information, including the individuals, organizations, and systems that collect, process, disseminate, or use information (JP 3-04).
>
> *See p. 2-1 for discussion of the information environment from a joint perspective.*

Nature of Information

A. Human Dimension (AODS7, p. 1-21.)

The human dimension encompasses people and the interaction between individuals and groups, how they understand information and events, make decisions, generate will, and act within an operational environment (FM 3-0). War is shaped by human nature and the complex interrelationship of cognitive (how people think) and psychological (mental or emotional state of a person) factors. Values and ethics are some of the factors that motivate both the cause for going to war as well as restrictions on the conduct of war. Fear, passion, camaraderie, grief, and many other emotions affect war participants' resolve. Emotions affect the behavior of combatants, how and when leaders decide to persevere, and when to give up. Individuals react differently to the stress of war; an act that may break the will of one enemy may only serve to stiffen the resolve of another. The will to act and fight emerges from the complex interrelationships in this dimension. Influencing these factors—by affecting perceptions, attitudes, and motivations—should underpin most, if not all, military objectives.

See p. 0-3 for factors that drive human behavior.

Informational Considerations

Informational considerations in the human dimension include identification of human relevant actors, analysis of how relevant actors tend to process information, and prediction of their likely behaviors.

> A **relevant actor** *(see p. 1-27)* is an individual, group, population, or automated system whose capabilities or behaviors have the potential to affect the success of a particular campaign, operation, or tactical action (JP 3-04). Understanding relevant actors and their relationships helps Army leaders to develop ways to influence their behavior through physical and informational means. Those relevant actors which the force intends to affect become **audiences** to inform or **target audiences** or **targets** for deception, physical attack, or other action.
>
> *See p. 2-64 for related discussion of audiences, targets, and target audiences.*

B. Information Dimension (AODS7, p. 1-21.)

The information dimension is the content and data that individuals, groups, and information systems communicate and exchange, as well as the analytics and technical processes used to exchange information within an operational environment (FM 3-0). Data and information are available globally in near real time. The ability to access data and information—from anywhere, at any time—broadens and accelerates human interaction across multiple levels, including person to person, person to organization, person to government, and government to government. Data and information may be highly controlled as in military classified information or may be publicly available. Publicly available information is information that has been published or broadcast for public consumption. This type of information is available to the public online, on request, through subscription or purchase, or could be seen or heard by a casual observer.

Informational Considerations

Informational considerations in this dimension include data and information used by relevant actors. Individuals, groups, and organizations record their perceptions in many formats ranging from spoken history to libraries, both physical and virtual. This body of information, collectively, can provide insights about how various groups, organizations, and countries might interpret Army operations. Although it is difficult to predict an individual's reaction to activities by Army forces, groups tend to be more consistent and predictable. A group, faction, or nation's prevailing narratives can provide a great deal of insight into how that group, faction, or nation might perceive Army operations. Threat doctrine is another example of relevant information that provides insight into how an enemy force may conduct operations and how it may respond to friendly actions. *(See facing page for further discussion.)*

Informational Considerations (in the Information Dimension)

Ref: ADP 3-13, Information (Nov '23), pp. 1-8 to 1-9.

Narratives. Identifying and understanding narratives in an OE are important informational considerations. Narratives—complementary, neutral, and hostile to the friendly force—are essential parts of any OE. A narrative is a way of presenting a situation or events that reflect a particular point of view with reasonable or believable logic. Individuals, groups, organizations, and countries all have narratives with many components that reflect and reveal how they define themselves. Political parties, social organizations, and government institutions, for example, have stories bound chronologically and spatially. Trust in the Army and Army forces is based upon demonstrated performance over time. Themes and messages complement friendly narratives. A theme is a distinct, unifying idea that supports a narrative. There may be multiple themes developed to support a narrative. A message is a narrowly focused communication directed at a specific audience to support a specific theme (JP 3-61).

(*See p. 3-15 for discussion of narrative, themes, and messages.*)

When considering the use of information by friendly, neutral, and threat actors, analysis should identify both **malign and benign information** including its content, purpose, source, and associated systems and processes. This includes identifying and analyzing misinformation, propaganda, disinformation, and information for effect.

Misinformation. Misinformation is unintentional incorrect information from any source. Misinformation is disseminated through ignorance or with the belief that the incorrect information is correct. Misinformation has no malicious intent. Identifying misinformation enables Army leaders to correct the record concerning misinformation about Army and friendly forces, operations, and other activities in an OE.

Propaganda. Propaganda is information that is biased or misleading and designed to influence the opinions, emotions, attitudes, or behaviors of any group to benefit the sponsor. Both disinformation and information for effect are forms of propaganda.

Disinformation. Disinformation is incomplete, incorrect, or out of context information deliberately used to influence audiences. Disinformation creates narratives that can spread quickly and instill an array of emotions and behaviors among groups, ranging from disinterest to violence. Relevant actors employ disinformation to shape public opinion, attract partners, weaken alliances, sow discord among populations, and deceive forces. Disinformation has a malicious intent. Sources of disinformation often rely on people to promulgate the information unwittingly, unaware that the information is inaccurate. Analysis allows Army forces to determine the source and purpose of information; what may first appear as misinformation may be a result of disinformation.

Information for Effect. Information for effect is the use, publication, or broadcast of factual information to negatively affect perceptions and/or damage credibility and capability of the targeted group. Unlike misinformation and disinformation which is incorrect or intentionally misleading, information for effect is factual and released at a time, at a place, and via means that will generate the most intended effect. Malign use of information for effect sometimes includes images or videos of friendly forces conducting operations that resulted in collateral damage. Videos of improvised explosive devices striking friendly forces provide another example of the use of information for effect to undermine the perceived strength of friendly forces. Friendly forces may mitigate information for effect through timely documentation of friendly force activities and release of statements and audio-visual products showing actions and results in near real time.

Nature of Information

C. Physical Dimension (AODS7, p. 1-20.)

The physical dimension is the material characteristics and capabilities, both natural and manufactured, within an operational environment (FM 3-0). While war is a human endeavor, it occurs in the physical world conducted with physical things. Each of the domains is inherently physical. Terrain, weather, military formations, electromagnetic radiation, weapons systems and their ranges, and many of the things that support or sustain forces are part of the physical dimension. Operations and activities in the physical dimension create effects in the human and information dimensions.

Informational Considerations

Informational considerations in the physical dimension include the natural and man-made aspects of an OE that affect communications—human to human, human to machine, and machine to machine. Geography, distance, and weather directly impact the ability of man-made capabilities to exchange data and information. The electromagnetic spectrum—the entire range of frequencies of electromagnetic radiation— permeates all domains and serves as a vital link for exchanging data and information across networks and information systems. Physical characteristics of the land domain also affect person-to-person communications. For example, a mountain range that separates two populations or groups impacts contact and communications between them. Understanding physical aspects of the domains and the electromagnetic spectrum helps determine the impacts on friendly, threat, and neutral communications capabilities and courses of actions available to employ them.

Communications capabilities are important physical informational considerations. Analysis of friendly, threat, and commercial communications capabilities and their impact on operations guides communications planning, communications protection, and methods to degrade threat communications. Communications considerations include bandwidth—the maximum amount of data transmitted over a network connection in a given amount of time. Army leaders understand the physical limitation of data transport and exchange, and they develop communications plans that prioritize bandwidth use.

Important physical informational considerations include threat intelligence and information warfare capabilities. Analysis of threat collections, intelligence, cyberspace, and electromagnetic capabilities informs leaders on ways to protect friendly data and information as well as to help identify threat targets for attack.

See pp. 1-53 to 1-55 for an overview and discussion of threat information warfare.

Informational considerations also include the inherent informational aspects of operations. Everything Army forces do impacts an OE either intentionally or incidentally. Whether executing a feint or conducting resupply, all Army activities can be observed and have the potential to affect the behavior of relevant actors. The perceptions that relevant actors draw from observing operations and actions will likely drive their behaviors, potentially making them vulnerable to exploitation to include deception. As such, leaders consider the inherent informational aspects of operations and activities as well as their potential to reinforce, prevent, or change behaviors.

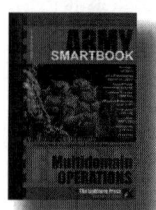

Refer to AODS7: The Army Operations & Doctrine SMARTbook (Multidomain Operations). Completely updated with the 2022 edition of FM 3-0, AODS7 focuses on Multidomain Operations and features rescoped chapters on generating and applying combat power: command & control (ADP 6-0), movement and maneuver (ADPs 3-90, 3-07, 3-28, 3-05), intelligence (ADP 2-0), fires (ADP 3-19), sustainment (ADP 4-0), & protection (ADP 3-37).

Chap 1 — Information Advantage

Ref: ADP 3-13, Information (Nov '23), chap. 2.

I. Information Advantage (Overview)

An **information advantage** is a condition when a force holds the initiative in terms of situational understanding, decision making, and relevant actor behavior. There are several forms of information advantage.

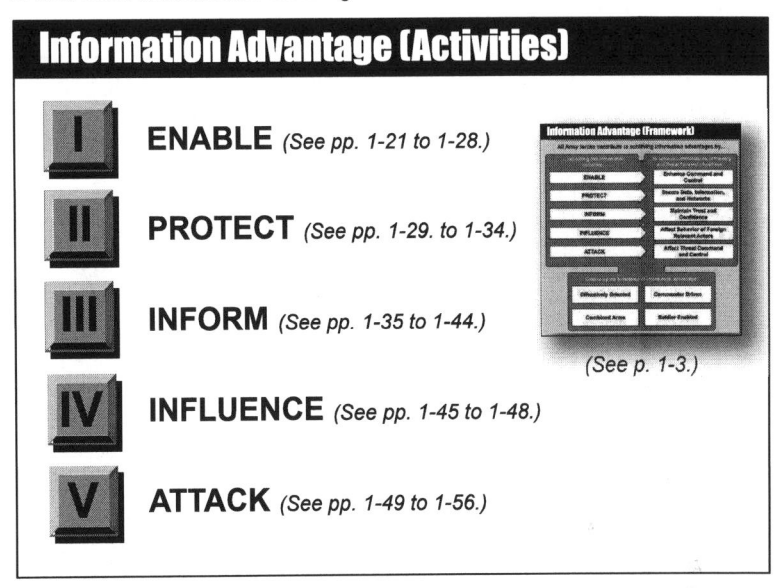

Information Advantage (Activities)

- **I — ENABLE** (See pp. 1-21 to 1-28.)
- **II — PROTECT** (See pp. 1-29. to 1-34.)
- **III — INFORM** (See pp. 1-35 to 1-44.)
- **IV — INFLUENCE** (See pp. 1-45 to 1-48.)
- **V — ATTACK** (See pp. 1-49 to 1-56.)

(See p. 1-3.)

When Army forces achieve an information advantage, they—
- Communicate more effectively than the threat.
- Collect, process, analyze, and use information to understand an OE better than the threat.
- Understand, decide, and act faster and more efficiently than the threat.
- Are resilient to threat information warfare, to include disinformation and information for effect.
- Maintain domestic support and the support of multinational partners.
- Degrade threat command and control (C2) by affecting the threat's ability to understand, make effective decisions, and communicate.
- Influence threats and other foreign relevant actors' behavior favorable to friendly objectives.

An information advantage can result from and exploit human and physical advantages or enable those advantages. Like human and physical advantages, information advantages are often temporary and change over time relative to the threat and changes in an OE. While friendly forces are seeking information advantages, threat forces are doing the same. As such, an information advantage is something to gain, protect, and exploit across as many domains as possible.

ADP 3-13, Information (Nov '23): Introduction & Overview

Information is central to everything we do—it is the basis of intelligence, a fundamental element of command and control, and the foundation for communicating thoughts, opinions, and ideas. As a **dynamic of combat power**, Army forces fight for, defend, and fight with information **to create and exploit information advantages**—the use, protection, and exploitation of information to achieve objectives more effectively than enemies and adversaries do.

Advancements in information technologies and increased global connectivity continue to shape how we interact with each other and how forces fight. These advancements accelerate and expand the ability of joint and Army forces to collect, process, analyze, store, and communicate information at a scale previously unimaginable.

> *** Editor's Note on <u>Changes in Information Terminology</u>**
>
> Based on changes to joint information doctrine, **Army forces will no longer use the terms** *information operations, information-related capabilities, or information superiority*.
>
> A significant joint doctrinal change is the transition from *joint information operations (IO)* to *operations in the information environment (OIE)*. Joint doctrine, however, retains the term *information environment*. The Army's new model of an operational environment established in FM 3-0 no longer includes an information environment. The term *informational considerations* aligns with the joint term *information environment*.
>
> The *Army is currently revising all its doctrine*, to include FM 3-13, to account for these changes and the Army's new information advantage framework. **As such, the INFO2 SMARTbook retains the original terminology as referenced from the original source, while recognizing this terminology is changing (marked with an asterisk).**
>
> - *ADP 3-13 (Nov '23) (See pp. 2-2 to 2-3 for joint doctrine changes.)*

ADP 3-13, Information, is the Army's first publication dedicated to information. It provides a framework for creating and exploiting information advantages during the conduct of operations and at home station. It represents an evolution in how Army forces think about the military uses of data and information, emphasizing that everything Army forces do, to include the information and images it creates, generates effects that contribute to or hinder achieving objectives. **As such, creating and exploiting information advantages is the business of all commanders, leaders, and Soldiers.**

ADP 3-13 operationalizes the two big ideas inherent in multidomain operations—combined arms and positions of relative human, information, and physical advantage. **We no longer regard information as a separate consideration or the sole purview of technical specialists. Instead, we view information as a resource that is integrated into operations with all available capabilities in a combined arms approach** to enable command and control; protect data, information, and networks; inform audiences; influence threats and foreign relevant actors; and attack the threat's ability to exercise command and control.

Army leaders are accustomed to **creating and exploiting relative advantages** through a combined arms approach that traditionally focuses on the human and physical dimensions of an operational environment. ADP 3-13 acknowledges that advantages in the information dimension complement and reinforce advantages in the human and physical dimensions. The advantages do not necessarily have to be great: small advantages exploited quickly help commanders gain and maintain the operational initiative. Combining these advantages slows threat decision making, increases its level of uncertainty, and allows Army forces to dictate the tempo of operations.

Information Advantage Framework

Fig. 2-2 depicts a framework for creating and exploiting information advantages—the use, protection, and exploitation of information to achieve objectives more effectively than the threat.

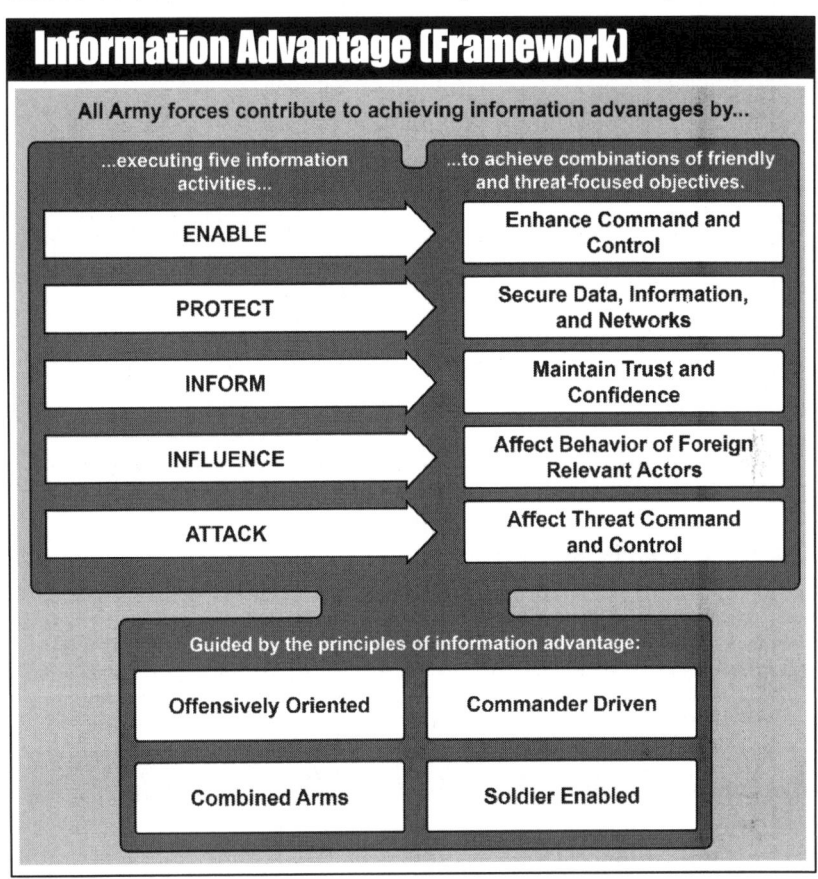

Information Advantage (Framework)

All Army forces contribute to achieving information advantages by...

...executing five information activities...	...to achieve combinations of friendly and threat-focused objectives.
ENABLE	Enhance Command and Control
PROTECT	Secure Data, Information, and Networks
INFORM	Maintain Trust and Confidence
INFLUENCE	Affect Behavior of Foreign Relevant Actors
ATTACK	Affect Threat Command and Control

Guided by the principles of information advantage:

- Offensively Oriented
- Commander Driven
- Combined Arms
- Soldier Enabled

ADP 3-13 is a new publication that represents an evolution in how Army forces think about military uses of data and information in competition, crisis, and armed conflict. It represents a change in mindset based on the recognition that all activities have inherent informational aspects that generate effects which contribute to or hinder achieving objectives. Accounting for advances in information technologies and threat information warfare capabilities, ADP 3-13 describes a combined arms approach to creating and exploiting information advantages to achieve objectives. ADP 3-13 incorporates the Army's operational concept of multidomain operations and related doctrine described in the FM 3-0. ADP 3-13 describes—

- A revised model for understanding an operational environment through the human, information, and physical dimensions.
- Information as a dynamic of combat power.
- Information advantages as a central component of multidomain operations.
- Considerations for how Army forces seek information advantages within the strategic contexts of competition below armed conflict, crisis, and armed conflict.

II. Information in Multidomain Operations
Ref: ADP 3-13, Information (Nov '23), pp. 2-2 to 2-3.

The Army organizes, trains, and equips its forces to conduct prompt and sustained land combat to defeat enemy ground forces and seize, occupy, and defend land areas. Trained and ready Army forces support joint force commanders in three strategic contexts: **competition below armed conflict, crisis, and armed conflict.** *(See pp. 1-16 to 1-20.)*

An **operation** is a sequence of tactical actions with a common purpose or unifying theme. Operations vary in scale of forces involved, duration, and level of violence. While most operations conducted by Army forces occur either below the threshold of armed conflict or during limited contingencies, the focus of Army readiness is on large-scale combat operations against a peer threat.

> **Multidomain operations** are the combined arms employment of joint and Army capabilities to create and exploit relative advantages to achieve objectives, defeat enemy forces, and consolidate gains on behalf of joint force commanders. Multidomain operations are the Army's contribution to joint campaigns during competition, crisis, and armed conflict. Below the threshold of armed conflict, multidomain operations are how Army forces accrue advantages and demonstrate readiness for conflict, deterring adversaries while assuring allies and partners. During armed conflict, Army forces use multidomain operations to close with and destroy the enemy, defeat enemy formations, seize critical terrain, and control populations and resources to deliver sustainable political outcomes.

The central idea of multidomain operations is the combined arms employment of all available joint and Army capabilities to create and exploit **relative advantages** to achieve objectives.

Relative Advantages

Relative advantages provide opportunities. A relative advantage is a location or condition, in any domain, relative to an adversary or enemy that provides an opportunity to progress towards or achieve an objective. During operations, small advantages can significantly impact the outcome of a mission, particularly when they accrue over time. Commanders seek and create relative advantages to exploit through action, and they continually assess friendly and enemy forces in relation to each other for opportunities to exploit.

Relative Advantages
- **Human**
- **Informational**
- **Physical**

Combined, these physical and information advantages can lead to a collapse of the enemy's morale and will—a human advantage. Army forces combine, reinforce, and exploit human, information, and physical advantages to achieve objectives across the competition continuum.

HUMAN Advantage

Human advantages are individual and group characteristics that provide opportunities for friendly forces. War is inherently a human endeavor—a violent struggle between multiple hostile, independent, and irreconcilable wills, each trying to impose its will on the other. Human will, instilled through commitment to a cause and leadership, is the driving force of all action in war. Army forces create and exploit human advantages throughout the conduct of operations. Combined with physical and information advantages, human advantages enable friendly morale and will, degrade enemy morale and will, and influence popular support.:

- Health, physical fitness, and toughness.
- Intelligence and intellect.
- Training.
- Leadership.
- Troop morale and will.
- Relevant actor trust.
- Positive relationships with foreign governments, populations, and forces.
- Cultural affinity and familiarity with indigenous populations and institutions.

INFORMATION Advantage *(See pp. 1-1 and 1-19.)*

An information advantage is a condition when a force holds the initiative in terms of situational understanding, decision making, and relevant actor behavior. There are several forms of information advantage. For example, a force that understands, decides, and acts more effectively than its opponent has an information advantage. A force that effectively communicates and protects its information, while preventing the threat from doing the same, is another form of an information advantage.

An information advantage can result from and exploit human and physical advantages or enable those advantages. Like human and physical advantages, information advantages are often temporary and change over time relative to the threat and changes in an OE. While friendly forces are seeking information advantages, threat forces are doing the same. As such, an information advantage is something to gain, protect, and exploit across as many domains as possible.

- Communicate more effectively than the threat.
- Collect, process, analyze, and use information to understand an OE better than the threat.
- Understand, decide, and act faster and more efficiently than the threat.
- Are resilient to threat information warfare, to include disinformation & information for effect.
- Maintain domestic support and the support of multinational partners.
- Degrade threat command and control (C2) by affecting the threat's ability to understand, make effective decisions, and communicate.
- Influence threats & other foreign relevant actors' behavior favorable to friendly objectives.

PHYSICAL Advantage

Physical advantages are most familiar to tactical forces, and they are typically the immediate goal of most tactical operations. Finding the enemy, defeating enemy forces, and seizing occupied land typically require the creation and exploitation of multiple physical advantages. These advantages include occupation of key terrain, the physical isolation of enemy forces, and the imposition of overwhelming fires. The exploitation of physical advantages reduces the enemy's capability to fight, which creates information and human advantages. Physical advantages implicitly communicate a message that can influence enemy forces' will to fight, sway popular support, and disrupt enemy risk calculus at all echelons.

- Geographic and positional advantages.
- Capabilities or qualitative advantages.
- Overall combat power, including numbers of systems and firepower.

Refer to AODS7: The Army Operations & Doctrine SMARTbook (Multidomain Operations). Completely updated with the 2022 edition of FM 3-0, AODS7 focuses on Multidomain Operations and features rescoped chapters on generating and applying combat power: command & control (ADP 6-0), movement and maneuver (ADPs 3-90, 3-07, 3-28, 3-05), intelligence (ADP 2-0), fires (ADP 3-19), sustainment (ADP 4-0), & protection (ADP 3-37).

III. Informational Power

Ref: ADP 3-13, Information (Nov '23), pp. 1-4 to 1-6.

Power—the capacity or ability to direct or influence the behavior of others—has many forms. **Informational power** is an ability to use information to support achievement of objectives and create information advantages. Informational power and physical power (strength or force) are interdependent and mutually supporting forms of power applicable below and above the threshold of armed conflict. An effective application of informational power to achieve objectives requires a whole of government, joint, and combined arms approach.

Information and the INSTRUMENTS OF NATIONAL POWER

Information is a vital resource for national security. From a U.S. government perspective, the informational instrument of national power is employed in combination with diplomatic, military, and economic power to advance national interests. Previously considered in the context of traditional nation states, the construct of information as an instrument of national power now extends to nonstate actors. Nonstate actors include terrorists, mercenary companies, and transnational criminal organizations—actors who use information to further their causes and undermine those of the U.S. government and its multinational partners. Nonstate actors can also include nongovernmental organizations and multinational corporations who can be supportive of U.S. interests.

The U.S. government employs informational power in three primary ways. First, it synchronizes its communications activities to influence the perception and attitudes of other governments, organizations, groups, and individuals deemed vital to strategic objectives. Second, the U.S. government coordinates efforts to secure cyberspace and critical infrastructure against information disruption. Third, the U.S. government provides information to bolster national will and resolve.

JOINT Informational Power (See pp. 2-4 to 2-5.)

For joint force commanders, the essence of informational power is the ability to exert one's will through the projection, exploitation, denial, and preservation of information in pursuit of military objectives. The joint force uses information to perform many simultaneous and integrated activities ranging from improving friendly understanding and decision making to affecting threat behavior. The joint force leverages the power of information to effectively expand the commander's range of operations. Joint force commanders apply informational power—

- To operate in situations where the use of destructive or disruptive physical force is not authorized or is not an appropriate course of action.
- To degrade, disrupt, and destroy threat C2.
- To prevent, counter, and mitigate the effects of external actors' actions on friendly capabilities and activities.
- To create and enhance the psychological effects of destructive or disruptive physical force.
- To create psychological effects without destructive or disruptive force.
- To confuse, manipulate, or deceive an adversary or enemy to create an advantage or degrade a threat's existing advantage.
- To prevent, avoid, or mitigate any undesired psychological effects of operations.
- To communicate and reinforce the intent of operations, regardless of whether those activities are constructive or destructive.
- To reinforce the will to fight in friendly forces and populations.
- To degrade the will to fight in threat forces and populations.

Information as a DYNAMIC OF COMBAT POWER *(AODS7, chap. 2.)*

Army forces create and exploit informational power similarly to the joint force through five information activities (enable, protect, inform, influence, and attack). Army forces also consider information as a dynamic of combat power employed with mobility, firepower, survivability, and leadership to achieve objectives during armed conflict. Combat power is the total means of destructive and disruptive force that a military unit/formation can apply against an enemy at a given time (JP 3-0). As a dynamic of combat power, Army forces fight for, defend, and fight with information.

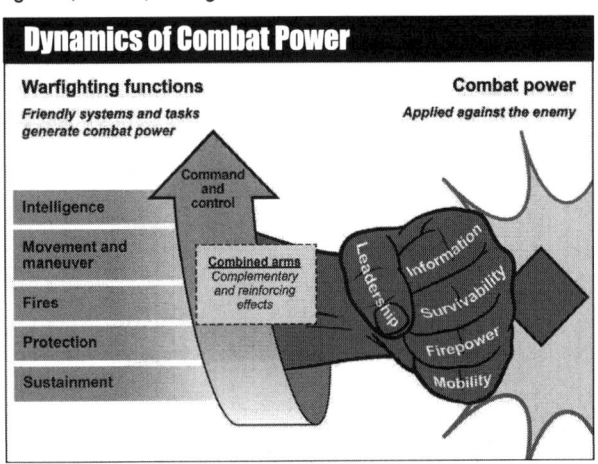

Dynamics of Combat Power

Warfighting functions
Friendly systems and tasks generate combat power

- Command and control
- Intelligence
- Movement and maneuver
- Fires
- Protection
- Sustainment

Combat power
Applied against the enemy

- Leadership
- Combined arms: Complementary and reinforcing effects
- Information
- Survivability
- Firepower
- Mobility

Combat power *is the total means of destructive and disruptive force that a military unit/formation can apply against an enemy at a given time (JP 3-0). It is the ability to fight. The complementary and reinforcing effects that result from synchronized operations yield a powerful blow that overwhelms enemy forces and creates friendly momentum.*

Army leaders at every level require and use information to seize, retain, and exploit the initiative and achieve decisive results. Army forces collect, process, and analyze data and information from all domains to develop understanding, make decisions, and apply combat power against enemy forces. Army forces fight for information about the enemy and terrain through reconnaissance and surveillance, and through offensive operations such as movement to contact or reconnaissance in force. Intelligence and cyberspace operations penetrate enemy networks and observe activities to gain and exploit information on the threat. Simultaneously, Army forces defend their own networks to secure friendly data and ensure secure communications. Friendly security operations, operations security, counterintelligence, and defensive cyberspace operations deny enemy access to friendly information and intentions.

Army forces fight with information to influence threat behavior. Creatively employing and concealing information can enable Army forces to achieve surprise, cause enemy forces to misallocate or expend combat power, or mislead enemy forces as to the strength, readiness, locations, and intended missions of friendly forces. Army forces also employ information as a means of amplifying the psychological effects of disruptive and destructive physical force to erode morale, impede decision making, and increase uncertainty among enemy forces. Army forces employ all relevant capabilities to attack threat data, information, and networks to hinder the threat's ability to exercise C2.

Refer to AODS7: The Army Operations & Doctrine SMARTbook (Multidomain Operations). Completely updated with the 2022 edition of FM 3-0, AODS7 focuses on Multidomain Operations and features rescoped chapters on generating and applying combat power: command & control (ADP 6-0), movement and maneuver (ADPs 3-90, 3-07, 3-28, 3-05), intelligence (ADP 2-0), fires (ADP 3-19), sustainment (ADP 4-0), & protection (ADP 3-37).

IV. Information Activities (Defined)

Ref: ADP 3-13, Information (Nov '23), pp. 2-3 to 2-6.

The information advantage framework *(see p. 1-3)* presents a framework for creating and exploiting information advantages. Within this framework, Army forces integrate all relevant military capabilities through the execution of five information activities (enable, protect, inform, influence, and attack).

> An **information activity** is a collection of tasks linked by purpose to affect how humans and automated systems derive meaning from, use, and act upon, or are influenced by, information. Each information activity incorporates several tasks and subtasks from the warfighting functions to achieve a variety of friendly and threat-based objectives.
>
> **Warfighting Functions**
>
> | Information Activities | Information Activities |
> |---|---|
> | Enable / Protect / Inform | Inform / Influence / Attack |
>
> Increase Effectiveness → Friendly Decision Cycle
>
> Threat Decision Cycle → Decrease Effectiveness
>
> Ref: ADP 3-13 (Nov '23), fig. 2-3.
>
> The enable and protect information activities increase the effectiveness of the friendly decision cycle as shown in Figure 2-3. The influence and attack information activities decrease the effectiveness of the threat's decision cycle. The inform information activity can increase the effectiveness of the friendly decision cycle while decreasing the effectiveness of the threat's decision cycle. The combined effects of enhancing the friendly decision cycle while degrading the threat's create a significant advantage for Army forces.

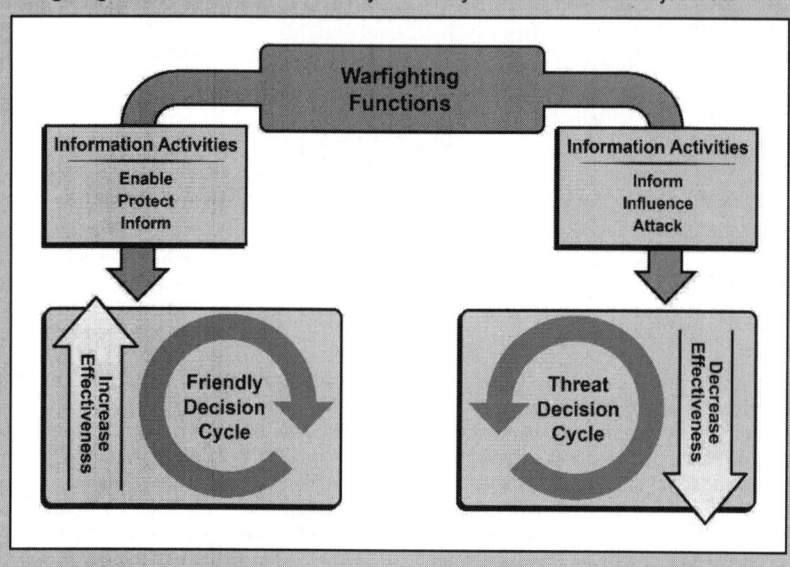

Information activities are interdependent. For example, both the protect and inform information activities help protect the force from malign influence. Guided by the principles of information advantage, Army leaders plan, prepare, execute, and assess information activities as part of the operations process.

I. Enable *(See pp. 1-21 to 1-28.)*

The enable information activity includes tasks that enhance friendly C2. The focus of this activity is to improve situational understanding, decision making, and communications. Throughout operations, Army forces collect, process, and analyze information to develop situational understanding that, in turn, facilitates effective decision making. Leaders then communicate their decisions to subordinates and direct actions. Based on feedback and assessment of the situation, Army leaders adjust operations as required. The network provides the backbone for exchanging data and information, enabling shared understanding and enhancing the exercise of C2 of Army forces. While all warfighting functions enhance the exercise of C2, the C2 and intelligence warfighting functions are the largest contributors. The C2 warfighting function tasks and system (people, processes, networks, and command posts) enable commanders to command forces and control operations. The tasks and systems in the intelligence warfighting function enable the exercise of C2 by facilitating the understanding of enemy, terrain, weather, civil considerations, and other aspects of an OE.

II. Protect *(See pp. 1-29 to 1-34.)*

The protect information activity includes tasks that secure friendly data, information, and networks. This activity focuses on denying threat access to friendly data and information while preserving friendly communications capabilities. Preventing threat access to friendly data and information not only protects the friendly force but hinders the threat's ability to accurately understand situations and make effective decisions. Defending friendly data, information, and networks enables friendly decision making and ensures functional communications. The protection and movement and maneuver warfighting functions are the largest contributors to the protect information activity.

III. Inform *(See pp. 1-35 to 1-44.)*

The inform information activity includes tasks that foster informed perceptions of military operations and activities among various audiences. The focus of this activity is to maintain the trust and confidence of internal (members of the U.S. Army, Army Civilians, contractors, and their family members) and external audiences (U.S. domestic and international audiences). The proactive release of accurate information to Army, domestic, and international audiences puts operations in context, facilitates informed perceptions about the Army, increases friendly force resiliency, and undermines threat disinformation activities. The C2 and intelligence warfighting functions are primary contributors to the inform information activity.

IV. Influence *(See pp. 1-45 to 1-48.)*

The influence information activity includes tasks that affect the thinking and ultimately the behavior of threats and other foreign audiences. This activity focuses on reinforcing or changing how individuals and groups think, feel, and act in support of objectives. Army forces influence threats to decrease combat effectiveness, erode unit cohesion, diminish morale and will, and deceive threat forces about friendly intent. Influence efforts toward other foreign audiences range from strengthening mutual support to discouraging an audience's support for an adversary or enemy. The inherent informational aspects of all warfighting functions contribute to influencing threats and other foreign audiences.

V. Attack *(See pp. 1-49 to 1-54.)*

The attack information activity includes tasks that affect the threat's ability to exercise C2. The focus of this activity is to affect threat data and the threat's physical capabilities used to communicate and conduct information warfare. This includes the data and communications between automated systems such as the communications between radars, fire control systems, and firing systems.

Threat C2 nodes (command posts, signal centers, networks, and information systems) and sensors (surveillance, target acquisition, and radars) are often high-payoff targets for Army forces. As part of the scheme of fires, Army forces attack these targets using physical destruction, electromagnetic attack, and offensive cyberspace operations to hinder the threat's C2 abilities. Army forces also attack threat data, information, and physical capabilities used to exchange data and information. Informed by intelligence, the fires and movement and maneuver warfighting functions are the primary contributors.

V. Information Activities and the Tenets of Operations

Ref: ADP 3-13, Information (Nov '23), pp. 2-12 to 2-14.

The tenets of operations are desirable attributes built into all plans and operations, and they directly relate to how Army forces should employ multidomain operations. Commanders use the tenets of operations to inform and assess courses of action throughout the operations process. The degree to which an operation exhibits the tenets provides insight into the probability for success. The tenets of operations are—

Tenets of Operations

- **Agility**
- **Convergence**
- **Endurance**
- **Depth**

Agility *(AODS7, p. 1-38.)*

Agility is the ability to move forces and adjust their dispositions and activities more rapidly than the enemy. Agility requires shared understanding, sound judgement, and rapid decision making often gained through the creation and exploitation of information advantages. Agility helps leaders influence tempo. Tempo is the relative speed and rhythm of military operations over time with respect to the enemy (ADP 3-0). It implies the ability to understand, decide, act, assess, and adapt more effectively than the enemy.

Information activities contribute to agility and tempo. The enable and protect information activities increase the effectiveness of the friendly decision cycle. The influence and attack information activities decrease the effectiveness of the threat's decision cycle. The inform information activity can increase the effectiveness of the friendly decision cycle while decreasing the effectiveness of the threat's decision cycle. The combined effects of enhancing the friendly decision cycle while degrading the threat's create a significant advantage for Army forces.

Convergence *(AODS7, pp. 1-39 to 1-42.)*

Peer threats employ adaptable and durable capabilities and formations over large geographic areas and multiple domains. They cannot be easily defeated in a single, decisive effort. Success requires Army forces to sustain attacks against multiple decisive points over time through convergence. Convergence is an outcome created by the concerted employment of capabilities from multiple domains and echelons against combinations of decisive points in any domain to create effects against a system, formation, decision maker, or in a specific geographic area. Its utility derives from understanding the interdependent relationships among capabilities from different domains and combining those capabilities in surprising, effective tactics that accrue advantages over time. Convergence occurs when a higher echelon (normally corps and above) and its subordinate echelons create effects from and in multiple domains in ways that defeat or disrupt threat forces long enough for friendly forces to effectively exploit.

Convergence requires the integration of relevant Army and joint capabilities and the synchronization of actions at multiple decisive points. Commanders integrate Army capabilities (communications, information collection, fires, electromagnetic warfare, and MISO) into subordinate formations by task organizing forces, establishing support relationships, prioritizing efforts, and delegating execution authorities. For joint informational capabilities, Army commanders generally coordinate for specific informational effects. This includes requesting effects from cyberspace operations, space operations, MISO, fires, maneuver, and special technical operations. The effective integration of joint capabilities into Army operations requires an understanding of multiple joint processes, especially the joint targeting process. To achieve convergence, Army commanders synchronize information activities into the concept of operations. Based on the commander's intent, operational approach, and planning guidance, planners synchronize information activities and associated tasks within the schemes of intelligence, information collection, maneuver, fires, protection, sustainment, and command and signal. Synchronizing information activities into a concept of operations enables a combined arms approach to operations necessary to achieve convergence.

Endurance *(AODS7, pp. 1-42 to 1-43.)*

Endurance enhances the ability to project combat power and extends operational reach. Endurance is the ability to persevere over time throughout the depth of an OE. Endurance is about resilience and preserving combat power while continuing operations for as long as necessary to achieve the desired outcome. Endurance stems from the ability to organize, protect, and sustain a force regardless of the distance from its support area and the austerity of the environment.

Information activities and their related tasks contribute to endurance. Robust and resilient networks help to maintain continuous communications across echelons. Securing and obscuring friendly data and information helps prevent threat collection and protects the force. Proactively informing internal and external audiences throughout the duration of a campaign helps to preserve will, friendly morale, and human resiliency. Influencing populations and other foreign relevant actors to support friendly objectives over time contributes to endurance. Degrading threat information warfare capabilities helps to preserve friendly combat power throughout the breadth and depth of an operational area.

Depth *(AODS7, p. 1-43.)*

Depth is the extension of operations in time, space, or purpose to achieve definitive results (ADP 3-0). While the focus of endurance is on friendly combat power, the focus of depth is on enemy locations and dispositions across all domains. Commanders achieve depth by understanding the strengths and vulnerabilities of the enemy's echeloned capabilities, then attacking them throughout their dispositions in simultaneous and sequential fashion. Leaders describe the depth they can achieve in terms of operational reach.

Operational reach is the distance and duration across which a force can successfully employ military capabilities (JP 3-0). Staffs assess operational reach based on available sustainment, the range of capabilities and formations, and courses of action compared with the intelligence estimates of enemy capabilities and courses of action. This analysis helps the commander understand the limits on friendly operations, risks inherent in the mission, and likely points in time and space for transitions.

Information activities contribute to depth. Army signal forces add depth by maintaining communications infrastructures throughout an area of responsibility and establishing networks in joint operations areas. These networks facilitate world-wide communications of both joint and Army forces. The joint force land component command creates depth by facilitating access to joint and Army capabilities, especially space, cyberspace, and electromagnetic warfare, to degrade enemy networks and systems in the extended deep area. Corps and division forces employ fires into deep areas to degrade enemy C2 and disrupt communications, to include communications between sensors and shooters. Inform information activities provide depth in friendly morale and will, public willingness to support operations, and reassurance of commitment to allies and partners.

Leaders enhance the depth of their operations by orchestrating effects in one dimension to amplify effects in the others. For example, a commander might decide to destroy an elite enemy formation first because it undermines the confidence of the enemy's other units.

VI. Principles of Information Advantage

Ref: ADP 3-13, Information (Nov '23), pp. 2-14 to 2-16.

A principle is a comprehensive and fundamental rule or an assumption of central importance that guides how an organization approaches and thinks about the conduct of operations.

Gaining and exploiting an information advantage involves four principles. These principles provide a starting point for thinking about the use of information and the employment of capabilities to create and exploit information advantages. The principles of information advantage include the following:

Principles of Information Advantage

- **Offensively Oriented**
- **Combined Arms**
- **Commander Driven**
- **Soldier Enabled**

Offensively Oriented

Offensively oriented suggests that offensive action, or maintaining the initiative, is the most effective and decisive way to achieve objectives. Any information advantage not sought or defended is potentially ceded to the threat. As such, Army leaders take the initiative to create, protect, and exploit information advantages in all domains.

> The force that anticipates better, thinks more clearly, decides and acts more quickly, and adapts more rapidly, stands the greatest chance to seize, retain, and exploit the initiative over an opponent.

No matter the echelon, the force that retains the initiative through offensive action forces its opponent to react. While it is necessary to defend or protect, Army leaders maintain an offensive mindset and anticipate events in pursuit of various information advantages throughout the conduct of operations. They aggressively collect and use information to understand, decide, and act while actively denying the threat from doing the same. In addition to passive measures to protect data and information, Army forces target and attack threat capabilities to hinder the threat's ability to collect and communicate information about friendly forces. Army leaders proactively release accurate and timely information to various audiences as opposed to just reacting to threat propaganda.

Combined Arms

The combined arms approach to operations is foundational to creating and exploiting information advantages. Combined arms is the synchronized and simultaneous application of arms to achieve an effect greater than if each element was used separately or sequentially (ADP 3-0). Leaders combine available organic, joint, and multinational capabilities in complementary and reinforcing ways to create and exploit information advantages in the same way they do with human and physical advantages. For example, Army forces precede or follow a massed artillery strike with surrender appeals through MISO.

All military capabilities can be employed for information advantage. For example, an infantry battalion can deceive an enemy by conducting a feint. A field artillery brigade can destroy enemy radars, communications, and CPs. The commitment of Army sustainment units to an area during competition can influence an adversary's decision making. Additionally, some Army, joint, and multinational units are specifically designed for the use, protection, denial, or manipulation of information. Signal, cyberspace, electromagnetic warfare, psychological operations, space, civil affairs, and public affairs units are examples. Commanders and staffs do not restrict their thinking to a select few specialized units or capabilities but consider all available capabilities in a combined arms approach to enable C2; protect data, information, and networks; and inform audiences, influence relevant actors, and attack threat C2.

Commander Driven

Commanders at every level require and use information to seize, retain, and exploit the initiative and achieve decisive results. Therefore, commanders must understand information, integrating it in operations as carefully as fires, maneuver, protection, and sustainment. Commanders think of information as a resource to achieve situational understanding, a tool to induce ambiguity and uncertainty in the threat, and the primary means to direct Army forces. Commanders direct the use of information and capabilities to penetrate threat decision-making processes, exploit information dependencies, achieve surprise, and disrupt the threat from within.

The decision to conceal or reveal information is a constant push and pull between advantages and disadvantages that inform risk. Throughout operations, commanders weigh the risks and benefits to revealing and concealing information. For example, revealing information about the friendly force as part of deterrence can also provide valuable intelligence to a threat force.

Commanders ensure information activities are integrated into the concept of operations through the operations process. This requires commanders to understand, visualize, and describe how they intend to use information and capabilities to create and exploit information advantages. Commanders direct information activities through orders while leading and assessing progress throughout operations. Based on changes in a situation that reveal opportunities and threats, commanders adjust information activities and related tasks as required.

Soldier Enabled

Informational considerations—those aspects of the human, information, and physical dimensions that affect how humans and automated systems derive meaning from, use, act upon, and are impacted by information—are not just for commanders, planners, and specialists. All Soldiers must protect information, help overcome their unit's disadvantages, and create and exploit information advantages. Developing and maintaining data literacy—the ability to derive meaningful information from data—is an important Soldier skill. Considerations such as OPSEC, physical security, noise and light discipline, and electromagnetic emission control apply to every individual Soldier and are critical to ensuring information advantage.

Every Soldier consumes, communicates, and relies on information to accomplish the mission. As representatives of the Army and the United States, Soldiers understand that their presence, posture, and actions always communicate a message that is open to interpretation. High visibility offers great opportunity as well as potential risk. Effective Soldiers at all levels understand the impact that their actions and messages communicate. This requires all Soldiers to understand the broader purpose of operations as communicated to them from commanders and other leaders. It also requires practicing OPSEC and disciplined communication through all forms of media—including personal media accounts—both in operations and while at home station.

VII. Warfighting Function Contributions

Information activities organize various tasks and capabilities from the six warfighting functions to help leaders visualize and describe how to create and exploit information advantages. A warfighting function is a group of tasks and systems united by a common purpose that commanders use to accomplish missions and training objectives (ADP 3-0). The warfighting functions are:

A. Command and Control *(AODS7, chap. 3.)*

The C2 warfighting function is the related tasks and a system that enable commanders to exercise authority and direction to accomplish missions. Tasks include command forces, control operations, drive the operations process, and establish the C2 system. The C2 system consists of the people, processes, networks, and command posts (CPs) that support the commander in the exercise of C2.

The C2 warfighting function significantly contributes to the enable information activity. The entire C2 system is designed to support commanders in their abilities to understand, visualize, describe, direct, lead, and assess faster and more effectively than their opponents.

B. Intelligence *(AODS7, chap. 5.)*

The intelligence warfighting function is the related tasks and systems that facilitate understanding the enemy, terrain, weather, civil considerations, and other significant aspects of the OE. The intelligence warfighting function synchronizes information collection with primary tactical tasks that support reconnaissance, surveillance, security, and intelligence operations. Intelligence is driven by commanders, and it involves analyzing information from all sources and conducting operations to develop the situation. Army forces execute intelligence, surveillance, and reconnaissance (ISR) through the operations and intelligence processes, with an emphasis on intelligence analysis and information collection.

The intelligence warfighting function contributes to the integration of all the information activities by providing relevant information and intelligence to decision makers. It directly contributes to developing situational understanding and informs decision making. The intelligence warfighting function contributes to understanding the human, information, and physical dimensions of an OE, to include identifying relevant actors, their relationships, and patterns of thinking.

C. Movement and Maneuver *(AODS7, chap. 4.)*

The movement and maneuver warfighting function is the related tasks and systems that move and employ forces to achieve a position of relative advantage in respect to the enemy. Maneuver directly gains or exploits positions of relative advantage. Commanders use maneuver for massing effects to achieve surprise, shock, and momentum. The movement and maneuver warfighting function directly contributes to the protect, influence, and attack information activities.

Movement and maneuver creates information advantages by placing units in positions that communicate an explicit or implicit threat to the enemy. Through reconnaissance and security, maneuver forces gain information on the enemy and terrain facilitating friendly decision making. They conduct security operations to protect friendly information and C2 nodes. Commanders also maneuver forces to secure areas to put network transport assets in place.

The movement and maneuver of forces has inherent informational aspects that create effects and must be accounted for during planning and execution. These include signaling intent, demonstrating capability, and driving tempo to cause confusion and disorder within the enemy system. The movement and maneuver warfighting function also contributes to information advantage through close combat that changes facts on the ground. Maneuver forces destroy enemy forces, to include their C2 systems and infrastructure, seize key terrain, and control physical areas. The moving and positioning of forces as part of deception contributes to confusing and influencing threat decision makers.

D. Fires (AODS7, chap. 6.)

The fires warfighting function is the related tasks and systems that create and converge effects in all domains against the threat to enable operations across the range of military operations. These tasks and systems create lethal and nonlethal effects delivered from Army, joint, and multinational forces. The fires warfighting function contributes to the enable, protect, influence, and attack information activities.

The fires warfighting function contributes to enabling the exercise of C2 through the targeting process. The delivery of fires contributes to protecting data and information, affecting threat C2 targets, and influencing target audiences. Through the targeting process, commanders identify, select, and prioritize targets and match the appropriate capability (or delivery platform) to targets to create desired effects. This includes identifying and attacking enemy C2 nodes, information systems, radars, ground control stations, and sensors to affect the enemy's decision cycle. Capabilities used to attack these targets range from cannons, rockets, and missiles to offensive cyberspace operations, electromagnetic attack, and offensive space operations.

E. Sustainment (AODS7, chap. 7.)

The sustainment warfighting function is the related tasks and systems that provide support and services to ensure freedom of action, extend operational reach, and prolong endurance. Sustainment contributes to all information activities by ensuring the friendly force is healthy, manned, equipped, maintained, and supplied. Sustainment activities also contribute to the influence information activity. Providing sustainment to relevant actors can reinforce or change their behavior. The position of sustainment forces and their activities can contribute to both deception and the communication of a will to fight.

F. Protection (AODS7, chap. 8.)

The protection warfighting function is the related tasks, systems, and methods that prevent or mitigate detection, threat effects, and hazards to preserve combat power and enable freedom of action. Protection encompasses the collective actions and measures required to preserve the potential of a force to be applied at the appropriate time and place. The protection warfighting function contributes to the protect information activities.

Protecting friendly data and information involves active and passive methods. Standard methods of protecting friendly information include signature management and OPSEC. Additionally, highly visible defensive measures are used to communicate messages of resolve to threats, while other less visible defensive measures are used to conceal, reduce, or eliminate friendly critical vulnerabilities. Survivability operations harden C2 facilities and information infrastructure and improve fighting positions, which protects combat power and preserves options for the commander. Physical security procedures help safeguard facilities and the information in them. Department of Defense Information Network (known as DODIN) operations, defensive cyberspace operations, and electromagnetic protection help protect the friendly network. OPSEC— a responsibility of all forces—helps to safeguard information and friendly intentions from threats, which in turn preserves options for the commander.

Refer to AODS7: The Army Operations & Doctrine SMARTbook (Multidomain Operations). Completely updated with the 2022 edition of FM 3-0, AODS7 focuses on Multidomain Operations and features rescoped chapters on generating and applying combat power: command & control (ADP 6-0), movement and maneuver (ADPs 3-90, 3-07, 3-28, 3-05), intelligence (ADP 2-0), fires (ADP 3-19), sustainment (ADP 4-0), & protection (ADP 3-37).

VIII. Information Advantages (Across Strategic Contexts)

Joint doctrine describes the strategic environment in terms of a competition continuum. Rather than a world either at peace or at war, the competition continuum describes three broad categories of strategic relationships: cooperation, competition below armed conflict, and armed conflict. Each relationship is defined as between the United States and another strategic actor relative to a specific set of policy aims. Within this competition continuum, Army forces support combatant commanders in achieving their objectives in three strategic contexts:

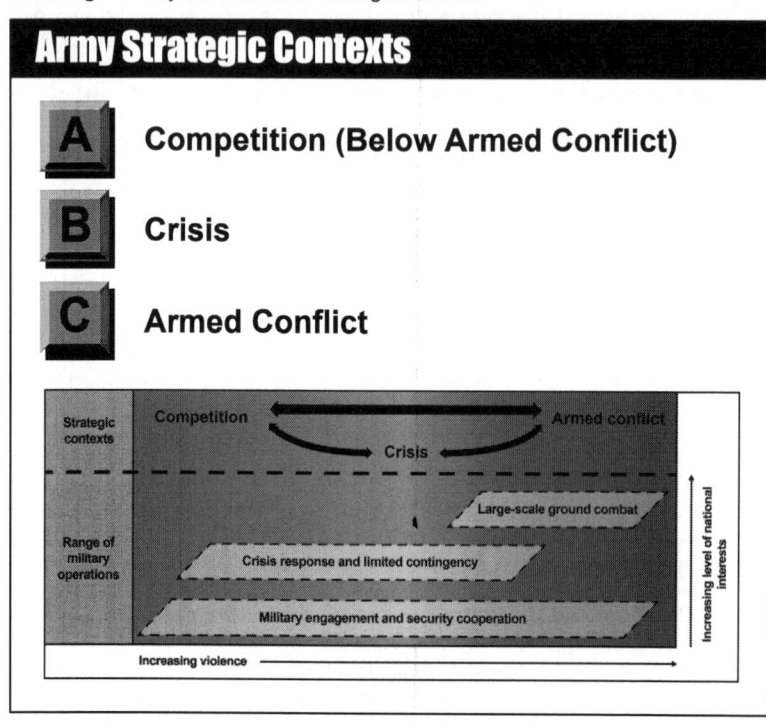

Army Strategic Contexts

A Competition (Below Armed Conflict)

B Crisis

C Armed Conflict

[Figure: Strategic contexts showing Competition, Crisis, and Armed conflict across range of military operations — Military engagement and security cooperation, Crisis response and limited contingency, Large-scale ground combat — with increasing violence and increasing level of national interests]

Ref: FM 3-0 (Oct. '22), fig. 1-3. Army strategic contexts and operational categories.

Whether in times of relative peace or periods of armed conflict, Army forces seek to create and exploit information advantages to achieve objectives.

Refer to AODS7: The Army Operations & Doctrine SMARTbook (Multidomain Operations). Completely updated with the 2022 edition of FM 3-0, AODS7 focuses on Multidomain Operations and features rescoped chapters on generating and applying combat power: command & control (ADP 6-0), movement and maneuver (ADPs 3-90, 3-07, 3-28, 3-05), intelligence (ADP 2-0), fires (ADP 3-19), sustainment (ADP 4-0), & protection (ADP 3-37).

A. Competition (Below Armed Conflict) *(AODS7, pp. 1-63 to 1-72.)*

Competition below armed conflict occurs when an adversary's national interests are incompatible with U.S. interests, and that adversary is willing to actively pursue those interests short of armed conflict. Operations during competition involve security cooperation and deterrence activities conducted under numerous programs within a combatant command. The combatant commander uses these activities to improve security within partner nations, enhance international legitimacy, gain multinational cooperation, and influence adversary decision making.

During competition below armed conflict, Army forces conduct operations and execute activities that support joint force campaigning goals, satisfy interagency requirements, and set the necessary conditions to employ Army combat power during crisis and armed conflict. Threat information warfare activities are continuous during competition. The theater army works with the joint force to thwart threat information warfare, communicate U.S. resolve, and achieve campaign plan objectives.

During competition, Army forces provide essential support to shaping foreign perceptions and behavior by—

- Using information to promote stability, cooperation, interoperability, and partnership among multinational partners as well as fostering legitimacy of U.S. and coalition efforts.
- Informing international audiences to create shared understanding, promote trust, mitigate malign information efforts, and enhance the legitimacy of U.S. and coalition operations and activities.
- Helping to develop and communicate a compelling narrative that influences foreign relevant actors to support friendly objectives or preempts the threat's messaging and malign information efforts.
- Executing MISO, participating in joint and combined exercises that demonstrate will and interoperability, maintaining readiness, and conducting security cooperation activities.

As part of competition below armed conflict, Army leaders engage and communicate with domestic audiences to maintain support at home and establish advantageous relationships with allies and partners abroad. Army forces help shape an OE by conducting security cooperation activities with partner nation armed forces and civilian agencies. These types of engagements, coordinated with applicable American embassies, help shape a credible narrative that builds trust and confidence by sharing information and coordinating mutually beneficial activities.

Shaping adversary behavior requires persistent engagement and the presence of sufficient Army forces to ensure alignment between stated objectives and subsequent actions. Physically demonstrating the scope and scale of capabilities necessary to compel desirable behavior is a critical component of influencing both adversary attitudes and behavior and assuring allies and partners. For example, a combined arms exercise with an allied nation's armed forces amplifies messages of resolve and reassurance that fosters positive perceptions and attitudes toward U.S. presence, posture, and objectives. This, in turn, builds confidence among allies and partners. Conversely, this same exercise can support conventional deterrence against an adversary.

During competition below armed conflict, Army forces protect information and remain vigilant against threat attempts to confuse situations and disrupt positive relationships among Army forces and partners. Army forces must expect threats to conduct disinformation campaigns designed to sow distrust or doubt among U.S. domestic audiences and among foreign partners. As such, Army leaders engage with and inform Army, domestic, and international audiences to put operations into context, build and maintain resiliency, and maintain the trust and confidence in the Armed Forces of the United States.

Information Advantage

B. Crisis (AODS7, pp. 1-77 to 1-86.)

A crisis is an emerging incident or situation involving a possible threat to the United States, its citizens, military forces, or vital interests that develops rapidly and creates a condition of such diplomatic, economic, or military importance that commitment of military forces and resources is contemplated to achieve national and/or strategic objectives (JP 3-0). A crisis may result from adversary actions, indicators of imminent action, or natural or human disasters. Success during a crisis is a return to a state of competition in which the United States, its allies, and its partners are in positions of increased advantage relative to the adversary. Should deterrence fail, Army forces are better positioned to defeat enemy forces during conflict.

During a crisis involving natural or human disasters, a rapid response is typically necessary. Such events frequently cause communications systems and networks to break down, disrupting the flow of information and potentially allowing a situation to worsen. Army communications capabilities can help prevent mass movements of people and unrest by disseminating information and influential messages. By providing relevant information about relief aid and the situation, Army forces help maintain calm and patience among affected populations. In addition, when working with local authorities to mitigate the effects of the event and alleviate suffering, Army forces can help bolster the legitimacy of and support for the indigenous government and its organizations.

During a crisis involving an adversary, opponents are not yet using lethal force as the primary means for achieving their objectives, but the situation potentially requires a rapid response by forces prepared to fight to deter further aggression. When directed, the Army provides a joint force commander with forces to help deter further provocation and sufficient combat power to maintain or reestablish conventional deterrence. The introduction of significant land forces demonstrates the will to impose costs, provides options to joint force and national leaders, and signals a high level of national commitment. As a crisis develops, the responsible joint force commander continues to employ capabilities from all Services to gain and maintain information advantages against the threat. Examples include—

- Extending communications capabilities to friendly forces within an operational area.
- Increasing or rapidly building technical, human, and procedural interoperability with allied and coalition partners.
- Increasing ISR across all domains.
- Executing MISO programs to complicate the decision making of threat leaders and reinforce desirable narratives with target audiences.
- Conducting public affairs and key leader engagements to promulgate information to U.S. domestic and international audiences in support of a friendly narrative, emphasizing the legitimacy of friendly goals and actions.
- Employing defensive cyberspace capabilities to increase protection of critical friendly systems.
- Employing offensive cyberspace, electromagnetic warfare, and offensive space operations to disrupt threat ISR, and communications.
- Planning and executing deception activities to mislead adversary decision makers and to set conditions for success should a crisis result in armed conflict.

OPSEC is vital to the success of operations during a crisis. Army units deploying to, or operating within, a joint operations area exercise strict OPSEC to protect friendly information and networks against cyberspace attacks.

Information Advantage (Examples)

Ref: FM 3-0, Operations (Oct. '22), chaps. 4 to 6.

Information Advantages (Competition) *(AODS7, p. 1-72.)*

Information activities play a key role during competition. They include Army support to the combatant command and unified action partner strategic messaging. Coordinating with interagency and other unified action partners helps to develop and deliver coherent messages that counter adversary disinformation. Army forces reinforce strategic messaging by maintaining and demonstrating U.S. Army readiness for operations. Examples of relative information advantages are—

- Identifying targets and conducting target development on threat capabilities.
- Setting the conditions for convergence by developing methods to penetrate adversary computer networks.
- Discrediting adversary disinformation by helping the JFC inform domestic and international audiences through Army and joint information activities.
- Promoting the purpose and outcomes of multinational exercises and training events.
- Continuously monitoring the operational environment to detect changes to adversary methods or narratives.

Information Advantages (Crisis) *(AODS7, p. 1-82.)*

Two key information activities are protecting friendly information and degrading the threat's ability to communicate, sense, make effective decisions, and maintain influence with relevant actors and populations. An example is the use of strategic messaging to undermine the credibility of an adversary by exposing violations of international law and showing that adversary narratives are false. Achieving information advantages is a commander-driven, combined arms activity that employs capabilities from every warfighting function. During crisis, commanders lead their staffs to refine information activities based upon plans and processes developed during competition. Examples include commanders and staffs focusing on the challenges and tasks of establishing a mission-partner environment, building or modifying an intelligence architecture, and creating or refining common operating procedures with allies and other partners.

Information Advantages (Armed Conflict) *(AODS7, p. 1-89.)*

Information advantages invariably overlap with and emanate from physical and human advantages. To gain an information advantage, units first require a physical or human advantage. Army forces create and exploit information advantages by acting through the physical and human dimensions of an operational environment. Leaders combine information advantages with other advantages to understand the situation, decide, and act faster than enemy forces. Examples of information advantages during armed conflict include—

- The ability to access enemy C2 to disrupt, degrade, or exploit enemy information.
- Opportunities created by deception operations to achieve surprise and thwart enemy targeting.
- The ability to mask electromagnetic signatures.
- The ability to integrate and synchronize friendly forces in denied or degraded environments through use of redundant communications.
- The ability to rapidly share information with domestic and international audiences to counter enemy malign narratives.
- The ability to inform a wide range of audiences to maintain legitimacy and promote the friendly narrative.
- The ability to rapidly share and analyze information among commanders and staffs to facilitate decisions and orders.

C. Armed Conflict (AODS7, pp. 1-87 to 1-126.)

The employment of lethal force is the defining characteristic of armed conflict, and it is the primary function of the Army. Lethality's immediate effect is in the physical dimension—reducing the enemy's capability and capacity to fight. Lethality extends into information and human dimensions where it influences enemy behavior, decision making, and will to fight.

> ### Information Advantage during Large-Scale Combat
>
> Leaders at all echelons placed major emphasis on creating and exploiting information advantages during OPERATION DESERT SHIELD and OPERATION DESERT STORM. Commanders integrated operations security, military deception, tactical deception, military information support operations (formerly called psychological operations), and electromagnetic warfare efforts to set conditions for successful large-scale ground combat
>
> During planning, senior leaders identified Iraqi command and control (C2) as a critical vulnerability, that if degraded, would significantly enhance friendly success. As such, Army forces targeted Iraq national-level C2 throughout the campaign. Simultaneously, coalition electromagnetic warfare and interdiction selectively blinded enemy reconnaissance and surveillance, protecting friendly force movements and operations from enemy detection. Deception operations continued to reinforce erroneous enemy perceptions of coalition intentions. Electromagnetic warfare and precision air strikes against operational and tactical C2 targets disorganized and isolated Iraqi forces. Simultaneously, Iraqi ground forces were targeted with messaging to reduce their morale and encourage surrender. When the ground attack commenced, Iraqi forces were scattered with numerous formations unable to coordinate their efforts. Successfully denying Iraqi forces the ability to collect information and to exercise C2 created significant advantages for coalition forces. These advantages reduced friendly casualties and significantly reduced the time required to achieve coalition objectives.

During armed conflict, Army forces continue to develop situational understanding, protect friendly information and data, and inform audiences while attacking the enemy's ability to exercise C2. Army forces use all military capabilities, including physical destruction, to gain and exploit information advantages. Army forces seek to—

- Affect (degrade, deny, disrupt, corrupt, and destroy) the ability of the enemy to exercise C2.
- Defend, counter, and mitigate the enemy's efforts to affect friendly C2 capabilities.
- Amplify the psychological effects of a destructive or disruptive force.
- Confuse, manipulate, or deceive enemy understanding and decision making.
- Communicate the intent of Army operations and activities to maintain legitimacy.

(Information Advantage) I. Enable

Ref: ADP 3-13, Information (Nov '23), chap. 3.

Information is the basis of C2, intelligence, and communication. Army forces collect, process, and analyze data and information to understand situations, make decisions, and develop plans. They communicate information to integrate, synchronize, and control operations. Information, in the form of feedback, enables Army leaders to assess progress and adjust operations as required.

The force that uses and exploits data and information to understand, make decisions, and act more effectively than its opponents has a significant advantage. The enable information activity contributes to this advantage through four related tasks: establish, operate, and maintain C2 systems; execute the operations process and coordinate across echelons; conduct the integrating processes; and enhance understanding of an operational environment (OE) as shown in Figure 3-1.

Enable

I. Establish, Operate, and Maintain Command and Control Systems

II. Conduct the Operations Process and Coordinate Across Echelons

III. Conduct the Integrating Processes

IV. Enhance Understanding of an Operational Environment

Purpose: Enhance Command and Control

Ref: ADP 3-13 (Nov '23), fig. 3-1. Tasks and purpose of the enable information activity.

I. Establish, Operate, and Maintain Command and Control Systems (AODS7, chap. 3.)

Command and control is the exercise of authority and direction by a properly designated commander over assigned and attached forces in the accomplishment of the mission (JP 1, Volume 2). Commanders cannot exercise C2 alone. Even at the lowest levels, commanders need support to command forces and control operations. At every echelon of command, each commander has a C2 system to provide that support. The command and control system is the arrangement of people, processes, networks, and command posts that enable commanders to conduct operations (ADP 6-0).

Information Advantage

A C2 system has many purposes, all of them information centric. Commanders organize their personnel, processes, networks, and command posts (CPs) to facilitate the exchange of information, inform their decision making, direct action, and control operations.

Organize People

The most important component of the C2 system is its people—commanders, seconds in command, and staffs. Coordinating, special, and personal staff sections are the building blocks for organizing a headquarters for the conduct of operations. Staff sections support commanders in making and implementing decisions. They provide relevant information and analysis, make running estimates and recommendations, prepare plans and orders, assess operations, and assist in controlling operations. Commanders consider the following when organizing the staff for operations:

- Staff sections.
- CP cells.
- Augmentation. *Typically, augmentation focused on enhancing information activities includes, such as those listed on p. 1-64.*
- Working groups and boards.

Organize Processes *(BSS7, pp. 5-5 to 5-6.)*

Commanders use various processes to enhance C2. The **operations process**—plan, prepare, execute, and assess—is the overarching process for the exercise of C2. Within the operations process, commanders and staffs conduct several **integrating processes** to enhance C2.

See facing page for an overview of the operations process and pp. 1-24 to 1-25 for an overview of the integrating processes.

The unit's **battle rhythm** synchronizes various processes, activities, and reports within the operations process.

Establish, Operate, and Maintain Networks

Headquarters at all echelons rely on networks to collect, process, store, display, and disseminate information. Without a robust, resilient, and protected network, commanders cannot effectively exercise C2. Networks provide infrastructure for voice, data, and video connectivity to support operations. Networks enable commanders to communicate with higher, lower, adjacent, supporting, and supported commands.

Organize Command Posts *(BSS7, pp. 5-7 to 5-18.)*

Effective C2 requires continuous, and often immediate, close coordination, synchronization, and information sharing across the staff. To promote this, commanders organize their staffs and other components of the C2 system into command posts (CPs) to assist them in effectively conducting operations.

CP survivability is vital to mission success. Survivability considerations make collaboration and coordination more difficult. Depending on the threat, CPs need to remain small and highly mobile— especially at lower echelons. CP survivability measures are closely related to the protect information activity because threats continuously seek to collect information about their composition and location.

Refer to AODS7: The Army Operations & Doctrine SMARTbook (Multidomain Operations). Completely updated with the 2022 edition of FM 3-0, AODS7 focuses on Multidomain Operations and features rescoped chapters on generating and applying combat power: command & control (ADP 6-0), movement and maneuver (ADPs 3-90, 3-07, 3-28, 3-05), intelligence (ADP 2-0), fires (ADP 3-19), sustainment (ADP 4-0), & protection (ADP 3-37).

II. Conduct the Operations Process and Coordinate Across Echelons

Ref: ADP 3-13, Information (Nov '23), pp. 3-5 to 3-9.

The Army's framework for exercising C2 is the operations process—the major command and control activities performed during operations: planning, preparing, executing, and continuously assessing the operation (ADP 5-0). Commanders use the operations process to drive the conceptual and detailed planning necessary to understand, visualize, and describe their OE and the operation's end state; make and articulate decisions; and direct, lead, and assess operations. An effective operations process depends upon the exchange of information and data.

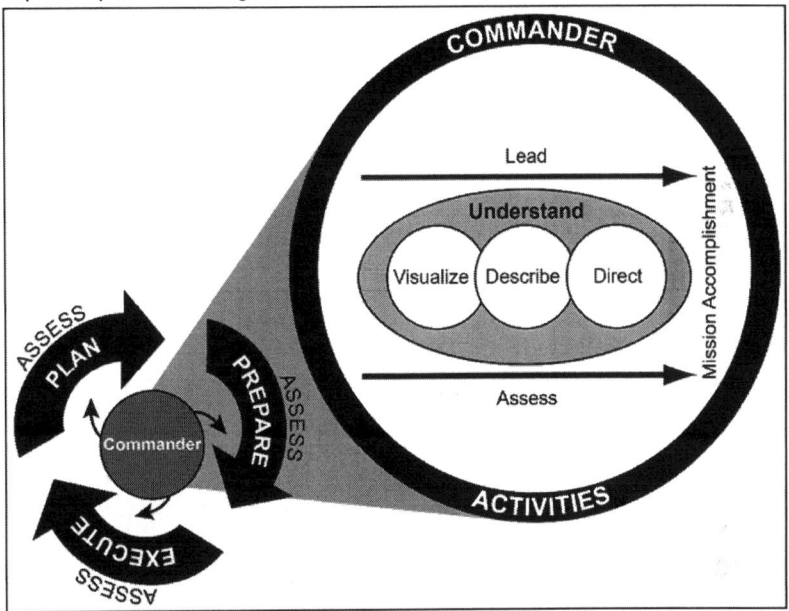

Ref: ADP 3-13, (Nov '23), fig. 3-3. The operations process.

The activities of the operations process are not discrete; they overlap and recur as circumstances demand. While planning may start an iteration of the operations process, planning does not stop with the production of an order. After the completion of the initial order, the commander and staff continuously revise the plan based on changing circumstances. Preparation for a specific mission begins early in planning and continues for some subordinate units during execution. Execution puts a plan into action and involves adjusting the plan based on changes in the situation and the assessment of progress. Assessing is continuous and influences the other three activities.

Commanders enhance the operations process and coordinate across echelons in many ways, to include—

- Prioritizing information requirements. *(BSS7, p. 1-23.)*
- Maintaining running estimates. *(BSS7, pp. 2-16 to 2-17.)*
- Creating, maintaining, and disseminating the COP.
- Establishing liaisons. *(BSS7, pp. 5-19 to 5-24.)*

III. Conduct the Integrating Processes

Ref: ADP 3-13, Information (Nov '23), pp. 3-9 to 3-11 and ADP 5-0, The Operations Process (Jul '19), pp. 1-15 to 1-17.

Within the operations process, commanders and staffs conduct several integrating processes to implement situational understanding, make decisions, and synchronize activities (to include information activities) into the concept of operations. An integrating process consists of a series of steps that incorporates multiple disciplines to achieve a specific end. Integrating processes begin in planning and continue during preparation and execution.—

Integrating Processes

- **Intelligence Preparation of the Operational Environment***
- **Information Collection**
- **Targeting**
- **Risk Management**
- **Knowledge Management**

Intelligence Preparation of the Operational Environment (IPOE)

(BSS7, pp. 3-3 to 3-52.*)

IPOE is the systematic process of analyzing the mission variables of enemy, terrain, weather, and civil considerations in an area of interest to determine their effect on operations. This includes informational considerations pertaining to the enemy, terrain, weather, and civil considerations. Continuous holistic IPOE contributes to creating and exploiting information advantages by improving understanding of an OE. A holistic approach—

- Describes the totality of relevant aspects of an OE that may impact friendly, threat, and neutral forces.
- Accounts for all relevant domains that may impact friendly and threat operations.
- Identifies windows of opportunity to leverage friendly capabilities against threat forces.
- Allows commanders to leverage positions of relative advantage at a time and place most advantageous for mission success with the most accurate information available.

Refer to ATP 2-01.3 for a detailed discussion of IPOE . *Doctrinally, this was formerly described/defined as intelligence preparation of the battlefield (IPB).*

Information Collection (BSS7, pp. 3-53 to 3-56.)

Information collection is an activity that synchronizes and integrates the planning and employment of sensors and assets as well as the processing, exploitation, and dissemination systems in direct support of current and future operations (FM 3-55). It integrates the functions of the intelligence and operations staffs that focus on answering CCIRs. Information collection includes acquiring information and providing it to processing elements. It has three steps:

- Plan requirements and assess collection
- Task and direct collection
- Execute collection

Information collection helps the commander understand and visualize the operation by identifying gaps in information and aligning reconnaissance, surveillance, security, and intelligence assets to collect information on those gaps. The "decide" and "detect" steps of targeting tie heavily to information collection.

Refer to FM 3-55 for a detailed discussion of information collection to include the relationship between the duties of intelligence and operations staffs.

Targeting *(BSS7, pp. 3-57 to 3-70.)*

Targeting is the process of selecting and prioritizing targets and matching the appropriate response to them, considering operational requirements and capabilities (JP 3-0). Targeting seeks to create specific desired effects through lethal and nonlethal actions. The emphasis of targeting is on identifying enemy resources (targets) that if destroyed or degraded will contribute to the success of the friendly mission. Targeting begins in planning and continues throughout the operations process. The steps of the Army's targeting process are—

- Decide
- Detect
- Deliver
- Assess

This methodology facilitates engagement of the right target, at the right time, with the most appropriate assets using the commander's targeting guidance. Targeting is a multi-discipline effort that requires coordinated interaction among the commander and several staff sections that together form the targeting working group. The chief of staff (executive officer) or the chief of fires (fire support officer) leads the staff through the targeting process. Based on the commander's targeting guidance and priorities, the staff determines which targets to engage and how, where, and when to engage them. The staff then assigns friendly capabilities best suited to produce the desired effect on each target.

Risk Management *(BSS7, pp. 3-71 to 3-74.)*

Risk—the exposure of someone or something valued to danger, harm, or loss—is inherent in all operations. Because risk is part of all military operations, it cannot be avoided. Identifying, mitigating, and accepting risk is a function of command and a key consideration during planning and execution.

Risk management is the process to identify, assess, and control risks and make decisions that balance risk cost with mission benefits (JP 3-0). Commanders and staffs use risk management throughout the operations process to identify and mitigate risks associated with hazards (to include ethical risk and moral hazards) that have the potential to cause friendly and civilian casualties, damage or destroy equipment, or otherwise impact mission effectiveness. Like targeting, risk management begins in planning and continues through preparation and execution. Risk management consists of the following steps:

- Identify hazards
- Assess hazards to determine risks
- Develop controls and make risk decisions
- Implement controls
- Supervise and evaluate

Knowledge Management *(BSS7, pp. 3-75 to 3-78.)*

Knowledge management is the process of enabling knowledge flow to enhance shared understanding, learning, and decision making (ADP 6-0). It facilitates the transfer of knowledge among commanders, staffs, and forces to build and maintain situational understanding. Knowledge management helps get the right information to the right person at the right time to facilitate decision making. Knowledge management uses a five-step process to create shared understanding. *Refer to ATP 6-01.1.*

Refer to BSS7: The Battle Staff SMARTbook, 7th Ed., updated for 2023 to include FM 5-0 w/C1 (2022), FM 6-0 (2022), FMs 1-02.1/.2 (2022), and more. Focusing on planning & conducting multidomain operations (FM 3-0), BSS7 covers the operations process; commander/ staff activities; the five Army planning methodologies; integrating processes (IPB, information collection, targeting, risk management, and knowledge management); plans and orders; mission command, command posts, liaison; rehearsals & after action reviews; operational terms & military symbols.

IV. Enhance Understanding of an Operational Environment

Success during operations demands timely and effective decisions based on applying judgement to available information and knowledge. As such, commanders and staffs seek to build and maintain situational understanding throughout the operations process. Understanding informational considerations of an OE bolsters this understanding. Several tasks assist commanders and staffs in understanding how information and information capabilities impact operations, to include—

- Analyze the operational and mission variables.
- Identify and describe relevant actors. *(See facing page.)*
- Identify likely behavior of relevant actors. *(See facing page.)*

A. Analyze the Operational and Mission Variables

The operational and mission variables are tools to assist commanders and staffs in developing situational understanding.

Upon receipt of a mission, commanders use the mission variables, in combination with the operational variables, to refine their understanding of the situation and to visualize, describe, and direct operations.

Operational Variables - PMESII-PT *(BSS7, p. 1-18.)*

Operational variables are categories of relevant information that commanders and staffs use to understand their OE. Commanders and staffs analyze and describe an OE in terms of eight interrelated operational variables known as PMESII-PT: political, military, economic, social, information, infrastructure, physical environment, and time.

Mission Variables - METT-TC (I) *(BSS7, p. 1-19.)*

METT-TC (I) represents the mission variables that leaders use to analyze and understand a situation in relationship to the unit's mission. The first six variables are not new. However, the increased reliance on information (military and private sector) to enable operations requires leaders to continuously assess the informational considerations on assigned missions. Because of this, the variable of informational considerations is added to the familiar METT-TC mnemonic. Within the mission variables, informational considerations are expressed as a parenthetical variable in that it is not an independent variable by itself, but it is an important consideration within each mission variable.

> **Informational considerations** are those aspects of the human, information, and physical dimensions that affect how humans and automated systems derive meaning from, use, act upon, and are impacted by information.

Refer to BSS7: The Battle Staff SMARTbook, 7th Ed., updated for 2023 to include FM 5-0 w/C1 (2022), FM 6-0 (2022), FMs 1-02.1/.2 (2022), and more. Focusing on planning & conducting multidomain operations (FM 3-0), BSS7 covers the operations process; commander/ staff activities; the five Army planning methodologies; integrating processes (IPB, information collection, targeting, risk management, and knowledge management); plans and orders; mission command, command posts, liaison; rehearsals & after action reviews; operational terms & military symbols.

B. Identify and Describe Relevant Actors

Ref: ADP 3-13, Information (Nov '23), pp. 3-12 to 3-13.

The analysis of human, information, and physical dimensions of an OE provides the context needed to understand how individuals, groups, populations, and automated systems operate.

See p. 2-64 for related discussion of audiences, targets, and target audiences.

Relevant Human Actors

Relevant human actors include individuals, groups, or populations whose behaviors have the potential to affect the success of a particular campaign, operation, or tactical action. Relevant actors may be friendly, neutral, or threat; military or civilian; and state or nonstate. Army forces use information combined with action to influence relevant actors in support of objectives.

When considering relevant human actors, staffs gain understanding by conducting two activities. First, commanders and their staffs describe the individuals, groups, and populations who can aid or hinder success of their missions. Some of these actors may exist outside the unit's area of operations. Second, the staff describes how the human, information, and physical dimensions affect each relevant actor.

Relevant Automated Systems

Automated systems are a combination of hardware and software that allow computer systems, network devices, or machines to function with limited human intervention. Automated systems with emerging artificial intelligence technologies can rapidly sort, collate, and identify trends, patterns, and vital information far faster and more efficiently than any human analysts can.

When considering relevant automated systems, staffs gain understanding by conducting the following two activities. First, commanders and their staffs remain aware that as automated systems become more sophisticated, they will have greater impacts on operations. Automated systems vary based on their degree of autonomy, intelligence, and sophistication. Additionally, their ubiquity makes it difficult to identify their presence and relevance among other actors. Conducting functional analysis as outlined in ATP 2-01.3 assists in identifying relevant automated systems within the area of operations. Second, staffs describe what effects informational and physical aspects of the environment have on each automated system.

C. Identify BEHAVIORS of Relevant Actors

Identifying current relevant actor behaviors helps planners formulate an operational approach to influence those behaviors. The staff helps the commander to develop a detailed understanding of the options available to affect relevant actor behaviors and assess which option might most strongly impact friendly operations.

- Identify current behaviors relative to impending friendly operations.
- Identify what relevant actor behaviors will likely affect operations.
- Describe how the selected behaviors of relevant actors may evolve over time.
- Describe how information and action can affect behavioral trends to yield outcomes favorable or unfavorable to friendly forces.
- Identify what broad actions friendly forces take to create effects in an OE that arrest or encourage behavioral trends.
- Identify potential second-and third-order effects of the operational approach.

Once the staff identifies relevant actors and their behaviors, the commander selects the appropriate means to affect behavior.

Considerations for Enhancing C2

Ref: ADP 3-13, Information (Nov '23), pp. 3-13 to 3-15.

Army doctrine provides numerous considerations to enhance C2 depending on the topic. For example, ADP 6-0 provides considerations for effective command and considerations for effective control. ADP 5-0 provides the principles of the operations process and considerations for effectively planning, preparing, executing, and assessing operations. FM 6-02 provides fundamental principles of signal support. In addition to the considerations above, Army commanders and staffs weigh several methods to enhance C2, to include—

Apply Principles of Mission Command *(BSS7, pp. 5-2 to 5-3.)*

Applying the principles of mission command significantly enhances friendly C2. Mission command is the Army's approach to command and control that empowers subordinate decision making and decentralized execution appropriate to the situation (ADP 6-0). Mission command is necessary because operations are inherently chaotic and uncertain. No plan can account for every possibility, and most plans must change rapidly during execution to account for changes in the situation. No single person is ever sufficiently informed to make every important decision, nor can a single person keep up with the number of decisions that need to be made during combat. Subordinate leaders often have a better understanding of what is happening during a battle than higher leaders do. Thus, subordinate leaders are more likely to respond effectively to threats and fleeting opportunities if allowed to make decisions and act.

Mission command facilitates subordinate ingenuity, innovation, and decision making to achieve the commander's intent when conditions change or current orders are no longer relevant. Subordinate decision making and decentralized execution appropriate to the situation help manage uncertainty and enable necessary tempo at each echelon during operations. Successful mission command is enabled by the principles of—

- Competence.
- Mutual trust.
- Shared understanding.
- Commander's intent.
- Mission orders.
- Disciplined initiative.
- Risk acceptance.

Ensuring Digital Readiness

Digital readiness is the ability of an individual and organization to maintain and operate information systems. Information systems help leaders process, store, and disseminate information concerning a warfighting function or functional area. Within a CP, there are multiple types of information systems that support C2, intelligence, fires, maneuver, air space control, air and missile defense, personnel management, transportation, and sustainment.

Developing and Maintaining Digital Literacy

Army and joint forces rely on the collection, processing, and analysis of data to understand situations, to make human and automated decisions, and for the exercise of C2. Soldiers and leaders must be equally informed about the potential limitations, threats, and risk associated with data. This requires improvements in data literacy—the ability to derive meaningful information from data so that it can be applied effectively to actions and outcomes. Data literacy includes the skills, knowledge, and attributes to read, manipulate, analyze, and communicate with data to effectively enable commanders to make accurate and timely decisions.

(Information Advantage)
II. Protect

Ref: ADP 3-13, Information (Nov '23), chap. 4.

All Army forces continuously provide protection. Protection is the preservation of the effectiveness and survivability of mission-related military and nonmilitary personnel, equipment, facilities, information, and infrastructure deployed or located within or outside the boundaries of a given operational area (JP 3-0). Achieving protection is a continuous endeavor, requiring Army leaders to apply a comprehensive, layered, and redundant approach in different contexts.

Protect

- **I. Secure and Obscure Friendly Information**
- **II. Conduct Security Activities**
- **III. Defend the Network, Data, and Systems**

Purpose: Secure Data, Information & Networks

Ref: ADP 3-13 (Nov '23), fig. 4-1. Tasks and purpose of the protect information activity.

I. Secure and Obscure Friendly Information

Securing information about Army forces is a responsibility of all Soldiers, Army Civilians, and contractors. For commanders and leaders, it means two things. First, leaders educate Soldiers, Army Civilians, contractors, and family members on the type and nature of data and information that threat forces seek. Second, leaders inculcate into their unit culture the imperative of securing friendly data and information.

> **Imperative of operations**: Account for being under constant observation and all forms of enemy contact. *(AODS7, p. 1-44.)*

Securing and obscuring friendly information begins with an understanding of what data exist relating to friendly forces. Army leaders must understand their own data and information signature from a threat's perspective. They must assume that threats constantly observe their formations from different domains and the electromagnetic spectrum.. Tasks that secure and obscure friendly information include—

- Implement operations security (OPSEC). *(See pp. 3-39 to 3-40.)*
- Conduct deception in support of OPSEC (DISO). *(See p. 3-28.)*
- Employ camouflage, concealment, and obscuration.

Information Advantage

II. Conduct Security Activities

Ref: ADP 3-13, Information (Nov '23), pp. 4-5 to 4-6.

All Soldiers share a responsibility for securing important information about Army forces. Some Army forces are manned, trained, and equipped to engage in specialized security activities, specifically conducted to deny threat forces relevant information. Security activities related to securing information include—

Security Activities

A. Conduct Security Operations
B. Implement Physical Security
C. Implement Personnel Security Program
D. Conduct Counterintelligence

A. Conduct Security Operations (AODS7, pp. 4-21 to 4-24.)

Commanders prevent threats from collecting information about friendly force activities in part by performing security operations. Security operations are those operations performed by commanders to provide early and accurate warning of enemy operations, to provide the forces being protected with time and maneuver space within which to react to the enemy, and to develop the situation to allow commanders to effectively use their protected forces (ADP 3-90). Security operations focus on the protected force or location. By denying threat actors a vantage point from which to observe friendly activities and dispositions, forces conducting security operations can protect friendly information against threat reconnaissance efforts. Army forces conduct four types of security operations:

- Screen.
- Guard.
- Cover.
- Area security.

Counterreconnaissance is a tactical mission task that encompasses all measures taken by a commander to counter enemy reconnaissance and surveillance efforts. It prevents hostile observation of a force or area and accounts for all the domains through which the threat can conduct reconnaissance in a particular situation. It involves both active and passive elements and includes combat action to destroy or repel enemy reconnaissance units and surveillance assets. Counterreconnaissance is not a distinct mission but an essential component to security operations.

Threat unmanned aircraft systems carry a variety of surveillance and reconnaissance capabilities, ranging from high-resolution video to infrared or electromagnetic reconnaissance. Unmanned aircraft systems carry a range of capabilities, to include surveillance, reconnaissance, targeting, electromagnetic attack, and air-to-surface weapons.

B. Implement Physical Security

Physical security consists of physical measures designed to safeguard personnel; prevent unauthorized access to equipment, installations, material, and documents; and safeguard them against espionage, sabotage, damage, theft, and terrorism. Army forces employ physical security measures in depth to protect personnel, information, and critical resources in all locations and situations against various threats. This total system approach is based on the continuing analysis and employment of protective measures, including—

- Physical barriers.

- Clear zones.
- Lighting.
- Access and key control.
- Intrusion detection devices.
- Biometrically enabled base access systems.
- Defensive positions.
- Nonlethal capabilities.

Refer to ATP 3-39.32 for additional information on physical security.

C. Implement Personnel Security Program

Personnel security plays an important role in protecting friendly information. Units should ensure that personnel in sensitive positions have the appropriate clearance, a need to know, and required certifications before granting access to critical network infrastructure. The clearance and sensitive position standard determines whether a person is eligible for access to classified information or assignment to sensitive duties. This standard evaluates if the person's loyalty, reliability, and trustworthiness for having access to classified information or assignment to sensitive duties is clearly consistent with the interests of national security.

Refer to AR 380-67 for more information about the Army personnel security program.

D. Conduct Counterintelligence (CI)

Counterintelligence is information gathered and activities conducted to identify, deceive, exploit, disrupt, or protect against espionage, other intelligence activities, sabotage, or assassinations conducted for or on behalf of foreign powers, organizations or persons or their agents, or international terrorist organizations or activities (JP 2-0). Counterintelligence (CI) is one of the Army's intelligence disciplines conducted by specially trained CI agents, technicians, and special agents. These specially trained personnel focus on detecting and identifying the FIE's intelligence collection activities targeting U.S. and multinational forces.

CI operations are broadly executed CI activities using one or more of the CI functions (investigations, collection, analysis and production, and technical services and support) that support a program or specific mission. The CI mission includes defensive and offensive activities conducted worldwide to protect Army forces, installations, and operations from the foreign intelligence collection threat. The CI mission encompasses four different mission areas:

- Counterespionage.
- CI support to force protection.
- CI support to research, development, and acquisition.
- CI-cyber.

CI relies on the Threat Awareness and Reporting Program (known as TARP) to identify systemic or personnel issues and to identify other inconsistencies that may indicate a vulnerability or incident of CI interest. The Threat Awareness and Reporting Program is an education, awareness, and reporting program to help identify incidents of potential CI interest. The program is a primary factor in obtaining information to initiate CI investigations in response to suspected national security crimes under Army CI jurisdiction. The Threat Awareness and Reporting Program education activities should be tailored to the supported unit based on the unit mission, unique foreign intelligence entities characteristics, and methods of operation.

Refer to ATP 2-22.2-1 for more information on Army CI.

Protect Considerations (AODS7, chap. 8.)
Ref: ADP 3-13, Information (Nov '23), pp. 4-10 to 4-13.)

Gaining and maintaining information advantage requires Army leaders to protect against threat attempts to access or affect friendly data, information, and communications. To do this, commanders and staffs incorporate protect information tasks into the operations process using a combined arms approach. To formulate plans for protecting friendly data, information, and networks, Army leaders must understand friendly vulnerabilities and be aware of what information the friendly force may be revealing to the threat. This understanding guides threat analysis and aids development of mitigation measures. Several fundamentals guide commanders and staffs in the planning and execution of tasks within the protect information activity:

See Yourself Physically and Virtually (AODS7, pp. 3-16 to 3-17.)

Protecting data, information, and networks begins with an awareness of the physical and virtual signatures (information footprint) the friendly force presents to the threat. Army leaders must understand how their headquarters and formations appear to a threat from each domain and from the electromagnetic spectrum. This understanding, combined with an understanding of threat information collection and attack capabilities, provides the basis for directing protect activities to mitigate friendly vulnerabilities.

> Observable signatures increase an enemy's likelihood of successfully detecting, collecting information about, and targeting Army formations and critical command and control nodes.

Friendly troop activities, gatherings of key leaders, patterns of life, social media posts, and physical signatures all help the threat paint a picture of the disposition, composition, and intent of the friendly force. For example, Soldier behaviors in garrison, out in the public, and online can indicate an upcoming deployment and provide insight into unit readiness. Troop movements and communications during crisis can highlight when forces deploy, where they deploy, what type of forces deploy, and for what purpose. During armed conflict, friendly forces emanate electromagnetic signatures that when aggregated can identify the size and composition of forces in a certain location.

- Electromagnetic signature.
- Personal electronic device.
- Visual signature.
- Radar signature.
- Infrared (heat) signature.
- Noise signatures.
- Social media signatures.

See the Threat and Account for Being Under Constant Observation (AODS7, pp. 4-2 to 4-3.)

In addition to understanding the signatures that friendly forces present, Army leaders must understand the threat's ability to collect on those signatures. The nine forms of contact provide a framework to account for threat observation.

> There are **nine forms of contact**: direct; indirect; non-hostile; obstacle; chemical, biological, radiological, and nuclear; aerial; visual; electromagnetic; and influence. *(AODS7, p. 4-3.)*

Air, space, and cyberspace capabilities increase the likelihood that threat forces can gain and maintain continuous visual, electromagnetic, and virtual contact with Army forces during competition, crisis, and armed conflict.

Peer threats possess a wide range of land-, maritime-, air-, and space-based intelligence, surveillance, and reconnaissance (ISR) capabilities that can detect and collect information on U.S. forces. During competition and crisis, **threat forces** employ multiple methods of collecting on friendly forces to develop an understanding of U.S. capabilities, readiness status, and intentions. They do this in and outside the continental United States. They employ space-based surveillance platforms to observe unit training and deployment activities. Threats can use this information to target Army forces during conflict. The proliferation of personnel using network-enabled electronic devices exacerbates this risk. Soldiers and their families must consider how their use of telecommunications, the internet, and social media makes them or their units vulnerable to adversary surveillance.

Leaders assume they are always under constant observation from one or more domains in all contexts, from home station through deployment, and while conducting operations. Leaders continuously assess and account for various information collection methods the threat uses to collect and exploit information that friendly forces generate. The intelligence process, intelligence preparation of the operational environment (IPOE), information collection, and the OPSEC process all facilitate understanding threat collection capabilities and methods.

Employ Combinations of Active and Passive Protection Measures

The threat seeks to gather, deny use of, and corrupt friendly data, information, and communications using many capabilities. Army leaders take a combined arms approach to counter threat information collection and attack methods by directing combinations of active and passive information protection measures.

When Army forces execute **active protection measures**, they act directly against threat information collection and information warfare capabilities to either deny or limit their use against Army forces. Active protection includes responding to ongoing attacks against Army data, information, and networks to limit the damage from threat information warfare activities. Army forces actively protect information while conducting operations in all the strategic contexts. Examples of active information protection include counterreconnaissance and cyberspace defense operations.

When Army forces execute **passive information protection measures,** they act to indirectly reduce the effectiveness of threat information collection and information warfare capabilities. While not all Army forces have the capabilities to take active information protection measures, all Army forces employ passive measures to disrupt the threat's ability to collect relevant information. Passive protection requires Soldiers actively considering what information they receive and what indicators their actions portray to the threat and other relevant actors. Examples of passive information protection include employing camouflage, noise, and light discipline; hardening critical infrastructure; and reducing electromagnetic emissions. Dispersion of CPs and having redundant capabilities are other examples.

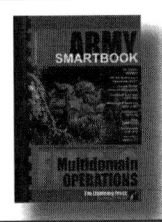

Refer to AODS7: The Army Operations & Doctrine SMARTbook (Multidomain Operations). Completely updated with the 2022 edition of FM 3-0, AODS7 focuses on Multidomain Operations and features rescoped chapters on generating and applying combat power: command & control (ADP 6-0), movement and maneuver (ADPs 3-90, 3-07, 3-28, 3-05), intelligence (ADP 2-0), fires (ADP 3-19), sustainment (ADP 4-0), & protection (ADP 3-37).

III. Defend the Network, Data, and Systems

Threat cyberspace and [electromagnetic warfare] capabilities jeopardize U.S. freedom of action in cyberspace and the electromagnetic spectrum. Because communications are a key command and control enabler, U.S. military communications and information networks present high-value targets.

Army and joint forces secure and defend the network to preserve the C2 system, protect friendly communications and network capabilities, and defeat threat cyberspace and electromagnetic reconnaissance. Various threats constantly attack Army networks. During competition, threats seek vulnerabilities and attempt to penetrate Army networks and systems to gather information and set conditions for future attacks. During armed conflict, threats attempt to destroy Army networks and corrupt and manipulate friendly information and data. Army forces conduct tasks to defend the network, data, and systems, including—

A. Conduct Cyberspace Security *(See p. 3-53.)*

Cyberspace security is actions taken within protected cyberspace to prevent unauthorized access to, exploitation of, or damage to computers and networks, including platform information technology (JP 3-12). The term "protected cyberspace" includes Department of Defense (DOD)-owned and -leased communications and computing software, data, security services, associated services, national security systems, and other relevant systems and services.

A threat's pathway into an Army network is often via a compromised individual-user device. All Soldiers conduct cybersecurity by knowing the techniques that threats use to penetrate Army networks. These techniques include unusual emails, odd attachments, compromised links, or other methods.

B. Conduct Defensive Cyberspace Operations *(p. 3-50.)*

Defensive cyberspace operations are missions to preserve the ability to utilize and protect blue cyberspace capabilities and data by defeating on-going or imminent malicious cyberspace activity (JP 3-12). Defensive cyberspace operations and cyberspace security share a common objective of a secure network.

Note. Joint doctrine list three categories of cyberspace: blue, red, and gray. The term "blue cyberspace" denotes U.S. cyberspace (areas in cyberspace owned or controlled by the United States Government or a U.S. person) and other areas of cyberspace the DOD is ordered to protect. The term "red cyberspace" refers to those portions of cyberspace owned or controlled by, or on behalf of, an adversary or enemy. The Army refers to blue, red, and gray cyberspace as friendly, threat, and neutral cyberspace respectively.

C. Conduct Communications Security

Communications security is actions designed to deny unauthorized persons information of value by safeguarding access to, or observation of, equipment, material, and documents with regard to the possession and study of telecommunications or to purposely mislead unauthorized persons in their interpretation of the results of such possession and study (JP 6-0).

D. Conduct Electromagnetic Protection *(See pp. 3-58 to 3-59.)*

Electromagnetic protection is a division of electromagnetic warfare involving actions taken to protect personnel, facilities, and equipment from any effects of friendly or enemy use of the electromagnetic spectrum that degrade, neutralize, or destroy friendly combat capability (JP 3-85). Electromagnetic protection measures eliminate or mitigate the negative impact resulting from friendly, neutral, enemy, or naturally occurring electromagnetic interference. Army forces execute a variety of electromagnetic protection tasks depending on threat capabilities and operational environments (OEs).

(Information Advantage)
III. Inform

Ref: ADP 3-13, Information (Nov '23), chap. 5.

See pp. 3-5 to 3-16 for related discussion of public affairs operations (JP/FM 3-61) and p. 2-19 for discussion of "inform" from a joint doctrine perspective (JP 3-04).

The U.S. Army has an obligation to inform. Army leaders keep internal audiences (Soldiers, Army Civilians, contractors, and family members) informed about organizational goals, priorities, values, and expectations. Army leaders keep external audiences (U.S. domestic and international) informed to maintain their trust and confidence. Within a larger national and joint narrative, Army leaders inform various international audiences to facilitate informed perceptions about military objectives and activities. Combined with demonstrated competence and professionalism, informing international audiences strengthens partnerships and alliances during competition, crisis, and armed conflict.

Army forces communicate accurate and timely information to internal and external audiences to gain an information advantage. The inform information activity contributes to this advantage through its related tasks: inform and educate Army audiences; inform U.S. domestic audiences; and inform international audiences as shown in Figure 5-1.

Inform

- **I** — Inform and Educate Army Audiences
- **II** — Inform U.S. Domestic Audiences
- **III** — Inform International Audiences

Purpose: Maintain Trust & Confidence

Ref: ADP 3-13 (Nov '23), fig. 5-1. Tasks and purpose of the inform information activity.

Note. The inform information activity relies on several public affairs terms. Public affairs are communication activities with external and internal audiences (JP 3-61). In public affairs, an audience is a broadly-defined group that contains stakeholders and/or publics relevant to military operations (JP 3-61). An audience can be internal or external. In public affairs, an internal audience is United States military members and Department of Defense civilian employees and their immediate families (JP 3-61). In public affairs, an external audience is all people who are not United States military members, Department of Defense civilian employees, and their immediate families (JP 3-61). External audiences are categorized as U.S. domestic and international audiences. In public affairs, a public is a segment of the population with common attributes to which a military force can tailor its communication (JP 3-61).

INFORM Considerations (See also pp. 3-13 to 3-13.)

Ref: ADP 3-13, Information (Nov '23), pp. 5-9 to 5-11.

Several fundamentals guide commanders in planning for and executing inform information activity tasks, to include—

Inform Considerations

- **Tell the Truth**
- **Timely Release of Information and Operations Security**
- **Compliance with Law and Policy**

Tell the Truth

The long-term success of a commander's communication strategy depends on maintaining the integrity and credibility of officially released information. Deceiving the public undermines trust in the Army. The accurate, balanced, and credible presentation of information leads to public confidence in the Army and the legitimacy of Army operations. Attempting to deny unfavorable information or failing to acknowledge its existence leads to media speculation, the perception of a cover-up, and the loss of public trust. Commanders and leaders, along with their public affairs officers, address issues openly and honestly as soon as possible.

Timely Release of Information and Operations Security

> *The First Amendment guarantees freedom of the press, but within the Department of Defense this right must be balanced against the military mission that requires operations security at all levels of command to protect the lives of US or multinational forces and the security of ongoing or future operations.*
> *- JP 3-61*

Other sources will fill an information vacuum created by Army leaders not communicating effectively. Some sources may provide accurate information, other sources will provide misinformation, and threats will spread disinformation—false information intentionally designed to undermine Army credibility and operations. Timeliness matters, as perceptions and opinions begin to develop with the first information received, whether that information is accurate or not.

Informing relevant audiences by releasing information can have a powerful effect on friendly and threat perceptions and decision-making cycles. Army forces that release more timely and accurate relevant information to relevant audiences achieve an advantage by forcing opponents to react rather than act. The following are considerations for releasing timely and accurate information:

- Early release of information sets the pace and tone for solving a problem or creating a problem for the opponent.
- Early release of information presents facts accurately from the beginning rather than attempting to correct the record later.
- Uncontrolled release (or leaking) of information jeopardizes trust and credibility.
- Information released as early as possible from the most accurate source increases effectiveness.
- Information released prompts meaningful dialogue and public involvement.
- Information released corrects misinformation and counters disinformation.

- Information released builds public trust and confidence in the command.
- Information released prevents perceptions of scandal or cover-up.

Commanders and staffs must continuously assess the timing and content of changes to public affairs guidance, press releases, social media posts, and other activities to ensure that they do not provide OPSEC indicators that a threat could exploit. When OPSEC concerns preclude the complete release of information, commanders determine methods to control the timing and tempo of messaging to balance risk to the force with friendly credibility with designated relevant actors and audiences. Commanders release timely, factual, coordinated, and approved information and imagery as part of the inform information activity. Commanders release public information consistent with security restraints in DODI 5200.01 and the principles of information outlined in DODD 5122.05 listed in Figure 5-3.

Department of Defense Principles of Information

It is the policy of the Department of Defense [DoD] to make available timely and accurate information so that the public, Congress, and the news media may assess and understand the facts about national security and defense strategy. Requests for information from organizations and private citizens will be answered in a timely manner. In carrying out the policy, the following principles of information will apply:

- Information will be made fully and readily available, consistent with the statutory requirements, unless its release is precluded by current and valid security classification. The provisions of Section 552 of Title 5, U.S.C., also known as the "Freedom of Information Act," will be supported in both letter and spirit.
- A free flow of general and military information will be made available, without censorship or propaganda, to the Service members and their dependents.
- Information will not be classified or otherwise withheld to protect the U.S. Government from criticism or embarrassment.
- Information will be withheld only when disclosure would adversely affect national security, threaten the safety or privacy of Service members, or if otherwise authorized by statute or regulation.
- The DoD's obligation to provide the public with information on its major programs may require detailed public affairs planning and coordination within the DoD and with other government agencies. The sole purpose of such activity is to expedite the flow of information to the public; propaganda has no place in DoD public affairs programs.

Ref: ADP 3-13 (Nov '23), fig. 5-3. Department of Defense principles of information.

Compliance with Law and Policy

Leaders keep the inform information activity congruent and synchronized with the other four information activities. All information dissemination, regardless of the communicator or medium, is intended to either inform or influence. The intent of the communication guides the commander's decision to either inform or influence a particular audience. Lack of synchronized inform activities can damage the credibility of Army forces' actions and undermine OPSEC.

To help coordinate and de-conflict inform tasks and influence tasks, commanders and staffs maintain awareness of U.S. laws and statutes guiding the use of information throughout the competition continuum, whether in garrison or a forward position. The tasks associated with the inform activity do not try to force a particular point of view on audiences. Soldiers provide facts so various audiences can increase their understanding and then make their own decisions.

I. Inform and Educate Army Audiences

Commanders establish and maintain a positive command climate—the characteristic atmosphere in which people work and live. Command climate is directly attributable to the leader's values, skills, and actions. A positive climate facilitates team building, encourages initiative, and fosters collaboration, mutual trust, and shared understanding. Commanders shape the climate of their organization no matter the size. Maintaining a positive command climate includes—

A. Inform Internal Audiences

Keeping internal audiences informed plays a crucial role in sustaining the morale and will of Army forces. Commanders and leaders keep internal audiences informed on organizational goals, priorities, values, and expectations, while encouraging feedback. They inform through various means ranging from conducting mission briefings to hosting town hall meetings.

An effective command information program combined with community engagement aids commanders in informing internal audiences. Command information is communication by a military organization directed to the internal audience that creates an awareness of the organization's goals, informs them of significant developments affecting them and the organization, increases their effectiveness as ambassadors of the organization, and keeps them informed about what is going on in the organization (JP 3-61).

Refer to DODI 5400.17 for policy concerning official use of social media.

Operations, particularly those involving armed conflict, are fraught with danger and hardship. Violence, fatigue, and fear characterize large-scale combat operations. To help build unit cohesion and maintain morale and will, commanders and leaders communicate to all Soldiers how their units' efforts and purpose fit into the overall purpose of the operations. Communicating "why we fight" helps Soldiers understand they are part of a larger team and effort. Shared understanding of purpose helps Soldiers reconcile their sacrifices toward a greater effort.

B. Educate Soldiers

Threats do not hesitate to employ a variety of influence techniques to weaken the resolve of Americans, especially members of the Department of Defense (DOD). Social media, internet-based communication, on¬line gaming, and other dynamic forms of communication allow threats to extend their reach in the human and information dimensions beyond what was previously possible. The potential for Soldiers to have contact with threat influence activities is high, even when in garrison. The chances of threat influence activities increase as Army forces initiate operations during crisis and conflict.

Soldiers must remain vigilant to recognize threat attempts to undermine their morale and will. The entire Army force is potentially subject to monitoring and threat influence activities through various mediums, to include the internet. Leaders train and educate Soldiers to maintain online awareness, to include identifying threats and applying operational security when posting information and images online. To provide Soldiers the ability to recognize and mitigate threat influence activities, as well as to withstand enemy influence attempts, leaders educate the force concerning—

- **Threat influence methods.** Threat influence methods can affect Soldiers directly or indirectly with the goal of influencing their thinking. The threat may employ direct influence activities, for example overt propaganda, or more subtle attacks like engaging in activities that amplify social or political differences.
- **The Army profession.** Understanding the role of the Army as articulated in ADP 1 can help defend against the effects of misinformation, disinformation, and information for effect. When Soldiers understand why the Army exists and their role in protecting their nation's interests, it becomes increasingly difficult for the threat to undermine their commitment to their duty.

Commander's Communication Synchronization (CCS)

Ref: ADP 3-13, Information (Nov '23), p. 5-1.

> The proactive release of accurate information to Army, U.S. domestic, and international audiences puts operations in context, facilitates informed perceptions about military operations, undermines threat propaganda, and helps achieve an information advantage.

For inform information activity tasks to be effective, Army words, images, and deeds must match. Commander's communication synchronization (CCS) is the process that helps do this. Commander's communication synchronization is a process to coordinate and synchronize narratives, themes, messages, images, operations, and actions to ensure their integrity and consistency to the lowest tactical level across all relevant communication activities (JP 3-61). Within this context, communication is the imparting or interchange of information, thoughts, and opinions by sending themes, messages, and facts through engagements and traditional and digital media platforms to designated audiences. CCS helps develop the commander's communication strategy that guides the execution of tasks that inform Army, U.S. domestic, and international audiences.

CCS is a top-down process starting at the U.S. government level and nesting down to Army tactical forces. At the national level, CCS focuses efforts for leaders to understand and communicate with key audiences to create, strengthen, or preserve favorable conditions to advance U.S. interests, policies, and objectives through coordinated programs, plans, themes, messages, and products synchronized with the actions of all instruments of national power.

Joint force commanders support the national security narrative by developing themes appropriate to their mission and authority. A theme is a distinct, unifying idea that supports a narrative. Sometimes commanders develop multiple themes to support a narrative. Operational-level themes are often created for each phase of an operation. Commanders continually nest operational themes with strategic themes and enduring national narratives to mitigate risks that phase-by-phase themes appear to give conflicting messages.

Messages support themes by delivering tailored information to a specific public audience. Commanders can also tailor messages for a specific time, place, and communications method. While messages are dynamic, they support enduring themes of higher headquarters and subordinate organizations. The dynamic nature inherent in messages provides joint force commanders and planners with agility in reaching various audiences.

Army commanders nest their communication strategy with the joint force. As the principal advisor for public information, command information, crisis communications, visual information, and community engagement, the public affairs officer manages the CCS process for Army commanders. Based on the commander's intent and planning guidance, the public affairs officer coordinates with other members of the staff to implement higher headquarters communication guidance to coordinate themes, messages, talking points, and images with Army operations. Part of this coordination includes determining the implications that a public message may have to non-primary audiences and recommended potential follow-on messaging. This continuous process requires deliberate planning in competition, crisis, and conflict.

See FM 3-61 for more information on public affairs and CCS. (See pp. 3-5 to 3-16.)

II. Inform United States Domestic Audiences

Federal laws and military instructions such as DODD 5122.05 and AR 360-1 require Army forces to inform domestic audiences of their operations, programs, and activities. The Department of the Army and Army commanders are responsible for informing the American people about the Army's mission and goals.

> Accurately informing the American people assists the Army in establishing conditions that lead to the public's understanding, trust, confidence, and support.

Army senior leaders ensure the operations and activities conducted by Army forces are aligned with the national security interests and values of the American people as articulated by various strategic documents. These documents include the National Security Strategy and the National Military Strategy. By informing the U.S. domestic audience, Army forces reassure the American public that they execute operations in accordance with national values. Commanders of Army formations inform domestic audiences primarily through public communication, which includes community engagement within the broader Army communication strategy.

Inform U.S. Domestic Audiences

- **A. Conduct Public Communication**
- **B. Conduct Community Engagement**
- **C. Correct Misinformation and Counter Disinformation Related to Army Forces or Operations**

Refer to FM 3-61 for more information on public communication and community engagement. (See pp. 3-5 to 3-16.)

A. Conduct Public Communication

Informing U.S. domestic audiences helps these audiences understand that Army operations align with American interests. This communication increases public trust and support through active engagements. Through public communication programs, commanders demonstrate they are community partners and responsible stewards of national resources.

Public communication includes the release of official information through news releases that encompass public service announcements, media engagements, town hall meetings, public engagements, and social networks. Public communication enables commanders to meet their obligations to keep the American people informed. Public communication objectives include the following:

- Increase public awareness of the Army's mission, policies, and programs.
- Foster good relations within the communities with which Army forces interact.
- Maintain the Army's reputation as a respected professional organization responsible for national security.
- Support the Army's recruiting and personnel procurement mission.
- Correct misinformation and counter disinformation.

B. Conduct Community Engagement

Community engagements are activities that support the relationship between military and civilian communities. Advised by public affairs personnel, commanders provide direction and purpose for engagement with civilian communities. These activities involve collaborating with groups of people affiliated by geographic proximity or special interests to enhance the understanding and support for the Army, Soldiers, and operations. Community engagement places special emphasis on two-way communications with public communities surrounding military installations. Effective relationships with key stakeholders must be enduring; trust cannot be built after a crisis occurs.

The Army relies on communities and regions surrounding its installations for direct and indirect support. Communities provide the Army access to resources needed to train and maintain readiness as well as extend support to families of mobilized or deployed Soldiers. Commanders recognize that a positive rapport between the Army and its host communities is mutually beneficial, supporting the Army as an institution as well as its individual Soldiers.

Members of the Army National Guard and United States Army Reserve live and work in the community and are integral members of their hometowns. A community's positive relationship with a local installation depends upon the command. A commander considers potential implications of every installation activity, operation, or major training activity. This is especially important during crisis management, mobilization, deployment, and redeployment operations, even if the installation or reserve unit is not directly involved. A commander also considers potential implications during national events concerning politically sensitive or controversial Department of the Army or DOD issues. Installation and reserve unit commanders and their staffs—assisted by their public affairs elements—need to implement effective programs that include the open, honest, accurate, complete, and timely released information their communities expect.

C. Correct Misinformation and Counter Disinformation Related to Army Forces or Operations

Threats use many methods in their attempts to attrit the U.S. domestic audience's trust and confidence in Army forces. Digital media platforms make it easy for threats to exploit information for effect, misinformation, and disinformation, but threats can use less technologically sophisticated methods as well. Intelligence, public affairs, and other staff members assist commanders in identifying misinformation and disinformation about the Army. Public affairs and leaders then correct the record as appropriate through official channels. All Army forces have a duty to be alert for information that erodes U.S. citizens' trust in Army forces and the Army as an institution.

> All Soldiers and leaders are authorized to correct misinformation about which they have personnel knowledge if the information is unclassified and releasable.

When countering misinformation and disinformation, Army forces preemptively and proactively tell the Army's story by providing public information and conducting community engagements. Critical to preempting misinformation or disinformation is a quick and active response by Army forces as soon as something negative occurs that might impact Army operations or perceptions about the Army. Public affairs specialists at all echelons assist the chain of command to get facts and context to the public audience as soon as possible. Soldiers refer potential information for effect and disinformation concerning policy, classified matters, or anything outside their scope of knowledge to their public affairs officer or chain of command.

Whether informing U.S. domestic or international audiences, effectively correcting misinformation and countering disinformation requires unity of effort. This includes coordination among the DOD, the military departments, Department of State, multinational partners, combatant commands, and Army forces. Annex J (Public Affairs) to operation plans and orders at all echelons provides direction and guidance that facilitate this unity of effort. Public affairs guidance, to include clear information and imagery release authority by echelon, enables Army commanders to rapidly correct misinformation and to counter disinformation.

Refer to FM 3-61 for more information on correcting misinformation and countering disinformation. (See pp. 3-5 to 3-16.)

III. Inform International Audiences

The presence of Army forces assures allies and partners while it deters threats. The presence of friendly forces reduces threats' perception of the benefit of aggression relative to restraint. The Army, as part of the joint force and whole of government, informs and assures allied and partner audiences about its activities and operations globally and within specific regions. The presence of Army forces provides the proof of a U.S. commitment that should be articulated by simple, clear, and synchronized messages conveyed from strategic to tactical levels.

Inform International Audiences

- **A. Conduct Community Engagement Outside the Continental United States**
- **B. Conduct Soldier & Leader Engagement**
- **C. Conduct Civil Affairs Operations**
- **D. Correct Misinformation and Counter Disinformation Related to Army Forces or Operations**

Overseas, the chief of mission, usually an ambassador, is the highest U.S. authority in a foreign country and has a U.S. government communication strategy within which all military engagements are nested. Army commanders conducting missions within a country coordinate with American Embassy personnel to ensure their inform efforts are synchronized with the U.S. government's communication strategy for that country.

Commanders employ public affairs activities to communicate with foreign audiences just as they communicate with domestic audiences when in the United States. Army forces maintain a scrupulous level of honesty and integrity when communicating to international audiences and should assume that anything communicated locally overseas is being communicated globally. As commanders conduct operations, physical activities and visible signatures from these actions confirm, reinforce, or contradict the information that commanders communicate through public affairs and other means. Avoiding contradictions is critical to maintaining information advantage.

When communicating to international audiences, commanders and staffs develop messaging appropriate to their mission and authority that support strategic themes developed by the National Security Council staff and Department of State, DOD, and other U.S. Government departments and agencies. Commanders ensure that their international themes and messages align with messaging to U.S. domestic audiences.

A. Conduct Community Engagement Outside the Continental United States

Community engagement is especially important for Army forces conducting operations overseas. These activities involve working collaboratively with, and through, groups of people affiliated by geographic proximity or special interests to enhance the understanding and support for Army forces and operations.

For Army forces conducting operations outside the United States, community engagement plays a critical role in enabling Army forces to achieve their objectives. In all operational contexts, Army forces themselves are the principal messaging tool that supports informing allies and partners. Commanders, advised by public affairs staff, include community engagement activities as a part of their operations, enhancing the inherent informational effect of Army operations and building the relationship between Army forces and the civilian communities near where they operate.

Since Army forces rely heavily on the support of civilian communities in the overseas regions in which they conduct operations to support strategic security objectives, community engagement is critical during competition. Community engagements build trust between Army forces and supporting communities. These engagements provide Army forces with a unique opportunity to understand concerns of the local population and to address and reduce concerns caused by Army operations. For both temporarily deployed forces and permanently forward-deployed forces, community engagement is critical to meet current or future operational requirements.

B. Conduct Soldier & Leader Engagement (SLE) *(See p. 3-72.)*

During operations in which Army forces are assigned an area of operations, commanders direct Soldier and leader engagements (SLEs) to inform audiences. Soldier and leader engagement is interpersonal interactions by Soldiers and leaders with audiences in an area of operations (ATP 3-13.5). Because Army forces conduct operations in and among populations, conducting SLEs effectively inform international audiences.

An SLE can be planned or unplanned. Planned SLEs include deliberate interpersonal interactions to provide specific information. Unplanned SLEs benefit from spontaneous interactions that allow greater understanding of an operational environment (OE) and attitudes of local populations, to include specific audiences, and process this understanding into time-sensitive feedback to other forces.

Leaders prepare their Soldiers before and during Army operations to conduct SLEs. Soldiers are invaluable in helping to understand specific audiences and their attitudes toward Army forces. They may be the sensors who report indications that Army operations are viewed positively or negatively. Behavior such as fear, hostility, indifference, or support of various audiences toward Soldiers enhances understanding and helps commanders assess whether their operations and words align. Because Soldiers understand the commander's engagement guidance, they can immediately engage audiences to mitigate negative effects of misinformation and disinformation. Soldier and leader behavior among foreign audiences is the most powerful means of informing them.

Refer to ATP 3-13.5 for more information on SLE.

C. Conduct Civil Affairs Operation *(See pp. 3-17 to 3-26.)*

Army forces conduct civil affairs operations principally to engage the civil component of an OE. Because civil affairs forces focus much of their operations on understanding local populations and their institutions, these forces can enhance a commander's understanding of the human dimension and associated civil considerations. Because civil affairs missions depend on engaging and developing relationships with relevant audiences, civil affairs operations substantially contribute to informing during the normal course of their duties.

Civil affairs forces assist commanders in informing international audiences, to include unified action partners and indigenous populations and institutions. Civil-military integration, a core competency of civil affairs forces, is essential to informing international audiences. Civil-military integration is the actions taken to establish, maintain, influence, or leverage relations between military forces and indigenous populations and institutions to synchronize, coordinate, and enable interorganizational cooperation and to achieve unified action (FM 3-57). Civil-military integration is essential to effectively integrate operations with commanders and unified action partners to achieve unity of effort. The establishment of a civil-military operations center, or other mechanisms, enables civil information sharing and integration.

Refer to FM 3-57 for more information on civil affairs operations.

D. Correct Misinformation and Counter Disinformation Within International Audiences

Threats use propaganda to attempt to gain a relative information advantage over the United States, its allies, and its partners. Propaganda is information that is biased or misleading and is designed to influence the opinions, emotions, attitudes, or behaviors of any group to benefit the sponsor. Threats use information for effect, misinformation, and disinformation as propaganda tools to influence international public opinions and to sow internal discord to fracture military alliances and partnerships.

> Disinformation is incomplete, incorrect, or out of context information deliberately used to influence audiences. Threats often rely on publics to promulgate disinformation that members of the public unwittingly believe to be correct information.

Information for effect, misinformation, and disinformation reside in the same outlets as factual, truthful information. Knowing where to search and being able to identify the types and tactics of threat information is critical to counter a threat's malign narrative. The staff's understanding of informational considerations of an OE helps that staff identify threat means, methods, and capabilities for using propaganda.

Commanders and staffs consider several factors when deciding if, how, and when to correct misinformation and deciding whether to, how, and when to counter disinformation. Not all misinformation or disinformation needs to be actively addressed. First, they consider what makes the disinformation believable to a specific audience. Understanding what makes information about an event newsworthy to a specific audience helps understanding the factors that enable a threat to effectively use disinformation.

Second, commanders and staffs consider the unity of effort necessary to counter disinformation. When coordinating information activities, commanders ensure unity of effort in countering disinformation. Commanders consistently communicate in an integrated and coherent manner regarding the actions and intentions of Army forces and their leaders to counter disinformation.

Speed is a third consideration when countering misinformation and disinformation. The first side that presents the information often sets the context and frames the public debate. Staffs work quickly to get accurate information and imagery out first, without rushing to failure by inadvertently releasing inaccurate or incomplete information.

Some audiences will not respond positively to friendly attempts to correct misinformation and counter disinformation. These types of international audiences may automatically assume any information released by Army forces to be disinformation. Commanders and staffs examine the informational considerations of an OE and focus on correcting misinformation and defeating threat disinformation among audiences willing to consider public information released by Army forces. In some cases, Army commanders coordinate for an external source to correct misinformation or to counter disinformation.

(Information Advantage)
IV. Influence

Ref: ADP 3-13, Information (Nov '23), chap. 6.

See p. 2-19 for discussion of influence from a joint doctrine perspective (JP 3-04).

To influence is to shape or alter the opinions, attitudes, and ultimately the behavior of threats and other foreign relevant actors. As a form of contact, Army forces influence threats to decrease their combat effectiveness, erode organizational cohesion, diminish will, and deceive threats about friendly intent. Army forces influence selected foreign audiences to increase support, decrease potential interference with Army operations, and undermine threat attempts to influence those same audiences.

The friendly force garners an information advantage by using information to influence the behavior of foreign relevant actors more effectively than an adversary or enemy does. The influence information activity contributes to this advantage through two related tasks: influence threat perception and behaviors and influence other foreign audiences as shown in Figure 6-1.

Influence

I   **Influence Threat Perception & Behaviors**

II   **Influence Other Foreign Audiences**

Note: U.S. audiences are not targets for military activities intended to influence.

Purpose: Affect Behavior of Foreign Relevant Actors

Ref: ADP 3-13 (Nov '23), fig. 6-1. Tasks and purpose of the influence information activity.

I. Influence Threat Perception and Behaviors

Influence information activities by Army forces, integrated into the combatant commander's campaign plan, support setting a theater, challenge threat activities, and facilitate campaign objectives. During competition and crises, influence efforts deter threat actions and erode threat cohesion and effectiveness.

During armed conflict, influence activities disrupt or corrupt enemy forces' understanding and decision making, decrease their combat effectiveness, erode command and control (C2), and degrade morale and will.

A commander's ability to integrate disparate capabilities and synchronize application in the human, information, and physical dimensions is critical to influencing threat behavior. Commanders understand their higher echelon commander's intent and concept of operations and so employ their joint and Army capabilities in ways that support making the threat act or react in a desired manner. Tasks specifically designed to influence threat perceptions and behavior include—

Influence Threat Perception and Behaviors

A. Conduct Deception Activities
B. Conduct Military Information Support Operations (MISO)

A. Conduct Deception Activities *(See pp. 3-27 to 3-32.)*

Surprise is a combat multiplier that amplifies the effects of the other principles of war and provides a relative advantage where none previously existed. Its effective use allows friendly units to strike at a time and place or in a manner for which the enemy is unprepared, which induces shock and causes hesitation. Surprise seldom lasts for long periods because enemies adapt, so rapidly exploiting the opportunities surprise affords is critical. Every echelon works to achieve surprise during an operation.

One way to achieve surprise is to use deception. Deception is the act of causing someone to accept as true or valid what is false. Army forces conduct deception activities to cause enemy decision makers to act or not act in ways prejudicial to themselves and favorable to achieving friendly objectives.

Army forces support or conduct three types of deception:

- **Military Deception (MILDEC).** Military deception is actions executed to deliberately mislead adversary military, paramilitary, or violent extremist organization decision makers, thereby causing the adversary to take specific actions (or inactions) that will contribute to the accomplishment of the friendly mission (JP 3-13.4).

- **Tactical Deception (TAC-D).** Army forces conduct TAC-D to cause the enemy to react or falsely interpret friendly operations. Tactical deception is a friendly activity that causes enemy commanders to take action or cause inaction detrimental to their objectives (FM 3-90). Properly planned and executed TAC-D helps Army forces to hide what is real and display what is false.

- **Deception in Support of Operations Security (DISO)** *(See p. 3-28.)*

B. Conduct Military Information Support Operations *(p. 3-33.)*

Military information support operations are planned operations to convey selected information and indicators to foreign audiences to influence their emotions, motives, objective reasoning, and ultimately the behavior of foreign governments, organizations, groups, and individuals in a manner favorable to the originator's objectives (JP 3-13.2). MISO can degrade enemy combat power, reduce civilian interference, minimize collateral damage, and increase a population's support for operations.

MISO focus on information and indicators to convey meaning and to influence specific target audiences—individuals or groups selected for influence. The Secretary of Defense approves all MISO programs submitted as part of combatant commander campaign and contingency plans. Combatant commanders plan and execute MISO in support of theater objectives. Within this framework, psychological operations (PSYOP) units execute MISO programs in support of combatant commanders, subordinate joint task forces, the theater special operations command, and Army forces. MISO programs directed at enemy forces focus on themes, such as—

- Degrading enemy combat power by encouraging surrender, desertion, and malingering.
- Reducing the will of the enemy to resist.
- Degrading the decision-making abilities and operational effectiveness of the enemy.
- Exploiting and amplifying friendly successes on the battlefield.
- Exploiting and amplifying enemy failures and actions on the battlefield.

Army PSYOP forces are trained and equipped to conduct MISO.

II. Influence Other Foreign Audiences

Within the framework of the combatant commander's campaign plan, Army forces seek to influence other foreign audiences. For one audience, efforts may be designed to increase mutual support and deepen existing relationships. For a different audience, influence efforts may focus on maintaining neutrality or changing from neutrality to supporting friendly positions.

Influence Other Foreign Audiences

A. Conduct Soldier and Leader Engagement
B. Conduct Military Information Support Operations (MISO)
C. Conduct Civil Affairs Operations

During crises, Army forces seek to keep neutral audiences from inadvertently obstructing operations and allowing Army forces to focus their attention as much as possible on frustrating threat attempts to achieve goals and objectives. Should neutral audiences purposely interfere with operations, analysis can determine if the reasons are related to issues specific to the group's circumstances rather than a perceived alignment (and realignment) with a U.S. adversary. A calculated, appropriate U.S. response to ensure continued neutrality and reduction in interference helps avoid pushing a neutral audience to align with a threat. The continued neutrality of unaligned audiences is vital during fluid situations. Planners and advisors integrate the influence activities to ensure Army forces minimize mistakes and quickly mitigate the effects of any that occur. At the same time, it is vital to position Army units so they can rapidly and effectively exploit the effects of enemy actions and mistakes, such as collateral damage, civilian casualties, and human rights violations. In doing so, Army forces can increase the likelihood of swinging neutral audiences to the friendly position in dynamic environments.

When Army influence efforts maintain an audience's neutral stance, the audience is less likely to become an impediment to operations or become an adversary asset. If a neutral audience can be persuaded into becoming a friendly audience, then it becomes a potential impediment to an adversary and can increase the dilemmas it potentially faces. Primary tasks to influence other foreign audiences include—

A. Conduct Soldier and Leader Engagement (See p. 3-72.)

SLE is interpersonal interactions by Soldiers and leaders with populations in an area of operations. It can occur as an unplanned face-to-face encounter on the street or a scheduled meeting. Engagements can also occur via telephone calls, video teleconferences, or other audiovisual mediums. SLE supports both inform and influence activities.

B. Conduct Military Information Support Operations (p. 3-33.)

In addition to influencing threats, PSYOP units conduct MISO to influence other foreign audiences and populations. PSYOP units and staffs in Army headquarters help commanders align actions and messaging that influences these audiences to align them more closely with friendly goals and objectives.

C. Conduct Civil Affairs Operations (See pp. 3-17 to 3-26.)

Civil affairs operations are integrated with other influence activities to affect the behavior of foreign relevant actors. In this way, civil affairs operations act as the tangible connection for the commander to produce desired effects in the civil component of an operational environment (OE).

Influence Considerations

Ref: ADP 3-13, Information (Nov '23), pp. 6-6 to 6-7.

> A force that uses information to deceive and confuse an opponent has an advantage. Using information to influence relevant actor behavior more effectively than an adversary or enemy is another information advantage.
> - FM 3-0

Influence activities align actions and messages with objectives to create desired, specific, and measurable changes in behavior that give commanders an advantage. Army leaders consider the following in planning and executing influence activities:

Deliberate Versus Incidental Influence

All warfighting functions contribute to the influence information activity because all military activities can influence threat behavior or the perceptions of a foreign audience. Regardless of the mission, Army forces consider the likely psychological impact of their operations and tasks on relevant actor perceptions, attitudes, and other drivers of behavior. The inherent informational aspects of operations produce cognitive effects on threats and other foreign relevant actors, including fear, anger, or confidence. These effects can erode, build, create, or negate other physical, human, and information advantages. Commanders and staffs prevent or minimize the negative consequences of undesired or unplanned effects by considering how operations and actions affect an OE and influence the people in it. They then plan actions and communicate messages to elicit desired behaviors and support the national and operational narratives.

Planners and subject matter experts advise the commander about potential unplanned effects of operations and actions. Collateral damage, fratricide of allied or partner nation forces, or bad behavior by friendly forces all can have serious negative consequences that require commanders and staffs to have contingency plans and staff battle drills in place to mitigate.

Language, Regional, and Cultural Expertise

Leveraging information for the purpose of affecting behavior of relevant actors requires an understanding of the drivers of human behavior. These drivers include cultural aspects of the population, like language, arts, customs, and religion, as well as geographical considerations. Planners assess these aspects to understand threat and other foreign relevant actors and develop plans to influence them. Language, regional, and cultural subject matter experts enable operations when they provide a thorough understanding and appreciation of local populations, government officials, partners, and allies.

Authorities

Influence activities can be lethal or nonlethal, may be attributable or nonattributable, may require specific permissions and authorities, may be politically and time sensitive, and are governed by policies and statutes. Influencing threats requires integration of influence considerations into the targeting process to ensure that staffs plan the most effective combination of lethal and nonlethal capabilities in context. Commanders and staffs employ as many legal and authorized potential means to influence their target audiences as required. Commanders and staffs must address authorities and approval in planning influence activities.

Authorities govern employment of joint and Army influence capabilities, to include civil affairs operations, MISO, MILDEC, cyberspace operations, and technical effects. U.S. and international law, DOD policies, status-of-forces agreements, treaty obligations, operation orders, and other binding documents may provide both authorities and limitations to conduct certain activities.

(Information Advantage)
V. Attack

Ref: ADP 3-13, Information (Nov '23), chap. 7.

See p. 2-19 for discussion of "attack" from a joint doctrine perspective (JP 3-04).

The threat is increasingly reliant on space, cyberspace, and the electromagnetic spectrum (EMS) for intelligence, surveillance, and reconnaissance (ISR); target acquisition; fire control; communications; and C2. Threat forces increasingly communicate (human to human, human to machine, and machine to machine) through the cyberspace domain. The cyberspace domain consists of the network and information technology infrastructures, resident data, the internet, telecommunications networks, computer systems, processors, and portions of the EMS that facilitate or inhibit them. Threats also employ information warfare capabilities through space, cyberspace, and the EMS to attack friendly data, information, and communications and to spread propaganda.

Affecting the threat's ability to use data and information to communicate, command, and control its forces or conduct information warfare provides the friendly force an advantage. The attack information activity contributes to this advantage through two related tasks: degrade the threat's ability to exercise C2 and affect threat information warfare capabilities as shown in Figure 7-1.

Attack

- **I. Degrade Threat Command and Control**
- **II. Affect Threat Information Warfare Capabilities**

Purpose: Affect Threat Command and Control

Ref: ADP 3-13 (Nov '23), fig. 7-1. Task and purpose of the attack information activity.

While both attack tasks affect the threat's use of data and information, each task has a different focus. Degrading threat C2 focuses on negatively affecting threat situational understanding, networks, and information systems. Affecting threat information warfare capabilities focuses on protecting friendly forces from threat cyber and electromagnetic attacks and contributes to a broader joint and national effort in attacking threat disinformation, propaganda, and legitimacy.

Information Attack Methods

Threat C2 nodes (command post [CP], signal centers, networks, and information systems); ISR sensors and systems; and fire control and target acquisition radars and systems are often high-payoff targets for Army forces. As part of the concept of operations and scheme of fires, Army forces attack these targets through a combination of methods: physical destruction, electromagnetic attack (EA), cyberspace attack, and offensive space operations.

See following pages (pp. 1-50 to 1-51) for an overview and further discussion.

Information Attack Methods

Ref: ADP 3-13, Information (Nov '23), pp. 7-2 to 7-5.

Army leaders combine available organic, joint, and multinational capabilities in complementary and reinforcing ways to create and exploit an information advantage.

Note. Additional classified capabilities, activities, and programs exist that can affect threat C2, networks, and systems. Technical effects are one or more capabilities, activities, or programs planned, coordinated, or executed that utilize classified means to accomplish an objective or enable military operations. Commanders requiring the execution of technical effects for an operation should understand that authorities and approvals generally reside at the combatant command or higher level and will often require long lead times for approval and execution.

Physical Destruction

In the context of information attack, physical destruction is the application of fires and maneuver to affect threat C2 and communications. Targets for physical destruction range from enemy CPs and communications centers to sensor and fire control systems. During armed conflict, commanders direct or coordinate for surface-to-surface fires, air-to-surface fires, and surface-to-air fires against threat C2, ISR, and information warfare targets. Commanders also direct maneuver forces to conduct raids and other offensive operations to seize or destroy enemy C2 nodes.

Physical destruction capabilities are inherent in combined arms formations and often provide more immediate results than employing other methods of attack. Depending on the echelon, organic indirect fires, to include mortars, cannons, rockets, and missiles, are well suited to destroy threat C2 nodes. Attack aviation and ground maneuver units can also execute physical destruction tasks focused on a threat C2 system. Army forces likewise nominate threat C2 and ISR targets to the joint force commander for physical destruction. Depending on priority, the joint force may attack these targets with fires or special operations forces.

Commanders and staffs consider rules of engagement, availability of assets and munitions, the potential for collateral damage, and the impact on escalation when directing physical destruction. At brigade and below echelons, physical destruction of the enemy's communications equipment can effectively create an advantage. At echelons above brigade, physical destruction is often combined with EA and cyberspace attacks to affect threat situational understating and the threat's ability to exercise C2.

Electromagnetic Attack (EA) *(See p. 3-56.)*

Threat forces rely on communications equipment using broad portions of the EMS to conduct operations. This equipment allows threats to talk, transmit data, provide navigation and timing information, and to exercise C2. Threat forces also collect signals in the EMS to build understanding and to target friendly forces and equipment. EA prevents or reduces an enemy's effective use of the EMS by employing jamming and directed-energy weapon systems against enemy spectrum-dependent systems and devices.

Electromagnetic attack is a division of electromagnetic warfare involving the use of electromagnetic energy, directed energy, or antiradiation weapons to attack personnel, facilities, or equipment with the intent of degrading, neutralizing, or destroying enemy combat capability and is considered a form of fires (JP 3-85). EA systems and capabilities include—

- Jammers.
- Directed energy weaponry.
- Radio frequency emitters.
- Technical means of deception.
- Antiradiation missiles.

Cyberspace Attacks *(See p. 3-54.)*

Cyberspace attacks are actions taken in and through cyberspace that create denial (i.e., degradation, disruption, or destruction) or manipulation effects in cyberspace and are considered a form of fires (JP 3-12). Cyber forces execute cyberspace attacks through defensive cyberspace operations-response actions (known as DCO-RA) and offensive cyberspace operations (known as OCO). Cyberspace attacks require coordination with other U.S. Government departments and agencies and careful synchronization with other lethal and nonlethal effects through the targeting processes.

Cyberspace attacks are executed under the authority of the Secretary of Defense. The effects from these attacks provide windows of opportunity Army forces can exploit. For example, the joint force commander times cyberspace attacks to affect threat air defense and fire control systems so that they do not interfere with joint and Army forces attacking in a specific area. Additionally, the joint force commander may provide direct offensive cyberspace operations support to corps and below Army commanders in response to requests via the joint targeting process.

Cyberspace attack actions create denial effects in cyberspace or manipulation in cyberspace to create denial effects in the physical dimension. In some cases, cyberspace attack actions can lead to physical destruction. Cyberspace attacks affect physical processes when they modify or destroy cyberspace capabilities that control the physical process. Some examples of effects created by a cyberspace attack include—

- Deny.
- Disrupt.
- Destroy.
- Manipulate.

Space Operations *(See pp. 61 to 3-70.)*

Space capabilities enable joint and Army operations. Space capabilities include space situational awareness; positioning, navigation, and timing; satellite communications; satellite operations; missile warning; environmental monitoring; space-based surveillance and reconnaissance; defensive space operations; and offensive space operations. Army space planners at all echelons advise commanders on the current space assessment and ways to coordinate for and integrate space capabilities and effects into operations.

Space operations enable freedom of action in the space domain for the United States and its allies. Offensive and defensive space operations, including navigation warfare, enable freedom of action in space and counter efforts to interfere with or attack space forces of the United States, allies, or commercial partners.

- **Offensive Space Operations.** Offensive space operations are actions taken to negate attacks against U.S. and friendly space assets and threat freedom of action. Measures include actions against ground, data link, and space segments or users to affect an enemy's space systems, or to thwart hostile interference on U.S. and multinational space systems:
    - Deceive
    - Disrupt
    - Deny
    - Degrade
    - Destroy.
- **Navigation Warfare.** Navigation warfare aims to ensure unimpeded access to the Global Navigation Satellite System for joint forces and multinational partners while denying it to the enemy. It encompasses various offensive, defensive, and support activities (such as surveillance, reconnaissance, and EMS management) to ensure unimpeded availability and integrity of positioning, navigation, and timing information.

I. Degrade Threat Command and Control

To degrade means to reduce or to lower. Army forces create and exploit every opportunity to degrade the threat's ability to exercise C2. As with the friendly forces, information is a central resource for the threat to exercise C2. Threats collect information, process and analyze it to understand, and use it to inform decisions. Before a threat actor can make a decision, an Army force aims to prevent, delay, or alter that threat's decisions by degrading its access to information, manipulating the information available, or overwhelming its systems and processes with large amounts of information. After a threat decision is made, an Army force aims to prevent, alter, or limit the threat force's ability to execute military actions by attacking threat C2 nodes, networks, and information systems. Limiting the information available to an enemy or adversary while also inhibiting the ability to exchange what information it does have thus provides significant military advantage.

The protect information activity contributes to degrading threat C2 by denying the threat's access to friendly data and information (see pp. 1-29 to 1-34). The influence information activity contributes to degrading threat C2 by affecting threat perceptions (see pp. 1-45 to 1-46).

The attack information activity degrades threat C2 by—

A. Affect Threat Understanding of an Operational Environment

Threat decision makers use information from a variety of sources to make decisions. Threat decision makers may rely on traditional intelligence sources—such as geospatial intelligence, human intelligence, and signals intelligence—as well as information gained through cyberspace reconnaissance, social media exploitation, and collection of publicly available information. Staffs often process and analyze this information by both technical and human means before it reaches the decision maker. Each source of information and each step in this information process represent an opportunity for Army forces to impact the threat's decision making.

Commanders should consider all ways and means to affect the threat's ability to build and maintain situational understanding. Within the attack information activity, commanders direct or coordinate for physical destruction, EA, cyberspace attack, and technical effects to—

- **Disrupt or deceive** sensors that provide threat actors with intelligence.
- **Disrupt or manipulate** data transmissions among threat sensors, analysis capabilities, and decision makers.
- **Deceive** threat decision makers about friendly intentions and capabilities.
- **Disrupt or manipulate** communication between threat decision makers and units.

Note. In some instances, Army commanders may want to deter threat actions by improving the threat's understanding of friendly capabilities and intent.

B. Affect Threat Networks and Systems

Army commanders use many military capabilities to affect threat networks and systems. The type of capabilities a commander employs depends on the objective, the type of target system, acceptable levels of risk, and the strategic context. Commanders carefully consider what parts and the duration of threat networks and systems they desire to affect. In some instances, commanders want the threat to see and communicate the activities of friendly forces. In other instances, they may want to degrade certain networks and systems for a specified time. Commands may focus attacks on disintegration by targeting. In these instances, units target key nodes within threat networks and systems (C2, ISR, and fires) for disruption, destruction, or manipulation.

When exploited for intelligence purposes, threat networks and systems can provide significant strategic insight as well as operational warnings. As commanders and staffs plan and execute actions to degrade these networks and systems, they weigh the relative operational value of degrading these systems against the potential loss of intelligence. For instance, if an Army force is collecting key warnings on enemy intent from a radio channel, a commander may forego jamming that radio channel during an operation even though such jamming might hinder enemy maneuver. Conversely, the commander may determine that such jamming will provide a great enough operational advantage to Army forces that it will offset the loss in intelligence insight into enemy intentions.

II. Affect Threat Information Warfare Capabilities

Adversaries and enemies have active and effective information warfare capabilities. Many of the information activities discussed in previous chapters are intended to reduce the effectiveness of threat information warfare capabilities and the impact of threat information warfare on Army forces and operations. While it may be preferable for Army commanders to be able to prevent or mitigate the effects of these threat actions, it is by no means ensured that prevention or mitigation is possible in all instances. In cases where threat actors have decided to conduct information warfare attacks against Army and friendly forces and those attacks are imminent or ongoing, Army commanders may need to degrade or defeat threat information warfare capabilities directly.

To degrade or defeat threat information warfare capabilities, Army forces may employ the same types of capabilities and use methods like those used in other Army information activities. The type of threat information warfare capability, as well as the strategic context of the attack, may narrow or broaden the list of appropriate military capabilities employed in response. For instance, an appropriate response to a threat cyberspace attack on an Army logistics system during competition below armed conflict might be limited to a cyberspace security response coordinated through joint and whole of government channels. However, the same type of cyberspace attack during crisis or armed conflict might warrant employment of Army cyberspace attack capabilities, EW capabilities, or a physical strike on threat cyberspace capabilities.

A technical or physical attack on threat information warfare capabilities might not always be required to functionally defeat the threat. For instance, friendly forces could diminish the impact of a threat disinformation campaign by blocking or removing the network nodes and communications systems used to promulgate it. Friendly forces could also achieve a similar effect by exposing the existence of the disinformation campaign and those who promulgate it. In this scenario, commanders would leverage the activities and capabilities associated with other information activities and tasks, such as public affairs and military information support operations (MISO). Exposing false threat narratives with truthful information, to include the authorized released of intelligence and imagery, directly attacks threat legitimacy. Coordinated at the strategic and operational levels, Army tactical commanders support these efforts as directed, to include providing intelligence, information, and imagery for release by higher authorities.

See following pages (pp. 1-54 to 1-55) for an overview and discussion of threat information warfare.

Threat Information Warfare

Ref: ADP 3-13, Information (Nov '23), pp. 1-12 to 1-14.

> The [People's Liberation Army] defines information attack as any [information warfare] activity intended to weaken or deprive the enemy of control of information. Information attack is the primary means by which information warfare is won, and it is the key to achieving information superiority.
>
> - ATP 7-100.3

A **threat** is any combination of actors, entities, or forces that have the capability and intent to harm United States forces, United States national interests, or the homeland (ADP 3-0). Threats may include individuals, groups of individuals, paramilitary or military forces, criminal elements, nation-states, or national alliances. A threat may be a nation-state with an authoritarian government or a nonstate actor that follows an extremist ideology. Threats operate outside and within the United States.

In general, a threat can be categorized as an adversary or enemy. An adversary is a party acknowledged as potentially hostile to a friendly party and against which the use of force may be envisaged (JP 3-0). An enemy is a party identified as hostile against which the use of force is authorized (ADP 3-0). An enemy is called a combatant and is treated as such under the law of armed conflict. The most dangerous threats to the United States are peer threats. A peer threat is an adversary or enemy with capabilities and the capacity to oppose U.S. forces across multiple domains worldwide or in a specific region in which it has a significant relative advantage.

In the context of the threat, information warfare refers to a threat's orchestrated use of information activities (such as cyberspace operations, electromagnetic warfare, psychological warfare, and influence operations) to achieve objectives from the strategic to the tactical levels of warfare. At the tactical level, threat information warfare consists of specifically planned and integrated actions taken to achieve advantages at critical points and times.

Note. There is not a single-source threat doctrine on information warfare. While similar in many ways, each threat nation and threat force will employ informational power based on their capabilities and their understanding of military art and science. TC 7-100.2, Opposing Forces, provides a base model for threat tactical information warfare.

Threat information warfare seeks to blur the divide between peace and war, control access to information, shape an OE with narratives and propaganda, and deny opponents information in armed conflict through systems confrontation and destruction. Peer threats use diverse means to conduct information warfare, which may include—

- Cyberspace operations.
- Psychological warfare.
- Influence operations.
- Movement and positioning of forces.
- Deception.
- Electromagnetic warfare.
- Physical destruction.
- Political and legal warfare.
- Active measures (espionage, sabotage, and assassinations).
- The use of proxies and nonstate actors.

The Peoples Republic of China's Three Warfares Strategy is an example of threat information warfare. The Russian Federation's information warfare concept of reflexive control is another example.

Information Advantage

Russian Activities in Ukraine 2014

Russia's annexation of Crimea in 2014 is a prime example of a peer threat's use of information warfare in operations. In February 2014, the pro-Russia Ukrainian government in Kyiv was ousted, leading to widespread protests and instability throughout Ukraine. In Crimea, a Ukrainian peninsula along the northern coast of the Black Sea in Eastern Europe, widespread protests occurred against the interim government and demonstrations by pro-Russian separatists.

Russia used the ensuing chaos to insert numerous troops into the region. First, Russian forces used physical attacks to cut fiber-optic communications lines, electromagnetic warfare to jam telephones and radios, and cyberspace attacks to severely degrade news outlets and websites, effectively creating an information blackout. Then Russian forces entered Crimea wearing no identifying insignia and took swift control of key government infrastructure. Rather than being identified as invading Russian forces, they were simply referred to as "little green men." Russian control over information sowed doubt and confusion, delayed the ability to communicate and make decisions, and prevented Ukrainian forces from organizing and resisting. In short order, a large-scale surrender of Ukrainian forces had occurred, and Russia had taken control of Crimea.

Russian activities prior to and after the invasion and annexation of Crimea included—
- Providing overt and covert support to Crimean separatists.
- Manipulating and controlling the flow and content of information.
- Promoting a Russian nationalist narrative in the region and around the world.

Other threats, such as Iran and nonstate actors like Hamas, use similar theories and concepts to gain positions of relative advantage in or through the human and information dimensions. These positions allow them to exploit vulnerabilities in their opponents and negate physical advantages. The use of information warfare by nonstate actors is akin to the historic use of guerrilla warfare and tactics to gain physical advantages over a stronger and larger force.

Tactical threat information warfare attacks surveillance and target acquisition sensors, C2 centers and nodes, decision makers, data and information, telecommunications systems and infrastructure, population groups, and relevant actors. Threats typically target information links, such as radio frequency receivers, communications devices, and information protocols. Tactical threat information warfare activities are employed to—
- Destroy or disrupt friendly C2.
- Destroy or deceive friendly reconnaissance, surveillance, and target acquisition.
- Deny friendly situational understanding.
- Isolate key elements of a friendly force, particularly allies and partners.
- Distort or deny information to relevant actors and audiences.

Refer to our OPFOR THREAT SERIES. Today's operational environment presents threats to the Army and joint force that are significantly more dangerous in terms of capability and magnitude than those we faced in Iraq and Afghanistan. Major regional powers like Russia, China, Iran, and North Korea are actively seeking to gain strategic positional advantage.

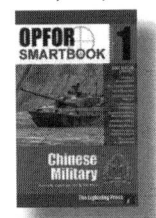

Attack Considerations

Ref: ADP 3-13, Information (Nov '23), p. 7-7.

The tasks that compose the attack information activity target threat data and information, information systems, communications, and information warfare capabilities. To be successful, Army leaders consider the following in planning and executing attack tasks:

Timelines for Preparatory Activities

Effective attacks require that commanders and staffs identify threat targets early during IPOE and continuously refine those targets throughout the operations process. In general, the more precise the required effect is, the more time it will take to analyze the target and prepare capabilities to create the effect. For example, coordinating for space and cyberspace capabilities often requires coordination and approval through the headquarters of several Army echelons, the combatant command headquarters, then to a supporting combatant command.

Effective units continuously conduct many preparatory activities during competition below armed conflict and through crisis and armed conflict. These activities include intelligence operations, cyberspace reconnaissance, electromagnetic reconnaissance, target audience analysis, and target development. During competition below armed conflict, units continuously conduct most of these preparatory activities at echelons above division. During crisis and armed conflict, units below corps may use organic assets and capabilities to continue these preparatory activities. However, units below corps will likely still rely on support from echelons above division to conduct cyberspace reconnaissance and augment division and below organic intelligence and EW capabilities. This reliance requires staffs to continuously coordinate requirements with higher echelon units throughout the operations process.

Precision and Scalability

Threat C2 systems present attractive targets. When friendly forces successfully attack, they render ineffective a large number of threat forces and weapon systems without directly attacking each individual system. When threat commanders cannot issue orders or direct their forces, it reduces the effectiveness of many or all that threat's capabilities. However, the fact that these C2 systems are so critical means that Army commanders must consider threat reactions and counteractions when attacking threat C2 systems.

To minimize the risk of unintended escalation during competition below armed conflict or in crisis, joint force commanders consider the precision and scalability of methods they select to attack threat C2 systems. Cyberspace attack, EW, space, and technical effects capabilities may vary widely in their ability to precisely affect a specific system or in their ability to scale effects on that system.

During armed conflict, Army commanders consider using lethal effects to degrade or destroy threat C2 systems. Delivering lethal effects through fires may be timelier than nonlethal technical effects, but it may also result in more collateral damage. Staffs develop and provide assessments of these tradeoffs to commanders during the targeting process.

Chap 1
(Information Advantage) Integration

Ref: ADP 3-13, Information (Nov '23), chap. 8.

I. Joint and Multinational Information Advantage

Gaining and exploiting information advantages is a whole of government, joint, and multinational effort requiring unified action. Unified action is the synchronization, coordination, or integration of the activities of governmental and nongovernmental entities with military operations to achieve unity of effort (JP 1, Volume 1). Unity of effort is the coordination and cooperation toward common objectives, even if the participants are not necessarily part of the same command or organization that is the product of successful unified action (JP 1, Volume 2). To facilitate unified action, Army commanders and supporting staff must understand the roles, capabilities, and processes of U.S. government, joint, and multinational organizations involved in creating and exploiting information advantages. (See paragraphs 1-13 through 1-20 for a discussion of informational power employed by the U.S. government.)

A. Information Joint Function *(See pp. 2-7 to 2-20.)*

The information joint function is the management and application of information to change or maintain perceptions, attitudes, and other drivers of behavior, and to support human and automated decision making. Combined with the other joint functions (command and control [C2], intelligence, fires, movement and maneuver, protection, and sustainment), the information joint function helps joint force commanders and staffs effectively use information during operations across the competition continuum. The primary joint tasks are—

- Understand how information impacts the operational environment (OE).
- Support human and automated decision making.
- Leverage information.

Joint Information Advantage *(See p. 2-1.)*

When the joint force successfully executes the tasks and subtasks associated with the information joint function, the joint force gains information advantages. Joint doctrine describes information advantage as the operational advantage gained through the joint force's use of information for decision making and its ability to leverage information to create effects in the information environment. The joint force applies information power to create and exploit information advantages in two primary ways:

- Planning and executing all operations, activities, and investments with deliberate intent to leverage its inherent informational aspects.
- Employing specially trained units to conduct joint operations in the information environment (OIE).

Leveraging the Inherent Informational Aspects of Operations *(See pp. 2-10 to 2-11.)*

Joint force action impacts the OE either intentionally or incidentally. All joint force operations, activities, and investments can affect the behavior of relevant actors. The conclusions that observers draw from interpreting joint force activities may drive them to act in ways to affect the joint force. Whether or not commanders consider this during planning, friendly activities do impact an OE and resonate in the operational area and potentially other operational areas.

Joint force commanders employ Army forces to achieve objectives while understanding the potential informational effects Army operations can have on an OE. Joint force commanders communicate to the senior Army headquarters how they intend to leverage these effects to accomplish various tasks, achieve joint information objectives, or support OIE. Army commanders nest the tasks and purpose they assign to subordinates to support the joint force commander's intent throughout the operations process.

Operations in the Information Environment *(See pp. 2-37 to 2-44.)*

Operations in the information environment are military actions involving the integrated employment of multiple information forces to affect drivers of behavior (JP 3-04). OIE affect drivers of behavior by informing audiences; influencing foreign relevant actors; attacking and exploiting relevant actor information, information networks, and information systems; and protecting friendly information, information networks, and information systems.

B. Multinational Considerations *(See p. 2-36.)*

Army forces integrate their information activities with multinational partners. Multinational operations is a collective term to describe military actions conducted by forces of two or more nations, usually undertaken within the structure of a coalition or alliance (JP 3-16). Multinational partners may contribute different or additional information forces to an operation. Each nation's force has unique capabilities and often operates with different authorities to employ key information capabilities in the various domains of an OE. Army forces anticipate and plan for most operations being multinational.

Multinational operations can present unique integration challenges. The differences between the various nations can affect how successfully a multinational force enables, protects, informs, influences, or attacks threat information capabilities to achieve objectives. Situational understanding affects how various commanders employ their assigned forces in support of achieving objectives. For example, if a multinational commander cannot anticipate and adjust for the various legal and regulatory restrictions for sharing classified information with allies and partners, subordinate commanders will probably understand the situation differently. Multinational national challenges include—

- National caveats on the use of respective forces.
- Doctrinal differences.
- Cultural and language barriers.
- Communications and procedural interoperability.
- Sharing of information and intelligence.
- Equipment interoperability limitations.
- Rules of engagement.

To help overcome these challenges, multinational commanders develop procedures to speed the exchange of relevant information to other nations, develop a standard lexicon that supports situational understanding of information and information capabilities, and ensure the staff is trained not to overclassify information. Participating in theater security cooperation activities helps Army forces appreciate partner capabilities and improves interoperability prior to conflict.

Refer to NATO's AJP-10.1 for allied joint doctrine on information operations. Refer to FM 3-13 for more information on multinational operations.

Army Forces and the Information Joint Function (See pp. 2-7 to 2-20.)

Ref: ADP 3-13, Information (Nov '23), pp. 8-4 to 8-5.

Although the Army's information framework differs from the joint information function, they both have the same goal: to create relative advantages that commanders can use to achieve objectives. To help understand the relationship between the Army's information framework and the information joint function, Figure 8-2 illustrates how Army information activities align with the joint information subtasks.

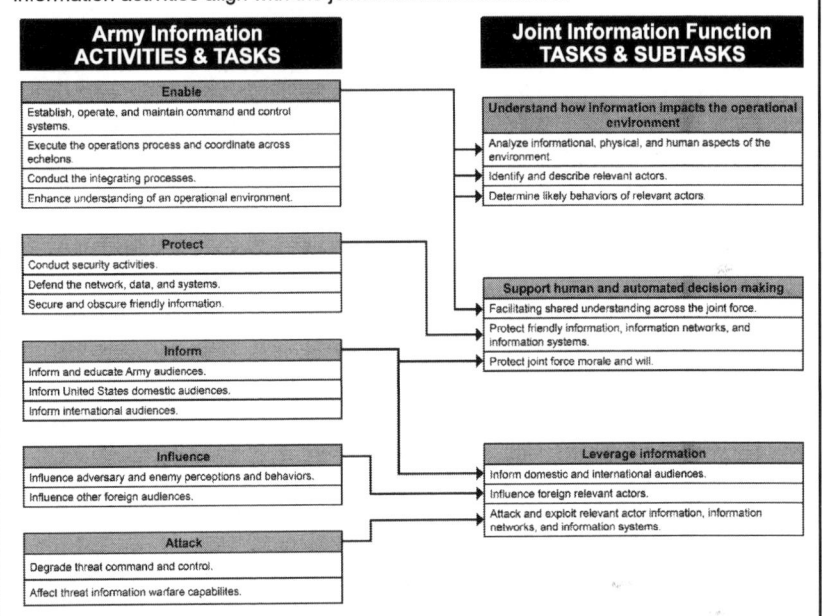

Army Information ACTIVITIES & TASKS	Joint Information Function TASKS & SUBTASKS
Enable	**Understand how information impacts the operational environment**
Establish, operate, and maintain command and control systems.	Analyze informational, physical, and human aspects of the environment.
Execute the operations process and coordinate across echelons.	Identify and describe relevant actors.
Conduct the integrating processes.	Determine likely behaviors of relevant actors.
Enhance understanding of an operational environment.	
Protect	
Conduct security activities.	**Support human and automated decision making**
Defend the network, data, and systems.	Facilitating shared understanding across the joint force.
Secure and obscure friendly information.	Protect friendly information, information networks, and information systems.
Inform	Protect joint force morale and will.
Inform and educate Army audiences.	
Inform United States domestic audiences.	
Inform international audiences.	
Influence	**Leverage information**
Influence adversary and enemy perceptions and behaviors.	Inform domestic and international audiences.
Influence other foreign audiences.	Influence foreign relevant actors.
Attack	Attack and exploit relevant actor information, information networks, and information systems.
Degrade threat command and control.	
Affect threat information warfare capabilites.	

Ref: ADP 3-13 (Nov '23), fig. 8-2. Army information activities relationship to joint subtasks.

Army forces use joint information doctrine when interacting with the higher joint headquarters but communicate with subordinate Army forces using Army doctrinal terms. Army information activities and tasks provide specificity while remaining aligned with the broader joint information tasks. The following echelons and types of units require a detailed understanding of both Army and joint information doctrine:

- Headquarters likely to be designated a joint task force.
- Theater armies.
- Headquarters likely to be designated as a land component command.
- Headquarters likely to be designated as a joint force land component command.
- Civil affairs units.
- Cyber units.
- EW units.
- Information operations units.
- Intelligence units.
- PSYOP units.
- Public affairs units.
- Space operations units.
- Special forces units.
- Security force assistance units.
- Digital liaison detachments.

II. Army Information Activities During Operations

Army commanders, supported by their staffs, integrate information activities into their concept of operations to create and exploit information advantages. Integration is the arrangement of military forces and their actions to create a force that operates by engaging as a whole. Subordinate commanders ensure they integrate their information activities with the higher commander's intent and concept of operations. Integration occurs at all echelons, with lower echelons relying on both the analytic capabilities, information capabilities, and experience of higher headquarters commanders and staffs to anticipate information requirements and synchronize the execution of tasks.

> ### Information Activities and the Operations Process
> To gain information advantages, the commander, supported by the staff, integrates information activities throughout the operations process. The operations process (plan, prepare, execute, and assess) is the major C2 activity performed during operations. Because nearly all military capabilities and actions can contribute to gaining or exploiting information advantages, the entire staff assists the commander in integrating information tasks during planning, preparation, execution, and assessment. The staff accomplishes this in the integrating cells (current operations, future operations, and plans); in working groups and boards; and through the integrating processes.

A. Planning *(See chap. 4.)*

Commanders, supported by their staffs, ensure information activities are fully integrated into plans and orders through the military decision-making process. This includes integrating information activities into the concept of operations and supporting schemes, to include schemes of intelligence, information collection, maneuver, fires, and protection. As part of planning, the staff should at a minimum—

- Consider how the informational considerations affect the warfighting functions' ability to contribute to mission accomplishment.
- Analyze the interaction of factors within the physical and human dimensions with those of the information dimension of an OE in the specific context of the operation being conducted.
- Identify unique employment considerations for information capabilities executing specific information tasks, such as MISO, offensive cyberspace operations, space operations, and EA.
- Anticipate and assess both the risk to the mission resulting from impacts and the signature resulting from unintentional inherent informational aspects of Army operations.
- Identify information tasks directed by higher headquarters required to accomplish the mission.
- Identify and consider ongoing or planned unified action partner information activities within an OE.
- Consider gaps in language, regional, cultural, or technical expertise required to understand an OE.
- Recognize potential undesirable information effects of friendly activities.
- Develop an assessment plan to monitor effects of information activities, including effects of other units, to enable adjustments as required.
- Identify required capabilities not organic to assigned forces and secure them from higher headquarters via contract or other means.
- Identify existing and required authorities needed to conduct information activities.

Integration of the Information Activities

Ref: ADP 3-13, Information (Nov '23), pp. 8-11 to 8-18 (table 8-1).

Each information activity and correlating subordinate tasks have staff leads. Information task leads assist the five information activity leads in integrating information tasks as depicted in Table 8-1. Most staff work occurs within the functional and integrating cells. The functional cells include intelligence, movement and maneuver, fires, protection, and sustainment. The integrating cells include current operations, future operations, and plans.

Activity Lead	Enable	Task Leads
Chief of staff	Establish, operate, and maintain C2 systems.	G-6 and KMO
	Execute the operations process and coordinate across echelons.	G-3
	Conduct the integrating processes.	Integrating process leads: G-2, G-3, chiefs of fires, chief of protection, and KMO
	Enhance understanding of an operational environment.	G-2
Activity Lead	**Protect**	**Task Leads**
Chief of Protection	Secure and obscure friendly information.	OPSEC officer
	Conduct security activities.	G-3
	Defend the network, data, and systems.	G-6
Activity Lead	**Inform**	**Task Leads**
PAO	Inform and educate Army audiences.	Army leaders and PAO
	Inform United States domestic audiences.	PAO
	Inform international audiences.	PAO
Activity Lead	**Influence**	**Task Lead**
G-3	Influence adversary and enemy perceptions and behaviors.	G-39
	Influence other foreign audiences.	
Activity Lead	**Attack**	**Task Lead**
G-3	Degrade threat command and control	Chief of Fires (DFSCOORD)
	Affect threat information warfare.	

C2   command and control   G-39   assistant chief of staff, information plans and operations
DFSCOORD   deputy fire support coordinator   OPSEC   operations security
G-2   assistant chief of staff, intelligence   KMO   knowledge management officer
G-3   assistant chief of staff, operations   PAO   public affairs officer
G-6   assistant chief of staff, signal

While most staff work occurs in the functional and integrating cells, successfully integrating the information activities into operations occurs when functional expertise from across the staff comes together in support of the commander's decision requirements. This occurs in integrating cells and when the commander directs temporary groupings of staff members in boards, working groups, and planning teams.

Army forces require authorities to conduct operations. Some information tasks—to include MISO, cyberspace operations, and some types of deception activities—illustrate activities requiring specific authorities. When execution authority is granted for these operations, the command may have to meet specific reporting requirements. The staff judge advocate verifies authorities required to execute required information tasks have been granted by the appropriate authority prior to execution.

The information task leads use established boards, working groups, and planning teams in conjunction with the integrating processes to incorporate the five information activities into the operations process. The operations assessment, plans synchronization, and targeting boards are examples of boards typically found within a unit's battle rhythm that help to integrate information tasks. The assessment, cyberspace electromagnetic activities, civil-military operations, information collection, knowledge management, protection, and targeting working groups exemplify working groups typically found within a unit's battle rhythm.

B. Preparing *(See chap. 5.)*

Preparation consists of those activities performed by units and Soldiers to improve their ability to execute an operation (ADP 5-0). Preparation creates conditions that improve friendly force opportunities for success. It requires commander, staff, and Soldier actions to ensure the force is ready to execute operations. Preparing to execute information tasks often requires Army forces to anticipate and account for requirements earlier than many other Army tasks. This is because some information tasks require additional coordination or lead times to create desired effects. Some types of preparation begin at home station during competition, for example, configuring and training various information systems. In other cases, units assigned to a combatant commander may already be conducting information activities to support the commander's campaign objectives during competition which will enable friendly operations in a crisis or during conflict.

Successful preparation enables leaders to—

- Improve situational understanding.
- Develop a common understanding of the plan.
- Train and become proficient on critical tasks.
- Task-organize and integrate the force.
- Maintain unit resiliency.
- Ensure forces and resources are positioned.
- Protect critical aspects of operations.

Preparation to execute information activities takes place within headquarters and by units across the Army. The staff executes various activities in preparation to integrate and assess information activities during execution. Some of these activities include—

- Continuing to revise and refine planned information tasks and support development of branches and sequels.
- Conducting external coordination and establishing liaison to integrate echelons and synchronize information tasks.
- Assessing ongoing information collection and updating information requirements.
- Tracking and monitoring the movement and integration of units executing specific information tasks.
- Coordinating for the necessary authorities to execute anticipated information tasks.
- Helping subordinate commanders to understand specified and implied information activities and tasks.
- Participating in rehearsals to ensure information tasks are synchronized with the concept of operations.
- Assessing and mitigating vulnerabilities created through inadvertent information signatures.

Some unit preparations include ensuring that friendly forces—

- Can identify misinformation and disinformation.
- Can identify threat information disruption or information attacks.
- Understand relevant actors within their assigned areas.
- Understand how to report relevant information.
- Understand what actions to take to reinforce the prevailing narrative.
- Understand how to reduce risks associated with friendly emission in the electromagnetic spectrum (EMS).
- Understand potential impacts of friendly use of EW capabilities on friendly communications.

C. Executing *(See chap. 6.)*

Execution is the act of putting a plan into action by applying combat power to accomplish the mission and adjusting operations based on changes in the situation (ADP 5-0). Commanders, staffs, and subordinate commanders focus their efforts on translating decisions into action. They direct action to apply combat power, of which information is a dynamic, to achieve objectives and accomplish missions.

The current operations integrating cell assesses the effects and performance of information activities during execution. Current operations integrating cell members compare the current situation to the plan as an operation progresses. These members modify information tasks as necessary to accomplish the mission. Common staff tasks related to executing information tasks include the following:

- Monitor information tasks and tasks intended to have effects in the information dimension.
- Nominate targets for attack.
- Update running estimates.
- Monitor networks and electromagnetic emissions.
- Deconflict information activities.
- Synchronize information tasks across warfighting functions with lower, higher, and adjacent headquarters, and with outside agencies when appropriate.

D. Assessing *(See chap. 8.)*

Information activities and tasks must be continually assessed to judge whether they achieve the desired outcome. Assessment is not a discrete step of the operations process. Assessing information activities and tasks is continuous and informs the other activities of the operations process. Staffs assess information activities and tasks while working in functional and integrating cells, and while participating in cross-functional meetings such as working groups and boards. The purpose of assessing information activities and tasks is to equip the commander with the analysis necessary to make better decisions.

Assessing information activities is an integral part of knowing if friendly forces have achieved various information advantages. Assessing information activities requires—

- Developing the assessment approach (planning).
- Developing and publishing the assessment plan (planning).
- Collecting information and intelligence (planning, preparation, and execution).
- Analyzing information and intelligence (planning, preparation, and execution).
- Communicating feedback/recommendations (planning, preparation, execution).
- Adapting plans or operations (planning, preparation, and execution).

Commanders typically use two types of indicators to assess. One type of indicator is referred to as a measure of performance. Units use a measure of performance to measure a friendly action tied to task accomplishment. It answers the question: Was the task accomplished? The echelon executing the information task is typically best able to assess whether it executed a specific task successfully. Another type of indicator is referred to as a measure of effectiveness. Units use measures of effectiveness to measure change in a system or in target behavior over time to assess if a force is achieving objectives or attaining the end state. Measures of effectiveness help to answer: Are the information activities and tasks Army forces execute contributing to achieving an objective or the end state? Because achieving objectives typically requires different units executing a variety of tasks, the echelon assigning the tasks can typically best measure effectiveness.

Refer to ATP 5-0.3 for further discussions on each step of the assessment process. Refer to JP 3-04 for more detail on assessing information advantage.

Information Training and Education

Ref: ADP 3-13, Information (Nov '23), pp. 8-11 to 8-18 (table 8-1).

The Army uses training and education to equip Soldiers and leaders with the knowledge and skills they need to compete and fight with information. All Soldiers receive informational training and education appropriate to their assigned specialty and when executing operations training aligned with a specific OE. Informational training and education help Soldiers develop skills they apply both at home station and while deployed. The training includes basic skills that every Soldier must know and apply to protect information and information systems, to understand and support narratives, and to recognize and become resilient when facing information disruption and malign behavior.

Common Training and Education

Army training and education provides Soldiers and leaders with an individual understanding of the information contest occurring during all three strategic contexts. Soldiers gain a better understanding of ways adversaries use information and technical means to undermine the United States on societal and global scales.

Training varies depending on the situation and a Soldier's experience. Examples of informational training common to all Soldiers include—

- Resiliency to information for effect, misinformation, and disinformation.
- Understanding of operational and strategic narratives.
- Soldier and leader engagement (SLE).
- Ability to protect friendly information and information systems.
- Ability to use information systems and access classified information.
- Ability to conceal visual signatures and patterns of life.
- Ability to limit personal electromagnetic signatures.

Technical Training and Education *(See chap. 3.)*

Technology and the threat's use of technology to collect information, protect information, manipulate information, shape attitudes and beliefs, mobilize mass actions, and hinder the commander's exercise of C2 continuously threaten Army forces. Adversaries armed with long-range precision weapons—and with the ability to integrate them with technical information capabilities for direct and indirect confrontation—pose operational challenges. To counter these challenges, the Army trains and educates technical specialists who possess the ability to counter and defeat threat activities. Examples of technical informational training for select information specialists include—

- Civil affairs operations.
- Counterintelligence (CI).
- Cyberspace operations.
- Electromagentic warfare (EW).
- Information management (IM).
- Intelligence and the various intelligence disciplines
- Knowledge management.
- Military deception (MILDEC).
- Military Information in Support of Operations (MISO).
- Network operations.
- Public affairs operations.
- Space operatlons.
- Operational law.

I. Joint Force Uses of Information

Ref: JP 3-04, Information in Joint Operations (Sept '22), chap. 2.

I. Military Operations and Information

Information is a resource of the informational instrument of national power at the strategic level. Information is also a critical military resource. The joint force uses information to perform many simultaneous and integrated activities. The joint force uses information to improve understanding, decision making, and communication. Commanders use information to visualize and understand the OE and direct and coordinate actions. The joint force leverages information to affect the perceptions, attitudes, decision making, and behavior of relevant actors. The joint force employment of information is of central importance because it may provide an operational advantage.

II. The Operational Environment (OE) and the Information Environment (IE)

An OE is the aggregated conditions, circumstances, and influences that affect the employment of forces and bear on the decisions of a commander. Each commander's OE is different from every other commander's OE.

Within the OE there exist factors that affect how humans and automated systems derive meaning from, act upon, and are impacted by information. We refer to the aggregate of social, cultural, linguistic, psychological, technical, and physical factors as the IE.

The IE is not distinct from any OE. It is an intellectual framework to help identify, understand, and describe how those often-intangible factors may affect the employment of forces and bear on the decisions of the commander.

The joint force plans and conducts activities and operations that have inherent informational aspects that will impact the factors that make up the IE. The joint force must account for those informational aspects so that joint force activities and operations affect the OE in a way that supports the JFC's objectives. Additionally, to ensure unity of effort among different commands, each JFC must consider and communicate how the informational aspects of their planned activities and operations may impact the factors that make up the IE to affect other OEs.

III. Information Advantage

Information advantage is the operational advantage gained through the joint force's use of information for decision making and its ability to leverage information to create effects on the IE. Commanders achieve this advantage in several ways: identifying threats, vulnerabilities, and opportunities along with understanding how to affect relevant actor behavior; obtaining timely, accurate, and relevant information with an ascribed level of confidence or certainty for decision making and the impact of decision making; influencing, disrupting, or degrading the opponent's decision making; protecting the joint force's morale and will; and degrading the morale and will of adversaries. The joint force exploits these advantages through the conduct of operations. For example, disabling an opponent's space-based assets might provide the joint force with the operational advantage of being able to communicate securely over long distances without interruption and of being able to move without being detected. The joint force could then exploit that advantage through an operation to destroy an enemy ground force. Likewise, gaining and maintaining sufficient goodwill among a

JP 3-04, Information in Joint Operations, Sept '22 (Summary of Changes)

Joint publication (JP) 3-04 guides how the joint force considers and uses information to support achieving its objectives. This JP identifies the operational significance of information in achieving commanders' objectives across the competition continuum. This publication is the result of a change in mindset based on the joint force's recognition that all activities have inherent informational aspects that impact the operational environment (OE) and can generate effects that may contribute to or hinder achieving commanders' objectives. The Department of Defense (DOD), in coordination with the other United States Government (USG) departments and agencies, supports the informational instrument of national power by using information to impact the way in which humans and systems behave or function. The joint force leverages information across the competition continuum to assure, deter, compel, and force relevant actor behaviors that support US interests.

Joint Force Transition from "Information Operations" (IO) to "Operations in the Information Environment" (OIE)

The establishment of the information joint function and the development of joint publication (JP) 3-04 on information in joint operations is driving changes across joint and Service DOTMLPF-P [doctrine, organization, training, materiel, leadership and education, personnel, facilities, and policy]. **One significant doctrinal change is the transition from joint information operations (IO) to operations in the information environment (OIE).** This transition is a substantial force development challenge requiring the joint force to evaluate how to organize forces and staffs to deliberately plan and execute OIE.

The Armed Forces of the United States are poised to fight and win the Nation's wars. Transregional, all-domain, and multifunctional threats require the joint force to conduct operations across the competition continuum to prevent armed conflict and set the conditions to prevail during armed conflict. To deter or defeat these threats and achieve strategic objectives, the joint force commander (JFC) should understand how information impacts the OE, use information to support human and automated decision making, and leverage information through offensive and defensive actions to affect behavior. Relevant actors include individuals, groups, populations, or automated systems whose capabilities or behaviors can affect the success of a particular campaign, operation, or tactical action.

The joint force can win tactical fights during armed conflict but has not always been able to translate victories into enemy behaviors that lead to intended, enduring, strategic outcomes. Defeat of an enemy, by whatever mechanism, is usually a psychological outcome. The enemy is not really defeated until they believe they are defeated. Even in operations without an enemy or adversary, such as foreign humanitarian assistance, successful outcomes hinge on the perceptions, attitudes, beliefs, and other drivers of behaviors of the affected population.

The joint force cannot rely on attrition or its ability to compel behavior through the use of destructive and disruptive lethal force. To support achieving the commander's objectives, the joint force deliberately leverages information through activities that inform audiences; influence foreign relevant actors; and attack and exploit information, information networks, and information systems.

JP 3-04, Information in Joint Operations CANCELS JP 3-13, Information Operations

This supersedes and cancels JP 3-13, Information Operations, 27 November 2012 Incorporating Change 1, 20 November 2014. Relevant material from JP 3-13 has been incorporated into the main body and appendices of this publication. Accordingly, JP 3-13, Information Operations, will be removed from the joint doctrine hierarchy.

Joint IO, as defined and practiced, had shortcomings that inhibited it from contributing to the commander's application of informational power. As defined, IO focused on the integration of information-related capabilities (IRCs) to affect the decision making of adversaries and potential adversaries, and effectively ignored other relevant actors that shape the strategic and operational environments. IO planning concentrated on the employment of those IRCs in support of broader joint force operations, ignoring planning for the inherent informational aspects of all activities.

JP 3-04 describes how the joint force applies informational power across the competition continuum. That application of informational power includes both the deliberate leveraging of the inherent informational aspects of activities as an imperative for all joint force operations, and the conduct of OIE. OIE are military actions involving the integrated employment of multiple information forces to affect drivers of behavior by: informing audiences; influencing foreign relevant actors; attacking and exploiting relevant actor information, information networks, and information systems. As such, OIE are distinct from, but complementary to, the joint forces' deliberate leveraging of the inherent informational aspects of military activities during all operations.

OIE calls for formations with the capabilities (i.e., the authorities and tools, as well as subject matter experts possessing in-depth skills, knowledge, and abilities to employ those tools) required to carry out actions that leverage information to affect behavior. Building and resourcing organizations with subject matter experts and tools is part of the joint and Service force development challenge.

JP 3-04, Information in Joint Operations, 14 September 2022, Active Terms and Definitions

Information Environment (IE). The aggregate of social, cultural, linguistic, psychological, technical, and physical factors that affect how humans and automated systems derive meaning from, act upon, and are impacted by information, including the individuals, organizations, and systems that collect, process, disseminate, or use information. Also called IE. (Approved for incorporation into the DOD Dictionary.)

Knowledge Management (KM). A discipline that integrates people and processes to create shared understanding, increased organizational performance, and improved decision making. Also called KM. (Approved for inclusion in the DOD Dictionary.)

Operations in the Information Environment (OIE). Military actions involving the integrated employment of multiple information forces to affect drivers of behavior. Also called OIE. (Approved for inclusion in the DOD Dictionary.)

Relevant Actor (RA). Individual, group, population, or automated system whose capabilities or behaviors have the potential to affect the success of a particular campaign, operation, or tactical action. (Approved for inclusion in the DOD Dictionary.)

Target Audience (TA). An individual or group selected for influence. Also called TA. (Approved for incorporation into the DOD Dictionary.)

Terms Removed from the DOD Dictionary

Supersession of JP Supersession of JP 3-13, Information Operations, 27 November 2012; Incorporating Change 1, 20 November 2014:

- Information operations
- Information operations intelligence integration
- Information-related capability
- Information superiority

IV. Informational Power

Ref: JP 3-04, Information in Joint Operations (Sept '22), pp. II-2 to II-5.

Informational power is the ability to use information to support achievement of objectives and gain an informational advantage. The essence of informational power is the ability to exert one's will through the projection, exploitation, denial, and preservation of information in pursuit of objectives. The joint force cannot achieve all of its strategic objectives by relying solely on attrition to coerce change in the behavior of an enemy or adversary. The joint force leverages the power of information as a means to support achievement of its objectives.

> **The joint force applies informational power in two ways.** First, the entire joint force plans and conducts all operations, activities, and investments to deliberately leverage their inherent informational aspects. Second, specially trained and equipped units conduct operations in the information environment (OIE). Leveraging the inherent informational aspects of activities in combination with OIE maximizes the effectiveness of all joint force activities.

The joint force can leverage the power of information to effectively expand the commander's range of options. The joint force applies informational power:

To operate in situations where the use of destructive or disruptive physical force is not authorized or is not an appropriate course of action (COA). The majority of joint force operations support campaigns and do not involve armed conflict. Leveraging information through operations that do not use destructive or disruptive force may be the only viable option to achieve the JFC's intent and objectives. Conducting noncombat operations and activities to communicate the purpose of joint operations, reinforced by information activities, may be the most effective way for the JFC to develop local and regional situational awareness, build networks and relationships with partners, shape the OE, keep tensions between nations or groups below the threshold of armed conflict, and maintain, enhance, and expand US global influence.

To degrade, disrupt, and destroy the C2 ability of an adversary or enemy. The joint force interferes with an adversary or enemy's ability to execute the decision cycle thus degrading their ability to make appropriate command decisions. This includes targeting intelligence, surveillance, and reconnaissance (ISR) and C2 systems to interfere with an enemy's ability to understand joint force operations and effectively control their forces.

To prevent, counter, and mitigate the effects of external actors' actions on friendly capabilities and activities. The joint force also uses information for defensive purposes. This includes denying an adversary or enemy access to friendly critical information that would allow them to impede joint force C2, understanding of the OE, movement and maneuver, and sustainment.

To create and enhance the psychological effects of destructive or disruptive physical force. The use of destructive or disruptive force creates psychological effects. Executing actions specifically to create desired psychological effects can elicit profound changes in behavior. Amplifying or manipulating certain features and details of these activities to emphasize the psychological effects of destructive or disruptive force can be a more effective way of achieving joint force objectives than relying on physical force alone to destroy or disrupt enemy capabilities.

To create psychological effects without destructive or disruptive force. The joint force conducts information activities to influence foreign relevant actors, in conjunction with other efforts (e.g., show of force, foreign military sales). In some of these activities, information is the main effort, supported by maneuver elements and the implicit threat of force.

To confuse, manipulate, or deceive an adversary or enemy to create an advantage or degrade the adversary or enemy's existing advantage. By leveraging information to confuse, manipulate, or deceive an adversary, the joint force has the potential to deter threats or induce actions favorable to the JFC. By doing so, the joint force may mislead adversary commanders as to the strength, readiness, locations, and intended missions of friendly forces, causing them to misallocate or waste combat power.

To prevent, avoid, or mitigate any undesired psychological effects of operations. This is particularly true in cases where civilians may be affected by armed conflict. This includes the potential consequences of physical harm, as well as the destruction of homes and key infrastructure. The joint force takes feasible precautions to protect civilians from harm and addresses civilian casualty incidents if they occur. These efforts include disseminating information to remove civilians from areas of risk, preparing deliberate public communication efforts to minimize reaction to the occurrence of any civilian casualties due to joint force operations, and providing releasable information on actions taken to minimize harm to civilians. More broadly, communication with the civilian population can allay their concerns during periods of increased tension or counter adversary efforts to stoke civil unrest.

To communicate and reinforce the intent of joint force operations, regardless of whether those activities are constructive or destructive. The JFC cannot assume audiences intuitively understand the intent of joint force operations and activities and behave in ways that support the JFC's objectives. Even when the joint force is engaged in constructive activities, audiences may misinterpret the JFC's intent. Planning activities to leverage information based upon an understanding of the intended and likely audiences will reduce the chance of misinterpretation. In some cases, competitors, adversaries, or enemies will attempt to use disinformation about the intent of joint force activities to undermine joint force credibility and freedom of action, or even take credit for the positive outcome of US activities. Planning and conducting joint activities and operations in ways that communicate the intent, supported by OIE, promotes understanding of the mission, enables initiative, and counters disinformation.

To prepare and support resilience in partner nations' populations. The imminent or perceived imminent threat of force can be as psychologically damaging as the use of force. Many PNs execute programs to instill and ensure resiliency in their populations to guard against the psychological effects of potential physical force as well as to guard against attempted influence by adversary informational activities.

Operations in the Information Environment (OIE)

OIE are military actions involving the integrated employment of multiple information forces to affect drivers of behavior by informing audiences; influencing foreign relevant actors; attacking and exploiting relevant actor information, information networks, and information systems; and protecting friendly information, information networks, and information systems. OIE are conducted in support of the JFC's operation or campaign objectives or in support of other components of the joint force. Joint forces continuously conduct OIE to remain engaged with relevant actors.

See pp. 2-37 to 2-34.

local population provides the operational advantage of joint forces being able to move more freely in the vicinity of the populace without the locals alerting insurgents to friendly force activities. The joint force could exploit that advantage by conducting operations to capture insurgents hiding in or near civilian populations and by conducting operations that facilitate the host nation (HN) delivery of services to the population.

V. Relevant Actors *(See p. 2-15.)*

Advantages are usually thought of in relation to an opponent. However, the joint force also recognizes that friendly and neutral actors also have the potential to positively or negatively impact the friendly mission. By understanding the importance of all the relevant actors and the relationships between them, the JFC develops operation plans (OPLANs) that effectively leverage information to support achievement of objectives. Those relevant actors that the joint force intends to affect then become audiences for inform tasks, target audiences (TAs) for influence tasks, or targets for joint fires or other action.

See p. 2-64 for further discussion of audiences, targets, and target audiences.

Relevant actors include individuals, groups, populations, or automated systems whose capabilities or behaviors have the potential to affect the success of a particular campaign, operation, or tactical action. The nature of information and how it impacts the OE will change how relevant an actor may or may not be to the success of joint force activities. Military operations have inherent informational aspects that can negatively or positively impact an actor and, ultimately, change how relevant they are to the joint force.

Automated systems are the sets of software and hardware that allow computer systems, network devices, or machines to function without human intervention. These automated systems detect and react to sensory inputs to make sense of their environment, act upon that sense making based upon programming or experience and receive feedback. These systems can be platform-based (e.g., satellite, robot) or may reside and act entirely in cyberspace (e.g., bots, malicious code). Depending upon purpose and required actions, these systems may have a varying degree of autonomy.

VI. Joint Force Use of Narrative

Narratives are an integral part of campaigns, operations, and missions. When two or more organizations' narratives are received by an actor, the narratives can be perceived as either competing or complementing. Competing and parallel narratives exist and are used by a broad range of actors (e.g., partners, allies, competitors, adversaries, enemies) to gain support for their efforts. The joint force strives to provide a compelling narrative that is integrated into OPLANs and resonates with relevant actors by fitting their frame of reference. An effective and integrated narrative can mitigate, undermine, or otherwise render competing narratives ineffective if it is accompanied by complementary actions.

The joint force uses narrative as part of campaigning to support understanding the purpose of military operations, link military activities with the activities of other USG departments and agencies, and reflect policy objectives. It provides an overarching expression of strategy and context to a military campaign, operation, or situation. A narrative provides internal and external audiences with the intended meaning of joint force operations, actions, activities, and investments. An effective narrative affects perceptions and attitudes to complement or compete with other narratives. While the joint force conducts all operations to achieve objectives, the narrative explains why the joint force is carrying out operations so the actions are planned and conducted in a way that complements the narrative and avoids a "say-do gap." Planning joint force missions to align with the narrative helps the joint force increase the probability that relevant actors will derive the intended meaning from joint force operations.

Refer to JP 3-04, App. A, Narrative Development.

II. Information (as a Joint Function)

Ref: JP 3-0, Joint Campaigns and Operations (Jun '22), chap. III.

A **joint function** is a grouping of capabilities and activities that enable JFCs to synchronize, integrate, and direct joint operations. A number of subordinate tasks, missions, and related capabilities help define each function, and some tasks and systems could apply to more than one function.

There are seven joint functions common to joint operations: **C2, information, intelligence, fires, movement and maneuver, protection, and sustainment**. Commanders leverage the capabilities of multiple joint functions during operations. The joint functions apply to all joint operations across the competition continuum and enable both traditional warfare and IW, but to different degrees, conditions, and standards, while employing different tactics, techniques, and procedures.

I. Information (as a Joint Function)

The elevation of information as a joint function impacts all operations and signals a fundamental appreciation for the military role of information at the strategic, operational, and tactical levels within today's complex OE.

> The **information function** encompasses the management and application of information to support achievement of objectives; it is the deliberate integration with other joint functions to change or maintain perceptions, attitudes, and other elements that drive desired relevant actor behaviors; and to support human and automated decision making. The information function helps commanders and staffs understand and leverage the prevalent nature of information, its military uses, and its application during all military operations. This function provides JFCs the ability to preserve friendly information and leverage information and the inherent informational aspects of military activities to achieve the commander's objectives. The information joint function provides an intellectual framework to aid commanders in exerting one's influence through the timely generation, preservation, denial, or projection of information.

All military activities have an informational aspect since most military activities are observable in the Information Environment (IE). Informational aspects are the features and details of military activities observers interpret and use to assign meaning and gain understanding. Those aspects affect the perceptions and attitudes that drive behavior and decision making. The JFC leverages informational aspects of military activities to gain an advantage in the OE; failing to leverage those aspects in a timely manner may cede this advantage to an adversary or enemy. Leveraging the informational aspects of military activities can support achieving operational and strategic objectives. The information function also encompasses the use of friendly information to influence foreign audiences and affect the legitimacy, credibility, and influence of the USG, joint force, allies, and partners. Additionally, JFCs use friendly information to counter, discredit, and render irrelevant the disinformation, misinformation, and propaganda of other actors.

The information joint function helps commanders and their staffs understand and leverage the pervasive nature of information, its military uses, and its application across the competition continuum, to include its role in supporting human and automated decision making. Information planners should consider coordination activities not only within the information joint function but also among all other joint functions. The information joint function organizes the tasks required to manage and apply information during all activities and operations.

II. Information Use Across the Competition Continuum

Ref: JP 3-0, Joint Campaigns and Operations (Jun '22), pp. III-24 to III-26.

COOPERATIVE Use of Information

During day-to-day activities, the joint force integrates information in SC and FHA activities by:
- Assuring and maintaining allies, widening/publicizing combined exercises and other PN cooperation activities, encouraging neutral actors that the joint force is the partner of choice or that they should remain neutral, and reminding partners of benefits to maintain their support.
- Informing enemies and adversaries of benefits to friendly multinational force membership and collective defense, informing enemies and adversaries that the joint force is committed to its allies and security agreements, and concealing investment priorities and costs.

COMPETITIVE Use of Information

During competition, the joint force conducts activities against state or non-state actors with incompatible interests that are below the level of armed conflict. Competition can include military operations such as CO, special operations, demonstrations of force, CTF, and ISR and often depends on the ability to leverage the power of information through OIE. Expect additional time to coordinate and obtain approval from DOD or other USG departments and agencies to use information due to increased risk. Specific information tasks may include:
- Informing allies and partners of malign influence and antagonistic behavior.
- Declassifying and sharing images that reveal or confirm enemy or adversarial behavior, recommending allies and partners communicate to relevant audiences within their areas of influence, and educating the joint force and allies about online disinformation activities to build understanding and resilience against propaganda.
- Influencing adversary's audiences to prevent escalation to armed conflict by demonstrating joint force resolve, strength, and commitment, as well as the costs and expectations of response actions.
- Targeting adversarial information, networks, and systems by temporarily denying communication or Internet access, disrupting jamming of Internet access to its internal population, and partnering with private-sector communication companies to remove inappropriate enemy and adversarial recruiting and fundraising advertisements.

Use of Information in ARMED CONFLICT

In addition to the above tasks, the joint force can use information defensively or offensively. JFCs can employ information as independent activities, integrated with joint force physical actions, or in support of other instruments of national power. Many of these information activities require additional authorities as they present larger strategic risks or risks to the joint force, though capabilities like PA, which has the preponderance of public communication resources and rarely requires additional authorities in armed conflict.

Defensive Purposes. Basic defense activities include protecting data and communications, movements, and locations of critical capabilities and activities. PA can assist in countering adversary propaganda, misinformation, and disinformation. MILDEC can help mask strengths, magnify feints, and distract attention to false locations. DCO can defeat specific threats that attempt to bypass or breach cyberspace security measures. EW can protect personnel, facilities, and equipment from any effects of friendly, neutral, or enemy use of the EMS. The management of EM signatures can mask friendly movements and confuse enemy intelligence collectors.

Finally, well-coordinated communication and messaging activities not only minimize OPSEC violations but also increase the consistency and alignment of joint force words, actions, and images. Conflicting messages or remaining silent allows adversaries and enemies to exploit or monopolize the media and propagate their agenda.

Offensive Purposes. Offensive information activities decrease enemy and adversary effectiveness, increase Ally and partner support and effectiveness, and reduce interference from neutral audiences.

Exploit Informational Weaknesses of the Threat

Communicate and provide images. JFCs expose illegal or malign activities to international and enemy civilian audiences such as enemy human rights abuses, reveal funding sources of enemies or adversaries, and demonstrate other actions inconsistent with the law of war and the treaties and customary international law embodied in these principles.

- Expose enemy decisions to their populaces that result in significant loss to their resources, lives, and treasures.
- Increase exploitation of adversary rifts, beliefs, or perceptions by publicizing enemy tactical failures, poor equipment readiness, inconsistent logistics, enemy surrenders, populace skepticism, and other internal vulnerabilities that distract enemy leadership.
- Manipulate enemy messaging to confuse their supporters, allies, and partners.
- Conduct OCO to deny use or confidence in enemy communication networks, information systems, or weapon systems.
- Disseminate messages to relevant enemy audiences to create or increase ambiguity.
- Conduct MILDEC in support of friendly attacks to mislead adversaries or foreign intelligence about friendly attack capabilities, locations, methods, and timing.
- Conduct physical movements or fires that support MILDEC by targeting adversary communication, information, or weapon systems in support of feints, demonstrations, or ruses to create perceptions that a targeted area is a primary maneuver objective.
- Destroy or nullify selected adversary intelligence collection capabilities.
- Conduct an EA to prevent or reduce an enemy's effective use of the EMS via the employment of systems or weapons that use EM energy (e.g., jamming in the form of EM disruption, degradation, denial, and deception).
- Employ systems or weapons that use radiated EM energy (to include directed energy [DE]) as their primary disruptive or destructive mechanism.
- Conduct signature management to support OPSEC, MILDEC, and offensive or defensive activities.
- Disseminate information that can reduce civilian interference, minimize collateral damage, and help to reduce military and civilian casualties.
- Recommend targets and provide support to enable USG departments and agencies to increase economic pressure activities. Examples include freezing enemy finance support, exposing threat finance transactions, exposing illegal arms trading, and exposing third-party financial and resource support to enemy activities.
- Conduct KLEs with international media, allied counterparts, and other third-party communicators that echo the joint force narrative.
- Avoid targeting and messaging of cultural locations or issues that unite enemy leadership and its citizens.
- Anticipate setbacks and opportunities by synchronizing and preapproving senior leader response messages.

III. Joint Force Capabilities, Operations, and Activities for LEVERAGING INFORMATION

Ref: JP 3-0, Joint Campaigns and Operations (Jun '22), pp. III-16 to III-25.

In addition to planning all operations to benefit from the inherent informational aspects of physical power and influence relevant actors, the JFC also has additional means with which to leverage information in support of objectives. Leveraging information involves the generation and use of information through tasks to inform relevant actors; influence relevant actors; and/or attack information, information systems, and information networks.

See chap. 3, Information Capabilities, for further discussion. (See also p. 2-19 and 2-41.)

Key Leader Engagement (KLE) *(See p. 3-74.)*
Most operations require commanders and other leaders to conduct KLE with key local and regional leaders to affect their attitudes, gain their support, and cultivate them as sources of information. Building relationships to the point of effective military engagement and influence usually takes time. An organic or reliable indigenous language, regional expertise, and cultural capability are critical for the successful conduct and management of KLEs.

Public Affairs (PA) *(See pp. 3-5 to 3-16.)*
Enemies and adversaries will make determined efforts to discredit US military efforts. PA contributes to the achievement of military objectives by truthfully informing US domestic and international audiences about US military operations. PA ensures the clear communication of CCMD, and joint force messaging supports the strategic narrative and counters adverse disinformation, misinformation, and propaganda.

Civil-Military Operations (CMO) *(See pp. 3-17 to 3-26.)*
CMO facilitates unified action in joint campaigns and operations. They are activities that establish, maintain, influence, and exploit relationships between military forces, indigenous populations, and institutions. Effective CMO results in the integration of military and other instruments of national power to achieve commander's objectives and US interests.

Military Deception (MILDEC) *(See pp. 3-27 to 3-32.)*
Commanders conduct MILDEC to mislead enemy decision makers and commanders and cause them to take or not take specific actions. The intent is to cause enemy commanders to form inaccurate impressions about friendly force dispositions, capabilities, vulnerabilities, and intentions; misuse their intelligence collection assets; and fail to employ their combat or support units effectively.

Military Information Support Operations (MISO) *(See p. 3-33.)*
MISO are planned operations to convey selected information and indicators to foreign audiences to influence their emotions, motives, and objective reasoning and ultimately induce or reinforce foreign attitudes and behavior favorable to the originator's objectives. MISO may use all means of communication, distribution, and message delivery as appropriate.

Operations Security (OPSEC) *(See pp. 3-39 to 3-44.)*
OPSEC uses a process to preserve friendly essential secrecy by identifying, controlling, and protecting critical information and indicators that would allow enemies and adversaries to identify and exploit friendly vulnerabilities. The purpose of OPSEC is to reduce vulnerabilities of the US and multinational forces to enemy and adversary exploitation, and it applies to all activities that prepare, sustain, or employ forces.

Signature Management *(See p. 3-57.)*
Signature management encompasses JFC actions to adjust, modify, or manipulate signatures—the observable aspects of administrative, technical, and physical joint force activities. JFCs oversee signature management in concert with OPSEC to protect friendly force information, information networks, and systems and to deliberately affect relevant actor decision making and behavior.

Electromagentic Warfare (EW) *(See pp. 3-55 to 3-60.)*
EW is the military action ultimately responsible for securing and maintaining freedom of action in the EMS for friendly forces while exploiting or denying it to adversaries. EW is an enabler for other activities that communicate or maneuver through the EMS, such as MISO, PA, or CO.

Combat Camera (COMCAM) *(See pp. 3-14 to 3-15.)*
Imagery is one of the most powerful tools available for informing internal and domestic audiences and for influencing foreign audiences. COMCAM forces provide imagery capability to the JFC across the competition continuum.

Historians
Historical reading and understanding are vital tools for commanders. Maintaining a command history is a command responsibility. Military historians deployed into combat provide real-time support to commander decisions, spark critical imagination and adaptation necessary for command leadership, and complete critical documentation for future lessons on military operations.

Space Operations *(See pp. 3-61 to 3-70.)*
Space operations support joint operations throughout the OE by providing space offensive and defensive operations; space-based surveillance and reconnaissance; missile warning; environmental monitoring; satellite communications; space domain awareness; space-based positioning, navigation, and timing (PNT); spacelift; satellite operations; and nuclear detonation detection. Space operations integrate offensive and defensive actions to achieve and maintain freedom of action in space.

Special Technical Operations (STO) *(See p. 3-72.)*
Commanders should deconflict and synchronize other activities with STO. STO action officers at CCMD or Service component HQs can provide military and civilian leadership with detailed information related to STO and its contribution to joint force operations.

Cyberspace Operations (CO) *(See pp. 3-47 to 3-54.)*
CO employ cyberspace capabilities to achieve objectives in or through cyberspace. Most DOD CO are routine uses of cyberspace to complete assigned tasks but not necessarily one of the three CO missions. These uses include actions like e-mail or researching information using the Internet. These activities do not require special authorities for DOD personnel; however, they are the source of most vulnerabilities to the DODIN when cybersecurity policies are not followed. The Cyber Mission Force and other cyberspace forces conduct specific OCO, DCO, and DODIN operations missions.

Refer to JFODS6: The Joint Forces Operations & Doctrine SMARTbook, 6th Ed. (Guide to Joint Warfighting, Operations & Planning). JFODS6 is updated for 2023 with new/updated material from the latest editions of JP 3-0 Joint Campaigns and Operations (Jun '22), JP 5-0 Joint Planning (Dec '20), JP 3-33 Joint Force Headquarters (Jun '22), and JP 1 Volumes I and II Joint Warfighting and the Joint Force (Jun '20). Additional topics and references include Joint Air, Land, Maritime and Special Operations (JPs 3-30, 3-31, 3-32 & 3-05).

IV. Information Joint Function Tasks

Ref: JP 3-04, Information in Joint Operations (Sept '22), pp. II-6 to II-15.

The information joint function encompasses the management and application of information to change or maintain perceptions, attitudes, and other drivers of behavior and to support human and automated decision making. The information joint function is the intellectual organization of the tasks required to use information during all operations— understand how information impacts the OE, support human and automated decision making, and leverage information (see Figure II-1). JFCs and their staff perform these tasks during all operations to accomplish their respective missions.

Information Joint Function Tasks

- **A** Understand How Information Impacts the Operational Environment (OE)
- **B** Support Human and Automated Decision Making
- **C** Leverage Information

A. Understand How Information Impacts the Operational Environment (OE) *(See pp. 0-8 to 0-10.)*

This task helps the joint force identify threats, vulnerabilities, and opportunities in the IE. It provides a foundation for, and supports the continued refinement of, joint intelligence preparation of the operational environment (JIPOE) products to improve the commander's decision making during planning, execution, and assessment of operations. There are three steps to understanding how information impacts the OE: analyzing of the informational, physical, and human aspects of the environment; identifying and describing relevant actors; and determining the most likely behaviors of relevant actors. These steps are continuous and iterative because the OE is always changing. Planners use the JIPOE products and inputs from other subject matter experts (SMEs) to understand the interrelationships between the informational, physical, and human aspects within the context of operational objectives. This task requires fusion of multi-source data from across, and external to, the joint force to achieve and maintain an understanding of how information impacts the OE. Sources of internally produced data for this task include inputs from intelligence, public affairs (PA), civil affairs (CA), cyberspace forces, psychological operations units, and C2 systems. Sources of information external to the joint force include USG departments and agencies, businesses, and academic communities, as well as foreign governments, international organizations, nongovernmental organizations (NGOs), and various traditional and nontraditional media sources. This task also relies on language, regional, and cultural expertise to help avoid mirror-imaging and other forms of bias.

Analysis of the Informational, Physical, and Human Aspects of the Environment

Understanding how information impacts the environment and identifying how it can be used to affect behavior requires analysis of the increasingly complex and dy-

Tasks and Outcomes of the Information Joint Function

Ref: JP 3-04, Information in Joint Operations (Sept '22), fig. II-1. Tasks and Outcomes of the Information Joint Function.

The information joint function provides the intellectual organization required to use information during all operations to create advantage in and through the information environment.

Tasks

- Understand how information impacts the operational environment.
- Support human and automated decision making.
- Leverage information.

Subtasks

Understand:
- Analyze informational, physical, and human aspects of the environment.
- Identify and describe relevant actors.
- Determine likely behaviors of relevant actors.

Support:
- Facilitate shared understanding across the joint force.
- Protect friendly information, information networks, and information systems.
- Protect joint force morale and will.

Leverage:
- Inform domestic and international audiences.
- Influence foreign relevant actors.
- Attack and exploit relevant actor information, information networks, and information systems.

Outcomes

Understand:
- Joint force identifies threats, vulnerabilities, and opportunities in the information environment.
- The joint force commander (JFC) has a better understanding of which drivers of relevant actor behavior to affect and how to affect them to achieve objectives.

Support:
- The JFC has accurate and timely information available on which to base decisions and is able to communicate those decisions for action.
- Joint force is able to maintain its morale and will against malign influence.

Leverage:
- The joint force is able to affect the drivers of relevant actor behavior and, ultimately, the behavior of those relevant actors in support of the JFC's objectives and enduring outcomes.
- Joint force operations and activities are perceived as legitimate and justified by domestic and international audiences.

The JFC uses the abilities provided by the information joint function during all operations. The understand task provides the JFC with the ability to identify threats, vulnerabilities, and opportunities in the IE and provides a better understanding of which drivers of behavior to affect to achieve objectives. These activities facilitate the availability of timely, accurate, and relevant information necessary for joint force decision making. The leverage task provides the JFC with the ability to inform audiences; influence foreign relevant actors; and attack and exploit, information, information networks, and information systems in support of the JFC's objectives and enduring outcomes. The joint force operationalizes the information joint function through operational design in planning of operations that use information and deliberately leverage the inherent informational aspects of its activities, and by conducting OIE.

Using operational design to plan operations that deliberately leverage the inherent informational aspects of activities and operations. Everything the joint force does impacts the IE, either by intent or incidentally. All joint force operations, activities, and investments have the potential to affect the perceptions, attitudes, and, ultimately, the behavior of relevant actors. The conclusions that observers draw from interpreting joint force activities may drive them to act in ways that impact the joint force. Whether or not commanders consider this during planning, their activities will impact the IE and resonate in their operational area and potentially other operational areas.

(Joint Ops) II. Information (as a Joint Function) 2-13

namic relationship of the informational, physical, and human aspects of an environment. A systems approach, such as political, military, economic, social, information, and infrastructure (PMESII), that focuses on the interactive nature and interdependence of each of the aspects to characterize an environment, has been found to be a best practice. Analysis using the three aspects does not separate elements of the environment into "bins" for individual analysis. Instead, this systems approach is a way of describing the different characteristics of objects, activities, or relevant actors; their informational, physical, and human aspects, and the context in which they exist. These results are included in the information staff estimate and help identify the relevant actors the joint force needs to affect, how to use information to effectively impact those relevant actors, and what friendly information the joint force needs to protect. The running estimate integrates intelligence and other information that characterizes the informational, physical, and human aspects of the environment against the established baseline to identify threats, vulnerabilities, and opportunities.

Informational aspects reflect the way that individuals, information systems, and groups communicate and exchange information. Informational aspects are the sensory inputs (e.g., content, medium, format, and context) of activities that a receiver interprets and uses to assign meaning. The content of communication can be verbal and nonverbal. If nonverbal cues do not align with the verbal message, ambiguity is introduced and uncertainty is increased. Medium refers to the system used to communicate (e.g., radio, television, print, Internet, telephone, fax, and billboard). The details of the medium can be described in as little or much detail as necessary. Format is how the information is encoded, such as what language is used, style of delivery (e.g., poetry, songs, imagery), tone, and volume. Context refers to the environment in which the communication happens (e.g., face-to-face, over the phone). Format and context can affect the content of a communication. For example, a text message may contain different content than the same communication delivered face-to-face. Actions are a form of nonverbal communication that have inherent informational aspects and are generally more impactful.

Physical aspects are the material characteristics, both natural and manufactured, of the environment that may inhibit or enhance communication. Physical aspects may create constraints and freedoms on the people and information systems that operate in it. Physical aspects are critical elements of group identity and impact how groups form, behave, or might be disrupted or cease to exist. For example, groups may be formed by the people inhabiting an island or an isolated jungle habitat. Similarly, a community might be disrupted by the building of a highway that divides a neighborhood and causes the creation of new, separate, and distinct communities. How information is exchanged is where the interplay between the informational and physical aspects is most apparent. As an example of this interplay, an isolated community without access to modern communications technology will likely have a stronger group identity and be more likely to communicate face-to-face compared to residents of a large modern city.

Human aspects are the interactions among and between people and the environment that shape human behavior and decision making. Those interactions are based upon the linguistic, social, cultural, psychological, and physical elements. Human aspects influence how people perceive, process, and act upon information by impacting how the human mind applies meaning to the information it has received. Individuals have distinct patterns of analyzing a situation, exercising judgment, and applying reasoning skills impacted by their beliefs and perceptions. Character and tradition are aspects that suggest how humans perceive a situation and how they might behave under particular circumstances in the future. For example, individual and group identity is often closely related to a geographical area, which can impact how individuals and groups in that region relate to one another and communicate along with the forms that communication may take. Describing these inextricably linked aspects will provide insight into relevant actors' worldviews that frame the perceptions, attitudes, and other elements that drive behaviors.

Identify and Describe Relevant Actors

The analysis of informational, physical, and human aspects of the OE provides the context needed to understand how individuals, groups, populations, and automated systems operate and makes it possible for the joint force to identify who or what is a relevant actor based upon the joint force mission and objectives. The staff conducts this analysis as part of the intelligence directorate of a joint staff (J-2)-led JIPOE process. Equipped with a thorough understanding of its objectives and the general context of the OE, the joint force undertakes deliberate steps to determine the environment in which the relevant actor exists. These efforts include the conduct of intelligence operations and communications with partners to improve knowledge of friendly, neutral, and threat actors and their social, cultural, political, economic, informational, cyberspace, and organizational networks. Intelligence's JIPOE and target systems analysis, psychological operations unit's target audience analysis (TAA), North Atlantic Treaty Organization (NATO) Strategic Communications Division's IE assessment, and CA's area studies and area assessments are analytical products and processes that can help identify and describe relevant actors. Other analysis products may be available from interagency and multinational partners.

In determining who or what is a relevant actor, the joint force considers the particular function and role of systems, individuals, groups, networks, and populations, while attempting to discern the affiliations and connections among them. Insight into institutions and their processes is often needed to comprehend the roles and relationships among actors. This includes a description of how relevant actors receive information and the factors that will impact the processing and interpretation of that information. The joint force should recognize that mission partners may be relevant actors that need to be understood to ensure unity of effort.

Identifying relevant actors goes beyond just listing entities of the friendly and enemy order of battle. It also includes a range of nonmilitary actors in the environment (e.g., local authorities, civilian supervisory control and data acquisition systems, religious leaders, community figures). Some potential relevant actors may exist far outside the geographic boundaries of an operation.

This is an iterative process where the staff continuously reassesses the relevance of actors and prioritizes them in regard to the commander's objectives and approach to mission accomplishment. The analysis and description of relevant actors will differ based upon whether the relevant actors are human or automated systems.

As part of JIPOE, network engagement and its associated analyses helps the JFC to identify and understand relevant actors and their associated links with others within a network.

For more information on network engagement, see JP 3-25, Joint Countering Threat Networks.

Identify Likely Behavior of Relevant Actors

This final step builds upon the previous steps to develop a detailed understanding of the range of available behavior options and assess which of those behaviors are most likely to have the greatest impact on the joint force. This is similar to traditional military planning where commanders and their staffs evaluate an enemy's most likely and dangerous COAs.

Identifying the likely behavior of relevant actors also helps the JFC and staff determine which relevant actor COAs in a given time and space will be advantageous or disadvantageous to friendly operations. This leads to the joint force being able to plan for activities that affect the drivers of behavior in support of achieving objectives.

Efforts to anticipate relevant actor reactions and decisions based upon joint force or other actions will be imperfect. Information will frequently be incomplete, imprecise, or flawed. Nevertheless, joint forces make use of the best information available.

Once the range of potential behaviors has been determined, the joint force is better able to select appropriate methods to affect future behavior, while considering intended and potential unintended effects. These predictions become inputs to identify initial collection requirements. Once collected and analyzed, the analysis will reveal which COA the relevant actor has adopted.

B. Support Human and Automated Decision Making

This task includes facilitating shared understanding across the joint force; protecting friendly information, information networks, and information systems; and protecting joint force morale and will. These activities help ensure the availability of timely, accurate, and relevant information necessary for joint force decision making.

Facilitate Shared Understanding

This task of the information joint function is related to building shared understanding in the C2 joint function and includes collaboration, knowledge management (KM) and information management (IM), and information and intelligence sharing. Although these are typically staff and organizational tasks conducted during daily operations, they are more critical and challenging in today's security environment given the exponential growth in the volume of information the joint force needs to analyze and share.

See facing page.

Protect Friendly Information, Information Networks, and Information Systems

This task helps ensure joint force C2 by protecting information and the systems and networks on which it resides from loss, manipulation, or compromise. Protection tasks are conducted during daily activities and are implied, if not specified, tasks for all units during all operations.

See following page.

Build, Protect, and Sustain Joint Force Morale and Will

Activities to build joint force morale and will reinforce the baseline strengths the Services have developed in their members to create a cohesive joint force and increase awareness of, and resistance to, malign influence and the demoralizing effects of operations to assure the joint force. Activities to protect joint force morale and will support the force's resiliency against trauma; deployment length; isolation; and propaganda, misinformation, disinformation, deception, persuasion, and dissuasion. As commanders build and protect the forces' resiliency, they prepare to sustain those gains. As conditions change in the OE, the force can be affected in a variety of ways. Sustaining resilience requires commanders to adapt to these changes. Examples of proactive measures and of countermeasures to build, protect, and sustain joint force morale and will include preparing Service members for the psychological effects of loss of life and mitigating those effects when they occur, conducting command information activities, facilitating shared understanding, authenticating trustworthy sources of information, establishing reliable and secure communications, conducting counter-deception and counter-propaganda activities, as well as conducting religious support and command psychologist activities, and facilitating face-to-face communication between command teams and Service members at the lowest echelon. The protection of information, information networks, and information systems task supports the protection of joint force morale by maintaining the integrity of information sent to and received from authenticated and reliable sources. Protecting information contributes to protecting joint force morale and will because it prevents adversaries from accessing or manipulating data and information to incite and spread dissension, confusion, and disorder.

Facilitate Shared Understanding
Ref: JP 3-04, Information in Joint Operations (Sept '22), pp. II-11 to II-12.

This task requires automated tools that manage and organize large quantities of disparate, structured, and unstructured data required for decision making. These tools, combined with people and processes, ensure the effective and timely transfer of knowledge to provide an operational advantage to commanders and other decision makers.

Collaboration
Collaboration includes activities such as sharing data across the joint force in real time; building situational awareness views; conducting collaborative planning and decision making; execution, coordination, and deconfliction of missions in near real time; and enabling IE visualization. That collaborative environment is enabled by communications systems and applications that improve long-distance, asynchronous collaboration among dispersed forces to enhance planning, execution, and assessment of joint operations. These systems and applications improve efficiency and common understanding during periods of routine interaction among participants and enhance effectiveness during time-compressed operations. Collaboration also requires information and intelligence sharing. Commanders at all levels should determine and provide guidance on what information and intelligence needs to be shared with whom and when. Standard operating procedures should include sharing information to the maximum extent allowed by US law and DOD policy.

Knowledge Management (KM) and Information Management (IM)
KM and IM facilitate understanding and decision making. KM is a discipline that integrates people and processes throughout the information lifecycle to create shared understanding, increase organizational performance, and improve decision making. KM identifies and fills knowledge gaps, minimizes or eliminates stovepipes, captures knowledge and transfers it to those who need to know, helps synchronize a battle rhythm, and cultivates a culture of sharing across multiple staff organizations. IM is the function of managing an organization's information resources for the handling of data and information acquired by one or many different systems, individuals, and organizations in a way that optimizes access by all who have a share in that data or a right to that information. IM provides a structure that supports and enables KM. Effective IM contributes to the KM tasks of knowledge creation and supports shared understanding for all unit members. Depending upon the size and mission of the command, the JFC may be supported in their information and KM responsibilities by various staff officers, including chief knowledge and information officers and supporting knowledge and IM officers.

Information and Intelligence Sharing
Sharing of information and intelligence with relevant USG departments and agencies, foreign governments, security forces, interorganizational participants, NGOs, and partner organizations in the private sector promotes interoperability and facilitates collaboration. The joint force shares information to the maximum extent necessary and allowed by US law and DOD policy (e.g., foreign disclosure law and policy). The public affairs officer (PAO), in coordination with the foreign disclosure officer, clears information for public release. While every country has its own sharing caveats, the United States often has additional responsibilities when leading an alliance or coalition. Risk to mission and risk to force related to sharing of information and intelligence is a consideration from the start of operational design in planning through execution. This includes consideration of the risk of not sharing information and intelligence.

Protect Friendly Information, Information Networks, and Information Systems

Ref: JP 3-04, Information in Joint Operations (Sept '22), pp. II-13 to II-14.

This task helps ensure joint force C2 by protecting information and the systems and networks on which it resides from loss, manipulation, or compromise. Protection tasks are conducted during daily activities and are implied, if not specified, tasks for all units during all operations.

Protect Information *(See pp. 1-29 to 1-34.)*

The protection of information includes passive and active measures to preserve information and prevent or mitigate competitor, adversary, and enemy collection, manipulation, and destruction of friendly information, to include attempts to undermine the trustworthiness of friendly information. Threats may attempt to manipulate or destroy friendly information to undermine the joint force's understanding, decision making, morale, and will. Activities that contribute to protecting information include intelligence, operations security (OPSEC), military deception (MILDEC), PA, IM, signature management, counterintelligence, cyberspace security procedures, and vulnerability assessments. Two categories of information relevant to joint operations are:

- **Classified information** is official information that has been determined to require, in the interests of national security, protection against unauthorized disclosure, and which has been so designated. Department of Defense Manual (DODM) 5200.01, DOD Information Security Program, identifies the procedures for classifying, marking, downgrading, declassifying, and safeguarding classified information.

- **Critical information.** Critical information is specific facts about friendly intentions, capabilities, and activities sought by adversaries and enemies to plan and act so as to thwart friendly mission accomplishment. Critical information may be unclassified information. For example, informational aspects can become signatures that divulge critical information by revealing intent or planned action. Department of Defense Directive (DODD) 5205.02E, DOD Operations Security (OPSEC) Program, directs personnel to maintain the essential secrecy of information that is useful to adversaries and potential adversaries to plan, prepare, and conduct military and other operations against the United States. This includes safeguarding critical information from unauthorized access and disclosure.

Protect Information Networks and Information Systems

Adversaries and enemies threaten joint forces through any vulnerability, to include joint force and partner networks, wireless apertures associated with weapons and C2 systems, and other processors and controllers. Through the cyberspace operations (CO) missions of Department of Defense information network (DODIN) operations and defensive cyberspace operations (DCO), cyberspace forces protect the DODIN and, when ordered, other friendly cyberspace capabilities from threats in cyberspace. This activity includes securing the DODIN from known vulnerabilities, educating DODIN users to recognize and thwart malicious cyberspace activity, implementing DOD cybersecurity policy, hunting for known or suspected threats in blue cyberspace, and engaging threats forward in gray and red cyberspace. These CO are informed by up-to-date knowledge about vulnerabilities in DODIN software and hardware, intelligence about malicious cyberspace activity, and counterintelligence analysis. Protecting the integrity and availability of friendly information helps support decision making.

See pp. 1-34 for related discussion from ADP 3-13 (Nov '23).

C. Leverage Information (See pp. 2-10 to 2-11 and 2-41.)

When commanders leverage information, they expand their range of options for the employment of military capabilities beyond the use of or threatened use of physical force. JFCs leverage information in two ways. First, by planning and conducting all operations, activities, and investments to deliberately leverage the inherent informational aspects of such actions. Second, by conducting OIE.

INFORM Domestic, International, and Internal Audiences

Inform activities are the release of accurate and timely information to the public and internal audiences, to foster understanding and support for operational and strategic objectives by putting joint operations in context; facilitating informed perceptions about military operations; and countering misinformation, disinformation, and propaganda. Inform activities help to ensure the trust and confidence of the US population, allies, and partners in US and MNF efforts; and to deter and dissuade adversaries and enemies from action. PA is the primary means the joint force uses to inform; however, civil-military operations (CMO), key leader engagement (KLE), and military information support operations (MISO) also support inform efforts.

See pp. 1-35 to 1-44 for related discussion from ADP 3-13.

INFLUENCE Relevant Actors

The purpose of the influence task is to affect the perceptions, attitudes, and other drivers of relevant actor behavior. Regardless of its mission, the joint force considers the likely psychological impact of all operations on relevant actor perceptions, attitudes, and other drivers of behavior. The JFC then plans and conducts every operation to create desired effects that include maintaining or preventing behaviors or inducing changes in behaviors. This may include the deliberate selection and use of specific capabilities for their inherent informational aspects (e.g., strategic bombers); adjustment of the location, timing, duration, scope, scale, and even visibility of an operation (e.g., presence, profile, or posture of the joint force); the use of signature management and MILDEC operations; the employment of a designated force to conduct OIE; and the employment of individual information forces (e.g., CA, psychological operations forces, cyberspace forces, PA, combat camera [COMCAM]) to reinforce the JFC's efforts. US audiences are not targets for military activities intended to influence.

See pp. 1-45 to 1-48 for related discussion from ADP 3-13.

ATTACK AND EXPLOIT Information, Information Networks, and Information Systems

The joint force targets information, information networks, and information systems to affect the ability of adversaries and enemies to use information in support of their own objectives. This activity includes manipulating, modifying, or destroying data and information; accessing or collecting adversary or enemy information to support joint force activities or operations; and disrupting the flow of information to gain military advantage. Attacking and exploiting information, information networks, and information systems supports the influence task when it undermines opponents' confidence in the sources of information or the integrity of the information that they rely on for decision making. Activities used to attack and exploit information include offensive cyberspace operations (OCO), electromagnetic warfare (EW), MISO, and CA operations. PA also contributes to this task by publicly exposing malign activities.

See pp. 1-49 to 1-56 for related discussion from ADP 3-13.

V. Key Considerations (Information Function)

Ref: JP 3-0, Joint Campaigns and Operations (Jun '22), pp. III-26 to III-27.

Intelligence Support to the Information Joint Function. Intelligence is critical to the effectiveness of information activities. Intelligence facilitates understanding the interrelationship of the informational, physical, and human aspects within the OE and the IE. By providing a society-centric, sociocultural understanding of the OE, intelligence can greatly assist the planning, integration, execution, and assessment of information activities to create desired effects. Intelligence in support of information activities may require longer lead times to establish behavior baselines for human decision making.

Information in the Targeting Process. Planners integrate information activities and capabilities into the targeting process during planning and execution to create and synchronize effects in support of objectives. Many information activities and capabilities have interagency coordination requirements. Targeting approval levels may increase the time required to plan, coordinate, and execute the process. Other USG departments and agencies lack trained personnel and procedures to satisfy interagency planning, execution, and assessment requirements. Fully analyzing and developing target sets for nonlethal action may also increase coordination time required. Some information activities such as STO may be compartmentalized. However, effective integration of information activities and capabilities in the targeting process results in an improved understanding of the entire joint and multinational force and increased opportunities to achieve JFC objectives.

Legal Considerations. US military activities, including OIE, must always be in compliance with US laws and policies. Planners deal with diverse and complex legal considerations. Legal interpretations can occasionally differ, given the complexity of technologies involved, the significance of legal interests potentially affected, and the challenges inherent for laws and policies to keep pace with the technological changes and implementation. Policies are regularly added, amended, and rescinded to provide clarity. As a result, legal restraints and constraints on information activities are dynamic. Multinational considerations further complicate them since each nation has its laws, policies, and processes for approving plans.

The complexity of the IE. The JFC will not be the only voice in the IE. Individuals, nongovernmental groups, or non-state actors can have outsize effects on JFC operations due to the rapid and comprehensive (i.e., multiple media formats) nature of traditional media and social media services. Consequently, the JFC's information plan should include social media services engagement as a form of information dissemination, along with traditional media outlets.

Leveraging relevant actors and the media they use to communicate is vital to establishing legitimacy, credibility, and influence. During mission analysis, the JFC and staff should identify relevant nonmilitary actors (e.g., indigenous formal and informal leaders and influencers) and their respective influences on the OE. Identifying and cultivating relevant actors and the media they use to communicate allows the joint force to develop ways to leverage their influence to accomplish the joint force mission. JFCs and staff should confirm or deny assumptions on relevant actors made during mission analysis through CMO, military engagement, and network engagement.

Proactive Communication Planning. The nature of the IE enables relevant actors to receive and react to information in the OE before JFCs can react. Some relevant actors create and leverage information to cause reactions. This reality leaves little time for the joint force to craft and disseminate communication to respond. Instead of constantly reacting to each negative or positive information event to gain an advantage in the IE, commanders ensure the proactive development and dissemination of information in line with communication plans. Additionally, the JFC needs to delegate information release decisions to the lowest possible level to enable timely action in the IE.

III. Unity of Effort (Information Forces)

Ref: JP 3-04, Information in Joint Operations (Sept '22), chap. 3.

Unity of effort is the coordination and cooperation toward common objectives, even if the participants are not necessarily part of the same command or organization. Unified action is the synchronization, coordination, and/or integration of the activities of governmental and nongovernmental entities with military operations to achieve unity of effort. It is essential to all DOD initiatives to achieve unity of effort through unified action with interagency partners, the broader interorganizational community, and multinational partners. The joint force collaborates with other USG departments and agencies and with multinational partners to effectively use and leverage information to achieve strategic objectives.

DOD's role in maintaining unity of effort in and through the IE is, for the most part, the same as it is for the physical domains. DOD establishes policies and sets the conditions for components and their staffs to identify adversarial and potential adversarial threats (including attempts to undermine US alliances and coalitions) and bring capabilities to bear in an effort to affect, undermine, and erode an adversary's or enemy's will. Additionally, DOD closely coordinates operations, activities, and investments with other USG departments and agencies to facilitate horizontal and vertical continuity of strategic themes, messages, and actions.

To facilitate unity of effort, the JFC and supporting staff should be familiar with the roles, expertise, and capabilities of individual and organizational stakeholders relative to the use of information and leveraging information to create relative advantage over an opponent. The JFC will need to understand what activities external organizations are currently doing to leverage information and whether the inherent informational aspects of their activities support or hinder the joint force objectives and mission. The JFC's challenge is how best to deconflict, synchronize, coordinate, and/or integrate activities to achieve unified action. This chapter describes the authorities of DOD related to information in joint operations, delineates various roles and responsibilities of organizations that support the joint force use and leveraging of information, describes DOD and interorganizational collaboration and multinational partner considerations regarding their contribution to OIE, and addresses legal considerations in the planning and execution of OIE.

Authorities

Military activities that leverage information frequently involve a unique set of complex issues. There are legal and policy requirements, including DOD directives and instructions, national laws, international laws (i.e., international treaties, the law of war), and rules of engagement, all which may affect these activities. Laws, policies, and guidelines become especially critical during peacetime operations and competition when international and domestic laws, treaty provisions, and agreements are more likely to affect planning and execution. Commanders should know who has the execution authority for the conduct of information activities. Many capabilities require separate and distinct execution authorities (e.g., MISO and some CO). Normally, the JFC is designated as the execution authority in the execute order (EXORD) but should consider requests for delegation of certain authorities down to the lower echelons to support tactical commanders. The exercise of operational authority over joint forces conducting information activities inherently requires a detailed and rigorous legal interpretation of authority and/or legality of specific actions. Legal considerations are addressed in more detail later in this chapter. Commanders will also need to know who has release authority for information. For example, release authority can be granted to the joint task force (JTF) PA for unclassified COMCAM products to expedite their release to the media.

The Department of Defense Strategy for Operations in the Information Environment established strategic initiatives for DOD to operate effectively in and through the IE, defend national interests, and achieve national security objectives. This strategy guides DOD support to the whole-of-government effort. It complements, and supports, other guidance documents, including the National Security Strategy of the United States of America, 2017 [short title: NSS]; 2018 National Defense Strategy of the United States of America: Sharpening the American Military's Competitive Edge [short title: NDS]; Department of Defense Cyber Strategy 2018; and the Department of Defense Strategy for Implementing the Joint Information Environment, which focuses on IT implementation. In general, the Department of Defense Strategy for Operations in the Information Environment describes operational-level objectives for OIE in which, through operations, actions, and activities in the IE, DOD has the ability to affect the decision making and behavior of adversaries and designated others to gain advantage across the competition continuum.

The Global Integrated Operations in the Information Environment [short title: GIOIE] EXORD directs the joint force to conduct globally integrated operations to maximize the cognitive impacts of combined informational power and physical force on an adversary and other relevant actor perceptions and decision making, to coerce behavior, communicate the costs of aggression, offer opportunities for updating alliances, and create new strategic partnerships to protect US interests. The GIOIE EXORD also addresses the need to improve OIE by adopting methods that deliberately align our analysis, decisions, investments, activities, operations, and relationships in time, space, and purpose.

Title 50, USC, Section 3093, states that any activity of the USG to influence political, economic, or military conditions abroad, where it is intended that the role of the USG will not be apparent or acknowledged publicly, is a covert action and is only authorized pursuant to a presidential finding. This is considered during the identification of attribution requirements and impacts any non-attribution or delayed attribution decisions. The law further states that traditional military activities fall outside of the statute.

Title 17, USC, governs the use of copyrights. The joint force uses a variety of multimedia formats and commonly incorporates music, symbols, graphics, and messages into its products. It is important to note these products are required to adhere to the copyright restrictions under Title 17, USC, that protect published and unpublished works in a variety of forms and formats.

Responsibilities

Information can have significant regional and global impacts that challenge the joint force with unanticipated threats, vulnerabilities, and opportunities. Effectively dealing with these challenges and communicating intended meanings to selected populations requires individuals and organizations across DOD and interagency partners to ensure coherency with, and align their policies and activities to, national strategic objectives. Unified command enables the synchronization, coordination, and/or integration of activities of governmental and nongovernmental entities with military operations to achieve unity of effort in support of an overall strategy. Senior leaders work with the other members of the national security community to promote unified action. A number of factors can complicate the coordination process, including various agencies' different and sometimes conflicting policies and overlapping legal authorities, roles and responsibilities, procedures, and decision-making processes for information activities. This section describes responsibilities of individuals and organizations related to achieving and maintaining unity of effort in the application of informational power.

I. Service Organizations
Ref: JP 3-04, Information in Joint Operations (Sept '22), pp. III-23 to III-27.

The Services man, train, and equip organizations to provide the joint force with the ability to leverage information during joint operations and to conduct OIE. Those Service organizations provide distinct specialized capabilities to the joint force (e.g., MISO, CMO, CO, PA, EW, COMCAM) or provide information commands composed of multiple specialized capabilities that focus on leveraging information and enable the joint force to create effects in the IE. Those Service-provided organizations that are trained and equipped to conduct OIE are referred to as OIE units.

United States Army
- Army Cyber Command (ARCYBER).
- 1st IOC [1st Information Operations Command] (Land).
- United States Army Special Operations Command (USASOC).
- United States Army Civil Affairs and Psychological Operations Command.
- The United States Army National Guard TIOG.

United States Navy
- United States Fleet Cyber Command (US FCC)/United States Tenth Fleet. US FCC reports directly to the Chief of Naval Operations as a Navy Echelon 2 command and is assigned to USCYBERCOM.
- United States Naval Information Forces (NAVIFOR). NAVIFOR mans, trains, and equips information warfare capabilities ashore and afloat.

USMC
- DC I [Deputy Commandant for Information].
- Marine Corps Forces Cyberspace Command (MARFORCYBER). MARFORCYBER is assigned to USCYBERCOM and conducts the full spectrum of CO.
- Marine Expeditionary Force Information Group (MIG). MIGs coordinate, integrate, and employ capabilities to ensure the MAGTF commander's ability to facilitate friendly forces maneuver and deny the enemy freedom of action in the IE.
- Marine Corps Information Operations Center (MCIOC). The MCIOC provides operational support to the Marine Corps forces and MAGTFs and provides OIE subject matter expertise in support of USMC OIE advocates and proponents to enable the effective integration of OIE into Marine Corps operations.
- Civil Affairs Group (CAG).

United States Air Force
- 16th AF [Sixteenth Air Force]/Air Force Cyber Command [AFCYBER]. 16th AF is responsible for developing, preparing, generating, employing, and presenting information warfare forces.
- 616th OC [616th Operations Center]. The 616th OC handles daily intelligence-gathering and offensive and defensive missions in the air, in cyberspace, and across the EMS.
- 16th AF Information Warfare Cell.

United States Space Force (USSF)
- Space Delta 6 CO. Space Delta 6, as part of Space Operations Command (SPOC), executes CO to protect space operations, networks and communications.
- Space Delta 8 Satellite Communications and Navigational Warfare. Space

II. Department of State (DOS) Organizations
Ref: JP 3-04, Information in Joint Operations (Sept '22), pp. II-7 to III-10.

DOS plans and implements foreign policy. DOS is led by the Secretary of State, who is the President's principal advisor on foreign policy and the person chiefly responsible for US representation abroad. DOS's primary job is to promote and communicate American foreign policy throughout the world. DOS interfaces with representatives of foreign governments, corporations, NGOs, and private individuals. A key DOS function is assembling coalitions to provide military forces for US-led multinational operations, as well as communicating the President's policies to other nations and international bodies. The following internal DOS offices have information-related duties and with whom DOD planners may need to coordinate OIE:

Under Secretary for Public Diplomacy and Public Affairs
The Under Secretary for Public Diplomacy and Public Affairs serves as the lead policy maker for DOS's overall public outreach and press strategies. The Under Secretariat team coordinates closely with the regional bureaus, functional bureaus, interagency partners, the private sector, and international partners to ensure DOS's public diplomacy and PA activities are consistent, forward-looking, supportive of US foreign policy, and grounded in research.

Bureau of Educational and Cultural Affairs
The Bureau of Educational and Cultural Affairs designs and implements educational, professional, and cultural exchanges and other programs that create and sustain the mutual understanding with other countries necessary to advance US foreign policy goals. The Bureau's programs cultivate people-to-people ties among current and future global leaders that build enduring networks and personal relationships and promote US national security and values sharing America's rich culture of performing and visual arts with international audiences.

Bureau of Global Public Affairs
The Bureau of Global Public Affairs serves the American people by effectively communicating US foreign policy priorities and the importance of diplomacy to American audiences and engaging foreign publics to enhance their understanding of and support for the values and policies of the United States. Some of the centers and offices in this bureau include:

- **Foreign Press Centers.** The Foreign Press Centers' mission is to deepen understanding of US policy and American values through engagement with foreign media. They provide clear and accurate understanding of policy and American values to global audiences via first-hand access.

- **Office of Global Social Media.** The Office of Global Social Media expands the reach of US foreign policy through new media and web-based communication technology. Working with the entire DOS, the team maintains the DOS's official blog, DipNote.

- **Office of Global Web Platforms.** The Office of Global Web Platforms oversees the DOS's use of websites to inform the public. State.gov delivers information about DOS, such as press releases, key policy information, and details about the US relationship with countries and areas of the world.

Office of International Media Engagement. The Office of International Media • Engagement creates and manages DOS mechanisms to ensure accurate coverage of US foreign policy priorities by major international media. The office oversees the DOS's six regional media hubs, which serve as overseas platforms for engagement of foreign audiences via the media. The office ensures DOS international media capabilities are integrated into the interagency press and PA planning and execution.

- **Office of Press Operations.** The Office of Press Operations supports the President and Secretary of State by explaining the foreign policy of the United States and the positions of DOS to domestic and foreign journalists. The office responds to press queries, conducts media interviews, monitors media for breaking international events, and coordinates special press briefings and conference calls.
- **Office of Public Liaison.** The Office of Public Liaison connects DOS to domestic audiences to advance the DOS's work at home and abroad. The Office of Public Liaison also responds to inquiries on foreign policy issues, handles requests for briefings from groups coming to DOS, and partners with organizations to sponsor major conferences and events.
- **Global Engagement Center.** The Global Engagement Center directs, leads, synchronizes, integrates, and coordinates efforts of the federal government to recognize, understand, expose, and counter foreign state and non-state propaganda and disinformation efforts aimed at undermining or influencing the policies, security, or stability of the United States and its allies and partner nations.

Office of Policy, Planning, and Resources

The Office of Policy, Planning, and Resources for Public Diplomacy and Public Affairs provides long-term strategic planning and performance measurement capability for public diplomacy and PA programs. It also enables the Under Secretary to better advise on the allocation of public diplomacy and PA resources, to focus those resources on the most urgent national security objectives and provide realistic measurement of public diplomacy's and PA's effectiveness.

US Advisory Commission on Public Diplomacy (ACPD)

The ACPD appraises USG activities intended to understand, inform, and influence foreign publics and to increase the understanding of, and support for, these same activities. The ACPD conducts research and symposiums that provide independent assessments and informed discourse on public diplomacy efforts across government. Supported by the Under Secretary of State for Public Diplomacy and Public Affairs, the Commission reports to the President, Secretary of State, and Congress.

Country Teams

The country team unifies the coordination and implementation of US national policy within each foreign country under the direction of the chief of mission (COM), working directly with the HN government, and consists of key members of the US diplomatic mission or embassy. Country teams meet regularly to advise the COM on matters of interest to the United States and review current developments in the country. The COM, as the senior US representative in each HN, controls information release in country. The CCMDs are the primary entry point for DOD personnel to coordinate with country teams in their AOR. CCMD staff coordinates all themes, messages, VI products, and press releases impacting a HN through the respective US embassy channels. The DOS foreign policy advisor (POLAD) at CCMDs can facilitate access to DOS and has reachback to resources for CCMD staff. The COM also directs the country team system, which provides the means for rapid interagency consultation and action on recommendations from the field (including US embassies, CCMDs with an AOR, and international programs) with a consistent USG voice and effective execution of US programs and policies. The CCDR and staff should establish habitual working relationships with relevant organizations before incidents occur that trigger planning and requests for military resources. As emergent events requiring planning develop, the normal flow of DOS and other agencies reporting from the field will increase significantly. Under the country team construct, USG departments and agencies are required to coordinate their plans and operations (including OIE) and keep one another and the COM informed of their activities (including activities that leverage information). The COM has the right to see all communication to, or from, mission elements, except those specifically exempted by law or executive decision.

III. Combatant Commanders (CCDRs)

Ref: JP 3-04, Information in Joint Operations (Sept '22), pp. III-12 to III-14.

The Unified Command Plan provides guidance to CCDRs, assigning them missions. CCDRs exercise combatant command (command authority) over assigned forces and are directly responsible to SecDef for the preparedness of their commands to perform, and their performance of, assigned missions. CCDRs are responsible for the implementation of strategy and US policy and the execution of assigned missions. One way they do this is by integrating, synchronizing, and employing forces to achieve effects in the IE that support achievement of operational objectives CCDRs also translate national strategic objectives into operational objectives that specify the desired behavior of relevant actors to support the attainment of enduring strategic outcomes.

CCDRs organize their staffs to best employ the information joint function. This may include standardizing organizational practices by aligning related capabilities into the same directorate, establishing routine working groups, and establishing a center with responsibility for the information joint function tasks, while maintaining PAO as their principal spokesperson, senior advisor, and a member of their personal staff. They also ensure all plans mitigate vulnerabilities, counter threats, and exploit opportunities in the IE. CCDRs develop and prioritize intelligence requirements that support leveraging information. CCDRs and subordinate JFCs develop, plan, program, and assess information activities during all phases of military engagement across the competition continuum; in coordination with the USD(P) and CJCS, identify and seek the appropriate delegated authorities required for leveraging information; integrate information guidance for theater planning and deliberate and contingency planning; develop interagency coordination requirements and mechanisms for each OPLAN; and ensure coordination and deconfliction of CCMD information and intelligence activities in all operational planning and execution. CCDRs guide the collaborative development of narratives for their assigned responsibilities and ensure actions across AORs or functional areas to align with that narrative.

The following CCDRs have additional responsibilities related to the information joint function:

Commander, United States Special Operations Command (CDRUSSOCOM)

CDRUSSOCOM is the designated joint proponent for MISO and CA, responsible for leading the collaborative development, coordination, and integration of MISO and CA capabilities across DOD. This responsibility is focused on enhancing interoperability and providing other CCDRs with MISO and CA planning and execution capabilities. CDRUSSOCOM also serves as the coordinating authority for MISO Web-Operations, and conducts transregional MISO with concurrence from applicable CCMDs. The USSOCOM Joint MISO Web Operations Center provides the JFC with a capability which facilitates and conducts MISO employing social media, mobile applications, websites, and other Internet-based capabilities and technologies to influence foreign audience behavior.

Commander, United States Cyber Command (CDRUSCYBERCOM)

CDRUSCYBERCOM is the coordinating authority for global CO. This responsibility includes planning, coordinating, integrating, synchronizing, and conducting OCO, DCO, and DODIN operations. CDRUSCYBERCOM conducts CO in support of national objectives and provides other CCDRs with CO planning and execution capabilities.

Commander, United States Strategic Command (CDRUSSTRATCOM)

CDRUSSTRATCOM's assigned responsibilities include strategic deterrence, nuclear operations, joint electromagnetic spectrum operations (JEMSO), global strike, global missile defense, and analysis and targeting. As the joint proponent for JEMSO, CDRUSSTRATCOM focuses on enhancing interoperability and providing other CCDRs with contingency EW expertise in support of their missions. This is in addition to the responsibilities shared by all CCDRs, in coordination with the USD(P) and through the CJCS. CDRUSSTRATCOM coordinates JEMSO.

Commander, United States Space Command (USSPACECOM)

Commander, USSPACECOM, plans and executes global space operations, activities, and missions. Space supports the flow of information and decision making. It may also serve as an information capability essential to the delivery of specific information in the IE. Space control consists of operations to ensure freedom of action in space for the United States and its allies and, when directed, deny an adversary freedom of action in space. The space control mission area includes defensive and offensive activities; supported by the requisite current and predictive knowledge of the space environment.

Commander, United States Transportation Command (USTRANSCOM)

Commander, USTRANSCOM, is responsible for mobility and joint enabling capabilities. One of USTRANSCOM's components is Joint Enabling Capabilities Command that provides mission-tailored capability packages on short notice for limited duration to assist the joint force plan, prepare, establish and operate joint force headquarters in globally integrated operations. The Joint Planning Support Element (JPSE) deploys expeditionary, mission tailored, joint SME across operations, plans, sustainment, intelligence, KM, and PA.

Refer to JP 3-08, Interorganizational Cooperation, for more information on the various organizations and their respective roles and responsibilities related to interorganizational cooperation.

Refer to JFODS6: The Joint Forces Operations & Doctrine SMARTbook, 6th Ed. (Guide to Joint Warfighting, Operations & Planning). JFODS6 is updated for 2023 with new/updated material from the latest editions of JP 3-0 Joint Campaigns and Operations (Jun '22), JP 5-0 Joint Planning (Dec '20), JP 3-33 Joint Force Headquarters (Jun '22), and JP 1 Volumes I and II Joint Warfighting and the Joint Force (Jun '20), Additional topics and references include Joint Air, Land, Maritime and Special Operations (JPs 3-30, 3-31, 3-32 & 3-05).

IV. The Joint Force

The JFC establishes and communicates command-specific guidance to ensure all joint force operations and activities are planned and executed to account for the effective management and application of information. This will include assigning responsibility for the tasks related to the information joint function. This may include standardizing organizational practices, establishing routine working groups, or establishing a center with responsibility for the information joint function tasks. Each of the directorates has responsibility related to information joint function tasks, but the JFC should assign overall responsibility and authority to a staff lead to ensure unity of effort. The JFC may choose to create additional staff or functional organizations to conduct or coordinate joint force activities related to the leveraging of information, coordinate with other organizations to obtain support, or synchronize activities with other organizations. This includes creating groups of specialized forces to conduct OIE. The JFC may choose to retain control of any newly created formation under the operations directorate of a joint staff (J-3) or create a separate task force. From this point forward, "OIE unit" will be used to represent a formation that conducts OIE. The JFC also identifies requirements for information planners to serve as OIE and capability SMEs and planners on the joint force staff or other headquarters staffs. During operational design and joint planning, the JFC provides planning guidance that describes the desired conditions that must exist in the IE to support mission accomplishment, how the joint force will leverage the inherent informational aspects of its activities to support the JFC's objectives, how information activities will support the scheme of maneuver, and the types and level risk that the JFC will accept in the IE. The JFC will also assign missions to OIE units.

A. JFC's Staff

The JFC's staff performs duties and handles special matters over which the JFC wishes to exercise close, personal control. JFCs and their staffs evaluate communication considerations with the interagency partners when planning joint operations. The staff advises the JFC on the inherent informational aspects of their activities, including how words and images will impact the JFC's operational areas. The staff also advises the JFC when their activities may have effects on the IE that impact other AORs. The chief of staff (COS) manages the staff. The staff group may include, but is not limited to, the PAO, staff judge advocate (SJA), KM officer, and POLAD.

- **Political Advisor (POLAD).** POLADs are senior DOS officers (often flag-rank equivalent) detailed as personal advisors to senior US military leaders and commanders, and they provide policy analysis and insight regarding the diplomatic and political aspects of the commanders' duties. Due to their status and contacts, they can enable interorganizational cooperation relationships and foster unity of effort. The POLAD provides USG foreign policy perspectives and diplomatic considerations and establishes links to US embassies in the AOR or joint operations area (JOA) and with DOS. They articulate DOS objectives relevant to the CCMD's theater strategy or JTF commander's plans.

- **Public Affairs Officer (PAO).** The PAO is the commander's principal spokesperson, senior PA adviser, and a member of the CCDR's personal staff. In that role, the PAO provides counsel to leaders, leads PA and communication activities, collaborates with other information planners to develop the narrative, supports the commander's intent, and supports community engagement and KLE. The PAO may also co-chair the JFC's information CFT.

- **Joint Force Staff Judge Advocate (SJA).** The joint force SJA, also titled the command judge advocate, is the principal legal advisor to the CCDR, with a focus on joint operational law issues pertaining to their commander's AOR.

Each joint staff directorate collaborates routinely, but to varying degrees, to plan, synchronize, support, and assess activities that leverage information.

Joint Organizations

Ref: JP 3-04, Information in Joint Operations (Sept '22), pp. III-14 to III-15.

The following joint organizations perform functions that support the joint force use and leveraging of information:

Joint Information Operations Warfare Center (JIOWC)

The JIOWC is a CJCS-controlled activity under the supervision of the JS Director for Operations. JIOWC enables the application of informational power at the strategic level and performs CJCS proponency responsibilities for joint enterprise information and information activities, MILDEC, and OPSEC, to create, enhance, or protect joint force advantages in the IE.

Joint Planning Support Element-Public Affairs (JPSE-PA)

JPSE-PA, a functional group within JPSE, plans, coordinates, and synchronizes PA activities with informational power activities to maximize support to campaign objectives and ensure execution of PA roles, responsibilities, and fundamentals. JPSE-PA provides ready, rapidly deployable, expeditionary joint PA capability to CCDRs to support joint operations, facilitate the rapid establishment of joint force headquarters, and bridge joint requirements supporting worldwide operations. JPSE-PA personnel assist development, planning, assessment, and synchronization of operational and mission narratives, themes, messages, PA and VI activities with the national narrative.

Joint Warfare Analysis Center

The Joint Warfare Analysis Center provides CCMDs, the JS, and other customers with effects-based analysis and precision targeting options for selected networks and nodes to carry out the national security and military strategies of the United States during peace, crisis, and war.

Joint Electromagnetic Warfare Center (JEWC)

JEWC integrates joint effects in the electromagnetic spectrum (EMS) by providing adaptive operational solutions and advocating for the coherent evolution of capabilities and processes to control the EMS during military operations. The JEWC assesses EW requirements, technology, and capabilities while conducting modeling, analysis, and EMS activity coordination between CCMDs and other USG departments and agencies. The JEWC also deploys EW experts, trains staffs, stands up forward planning cells, and delivers rapid warfighter support when required.

Joint Intelligence Support Element (JISE)/Joint Intelligence Operations Center (JIOC)

The JISE provides the JTF with tailored intelligence products and services with a continuous analytical capability. Capabilities of the element may include order of battle analysis, collection management, target intelligence, OIE analysis, a warning intelligence watch, and a request for information (RFI) desk. Alternatively, in a particularly large or protracted campaign, the JTF commander may decide to employ an operational-level JIOC. An operational-level JIOC incorporates the capabilities inherent in a JISE but is generally more robust. The JISE can provide population-centric, socio-cultural intelligence and physical network lay downs, including the information transmitted via those networks.

Defense Media Activity (DMA)

DMA is a mass media and training and education organization that creates and distributes DOD content across a variety of media platforms to audiences around the world.

B. Information Planning Cell & CFT

Ref: JP 3-04, Information in Joint Operations (Sept '22), pp. III-20 to III-23..

Information Planning Cell

JFCs may establish an information planning cell to provide command-level oversight and collaborate with all staff directorates and supporting organizations on informational considerations during planning and the conduct of operations. The information planning cell, composed of information professionals on the staff, serves as the focal point for planning how the joint force will leverage the inherent informational aspects of its activities and for planning OIE.

The information planning cell is a standing organization subordinate to the operations branch within the J-3 to provide command-level oversight on all aspects of leveraging information.

The information planning cell comprises personnel with subject matter expertise in OIE, specialized capabilities (e.g., CA, MISO, PA, EW, COMCAM, CO) and information activities (e.g., KLE, OPSEC) who serve as staff information planners. The J¬3 should tailor the composition of the cell as necessary to accomplish the mission. In cases where specialized capabilities have their own staff entities, SMEs may be assigned to the information planning cell as planners and serve as liaisons to their respective staff section (e.g., a PA planner assigned to the information planning cell would liaise with the PAO and PA staff; an EW planner assigned to the information planning cell would liaise with the JEMSO or joint electromagnetic spectrum operations cell [JEMSOC]).

The information planning cell members collaborate with all staff directorates and supporting organizations to ensure the joint force effectively leverages information as an element of maneuver in support of the JFC's objectives. Information planners provide subject matter expertise throughout operational design and the joint planning process (JPP) (see Chapter IV, "Operational Design and Planning"). The information planning cell supports the J-3 in the direction and control of operations to ensure the impacts, in and through the IE, of all activities support the JFC's objectives and enduring outcomes. Information planning cell members participate in staff joint planning groups (JPGs) or equivalent organizations and may be subtasked to serve as information planners in the JS J-5. The information planning cell chief heads the information CFT and may co-chair the information CFT with PAO. Information planning cell members comprise the core of the information CFT (see paragraph [5], "Information CFT") and is responsible for incorporating input from the information CFT into plans and overseeing execution of information activities.

The information planning cell collaborates with other staff sections to identify the inherent informational aspects of activities that should be included in those staff estimates. Additionally, they identify and maintain the information estimate.

Information Cross-Functional Team (CFT)

The information CFT is the JFC's forum for the development of a shared understanding of the IE and for the organization, coordination, and synchronization of joint force activities in and through the IE. The information CFT maintains situational awareness of the impact in and through the IE of operations, activities, and investments. As necessary, the information CFT develops and recommends alternatives or follow-on activities that support achieving the JFC's objectives.

The information CFT is comprised of members of the information planning cell and representatives from across the staff directorates, subordinate OIE units and information forces, and USG and other mission partners.

Members of the information CFT should establish ongoing communications with similar forums at the JS and other joint force, interagency, and multinational partners to ensure the joint force remains aware of the actions of others that may have impacts on those factors that make up the IE that will affect the JFC's OE. This awareness also helps identify threats, vulnerabilities, and opportunities in the IE.

Media Operations Center
A JFC may establish a media operations center to serve as the focal point for the interface between the military and the media during the conduct of military operations. The media operations center serves as a central meeting place for military personnel and media representatives and provides the media with a primary information source, a logistics support base, transmission capability, and a coordination base.

JEMSOC
The JEMSOC synchronizes and integrates the planning and operational use of electromagnetic support sensors, forces, and processes within a specific JOA to reduce uncertainties concerning the threat, environment, time, and terrain. The JEMSOC consolidates, prioritizes, integrates, and synchronizes the component electromagnetic spectrum operations (EMSO) plans and attendant EMS-use requests to produce a consolidated JEMSO plan. Joint force unity of effort in the EMS derives from the JEMSOC's integration of all joint force EMS actions across both the joint force's functional staff elements (e.g., signals intelligence, EMS management, EW, CO, fires) and the joint force's components.

KLE Cell
A KLE cell may be established to map, track, and distribute information about the key leaders within the JOA. The KLE cell should establish and maintain a human information database, recommend KLE responsibility assignment, deconflict KLE activities, conduct pattern analysis, develop a detailed background briefing on each key leader, suggest specific approaches for encouraging support for activities and objectives, ensure debriefs are conducted following engagements, and update the map with current information and intelligence and debrief information. The cell provides an updated map (with human information of the area), background information, and desired effects for KLE in the JOA to field units and staffs. The KLE cell coordinates subordinate command KLE activities to ensure a coherent effort across the JOA, gathering of debriefing information, and updating of the data base.

Counter Threat Finance (CTF) Cell
CTF cells are a central point to integrate threat finance intelligence into CTF operations and coordinate execution of CTF activities. The principal mission of a CTF cell is to identify and disrupt funding flows, financiers, and financial networks of terrorists, insurgents, and other relevant actors. CTF actions, activities, and operations are designed to deny, disrupt, destroy, or defeat the generation, storage, movement, and/or use of assets to fund activities that support an adversary's ability to negatively affect US interests.

Joint IM Cell
Depending on the size of the joint force and scope of operations, the COS may establish a joint IM cell within the joint operations center. The joint IM cell reports to the COS or joint operations center chief (or the J-3) and facilitates information flow throughout the JOA. The joint IM cell ensures the commander's dissemination policy is implemented as intended; takes guidance published in the commander's dissemination policy and combines it with the latest operational and intelligence information obtained from the joint operations center or joint analysis center; works closely with the joint network operations control center to coordinate potential changes in communications infrastructure to satisfy changes in the commander's information dissemination requirements; and coordinates the accurate posting of all current, approved commander's critical information requirements (CCIRs).

V. Information Forces
Ref: JP 3-04, Information in Joint Operations (Sept '22), pp. III-27 to III-29.

Information forces are those Active Component and Reserve Component forces of the Services specifically organized, trained, and equipped to create effects in the IE. These forces provide expertise and specialized capabilities that leverage information and can be aggregated as components of an OIE unit to conduct OIE. Information forces are available to the joint force through the RFF process.

Civil Affairs (CA)
CA provides expertise on the civil component of the OE. CA forces analyze and evaluate civil considerations for the commander and staff during mission analysis. CA forces promote the legitimacy of the mission by advising commanders on how to best meet their moral and legal obligations to the people affected by military operations. CA conducts civil reconnaissance and network engagement to help define the OE for the commander, to create options to influence the networks in support of US and joint forces information activities. CA coordinate, integrate, and synchronize plans and operations with the civil component. CA produce area studies, area assessments, and analysis that can help identify and describe civil considerations within the OE and refine the IE.

For additional guidance on CA and CMO, refer to JP 3-57, Civil-Military Operations.

Psychological Operations Forces
Psychological operations forces are trained and equipped to conduct MISO. Their primary task is to influence. They create effects in the IE, bringing significant human factors analysis, assessment, and capability to formulate MISO plans and programs that enhance the development and effectiveness of JFC's missions. MISO planners evaluate the psychological effects of military actions and advise the JFC and staff to maximize influence task effectiveness and minimize adverse impact and unintended consequences. The employment of psychological operations forces is governed by explicit policy and legal authorities that direct and determine how their capability is utilized. Synchronization of MISO with other actions precludes DOD messages or actions, and other agencies' messages and actions from contradicting or weakening each other.

For additional guidance on psychological operations forces and MISO, refer to JP 3-13.2, Military Information Support Operations.

Public Affairs (PA)
PA staffs are involved in planning, decision making, training, equipping, and executing operations, as well as integrating PA and communication activities into all levels of command and ensuring narrative alignment. PAOs and PA staffs also work with other planners to coordinate and deconflict communication activities. PA activities are divided into public information, command information, and community engagement activities, supported by research, planning, execution, and assessment to support the commander's intent and concept of operations (CONOPS). PAOs at all levels participate in planning, provide counsel to leaders and key staff members on the possible outcomes of military activities, lead development of the mission narrative, and identify the potential impact on domestic and international perceptions.

For more information on PA, refer to JP 3-61, Public Affairs.

Cyberspace Forces
Cyberspace forces comprise those personnel whose primary duty assignment is DODIN operations, DCO, or OCO. Cyberspace forces include the units of the Cyber Mission Force (CMF) as well as Service-retained units and various units assigned to CCMDs. CMF units operate under the tactical control of the supported CCDR or in direct support

or general support, depending upon the circumstances. Although it is possible for CO, including cyberspace-enabled OIE, to produce stand-alone tactical, operational, or strategic effects and thereby achieve objectives, commanders integrate most CO with other operations to create coordinated and synchronized effects required to support mission accomplishment. Cyberspace operations-integrated planning elements (CO-IPEs) integrate within CCDRs' CO support staff to provide CO expertise and reachback capability to USCYBERCOM. The CO-IPEs are organized from USCYBERCOM, JFHQ-DODIN and joint force headquarter-cyberspace personnel and are co-located at the supported CCMD.

For more information on CO, refer to JP 3-12, Joint Cyberspace Operations.

Electromagnetic Spectrum Operations (EMSO) Forces

JEMSO actions to exploit, attack, protect, and manage the electromagnetic environment rely on personnel and systems from EW, EMS management, intelligence, space, and cyberspace mission areas. EMSO personnel prioritize, integrate, synchronize, and deconflict all joint force operations in the electromagnetic environment, enhancing unity of effort. The result is a fully integrated scheme of maneuver in the electromagnetic environment to achieve EMS superiority and objectives.

For additional guidance on EMSO, refer to JP 3-85, Joint Electromagnetic Spectrum Operations.

(Combat Camera) COMCAM Forces

Imagery is one of the most powerful tools available for informing internal and domestic audiences and for influencing foreign audiences. COMCAM forces provide imagery support in the form of a directed imagery capability to the JFC across the competition continuum. COMCAM imagery supports capabilities that use imagery for their products and efforts, including MISO, MILDEC, PA, and CMO. COMCAM also provides documentation for sensitive site exploitation, legal and evidentiary requirements, battle damage assessment (BDA), operational assessment, and historical records.

For additional information on COMCAM, refer to CJCSI 3205.0I, Joint Combat Camera (COMCAM).

Space Forces

Space operations and activities that leverage information are mutually reinforcing. Space supports the flow of information and decision making. It may also serve as an activity essential to the delivery of specific information in the IE. Conversely, activities that leverage information to generate effects support achievement of space superiority. USSF Guardians on Service, CCMD, and other staffs ensure commanders and their staffs have a common understanding of space operations and how they should be integrated with other military operations to achieve unity of effort and meet US national security objectives. The Joint Combined Space Operations Center, on behalf of the Combined Forces Space Component Command, coordinates, plans, integrates, synchronizes, executes, and assesses space operations and facilitates unified action for joint space operations.

For more information, see JP 3-14, Space Operations.

VI. Interorganizational Collaboration

Interorganizational collaboration seeks to find common goals, objectives, and/or principles between diverse organizations to achieve unity of effort and, through planning and leveraging of cross-organizational capabilities, set the conditions to achieve unified action during execution. The relationship that the joint force establishes with relevant organizations helps it develop a more comprehensive awareness of the OE and understanding of the impact of information on the OE. Ultimately, these relationships help the joint force and the other organizations appreciate the impact of their activities and operations toward achieving shared objectives. This interdependency between the CCMD and other USG departments and agencies to achieve common objectives is a vital element of a whole-of-government effort. From an information planning perspective, unified action is particularly critical since the inherent informational aspects of activities resonate through the IE and may create desirable or adverse effects in the operational area. IT may also be leveraged to enable complex interorganizational coordination through collaborative virtual networking to facilitate reachback to the required SMEs. The JFCs and staff consider the capabilities and priorities of USG components, NGOs, and other interorganizational partners throughout the joint force's planning and execution of OIE.

A. Coordination and Synchronization

The deliberate coordination and synchronization of interorganizational efforts enables inclusion of the various perspectives, interests, and equities of each stakeholder; enhances friendly credibility and narrative; preserves legitimacy; mitigates the potential for conflicting messages; and improves the overall efficiency and effectiveness of whole-of-government efforts. To facilitate the working relationships among the stakeholders, the JFC will need to establish coordination and synchronization mechanisms to facilitate planning and execution with mission partners. The JFC establishes working relationships, specific organizational structures, and operational practices with external organizations to align activities and achieve unity of effort consistent with the overarching USG narrative. For example, the collaboration and synchronization of information activities can be accomplished through the establishment of cross-functional organizations (e.g., joint interagency task force [JIATF], JIACG) capable of leveraging information.

For more information, see JP 3-08, Interorganizational Cooperation.

B. USG Organizations

Effective integration of the appropriate USG organizations will enhance the overall success of joint force operations. There are a multitude of organizations inside and outside DOD that are relevant to the joint force's management and application of information. The JFC and staff, when appropriate, coordinate information activities and objectives for OIE with organizations that can impact the joint force's leveraging of information.

At the national level, the NSC, with its policy coordination committees and interagency working groups, advises and assists the President on all aspects of national security policy. OSD and the JS, in consultation with the Services and CCMDs, coordinate interagency support required to support the JFC's plans and orders. From an information joint function perspective, it is essential to coordinate activities that support creating the JFC's desired effects, with careful consideration of the inherent informational aspects of those activities. While a supported CCDR is the focal point for coordination of interagency supporting activities, interagency coordination with supporting commanders is also important. Prior to integrating interagency capabilities into their estimates, plans, and operations, JFCs should only consider those partners that can realistically commit their resources to the JFC's mission.

Any USG department or agency planning or conducting activities within the JOA, is considered a relevant organization. This is also true of private-sector entities and NGOs. JFCs and their staffs should consider how the capabilities of other USG components, NGOs, and members of the private sector (e.g., multinational corporations, academia, operational contract support) can be leveraged to assist in accomplishing their mission and broader national strategic objectives. JFCs should also consider the capabilities and priorities of interagency partners in planning and executing information activities. Such organizations do not necessarily need to have a physical presence in the JOA to have an impact. Joint planners need to account for these impacts.

In the case of international organizations, the JFC should determine the significance of their presence in the JFC's JOA and account for that presence in the JFC's planning and execution efforts.

C. Civil-Military Operations Center (CMOC) (See p. 3-24.)

The CMOC is a mechanism to coordinate CMO and can also provide operational- and tactical-level coordination between the JFC and other stakeholders. Horizontal and vertical synchronization among multiple CMOCs assists in unity of effort. The CMOC is the meeting place of stakeholders, providing a forum for military and other participating organizations. Sharing information is a key function of the CMOC. CMOCs receive, validate, and coordinate requests for support from NGOs, international organizations, indigenous populations and institutions, the private sector, and regional organizations. They also liaise and coordinate between joint forces and other agencies, departments, and organizations to meet the humanitarian needs of the populace. This level of interaction results in CMO having a significant effect on the perceptions of the local populace and improves understanding of the IE. Since this populace may include potential adversaries, their perceptions are of great interest to the information community. CMO can assist in identifying relevant actors; synchronizing communications media, assets, and messages; and providing news and information to the local population.

For more information on CMOCs, refer to JP 3-57, Civil-Military Operations.

D. Joint Interagency Coordination Group (JIACG)

The JIACG is an interagency staff group composed of USG civilian and military experts tailored to meet a validated CCDR's requirement. The primary role of the JIACG is to enhance interagency coordination. JIACGs facilitate unified action in support of plans, operations, contingencies, and initiatives. Members participate in planning and provide links back to their parent civilian departments and agencies to help synchronize JTF operations with their efforts. A JIACG provides the means to establish collaborative working relationships between civilian and military planners. For example, during joint operations a JIACG, as the bridge between the CCDR and interagency partners, provides the CCDR and subordinate commanders with an increased capability to coordinate and synchronize the joint force's leveraging of information with other USG departments and agencies. When augmented with other partners, such as international organizations, NGOs, and multinational representatives, the JIACG enhances the capability to conduct interorganizational cooperation.

E. Joint Interagency Task Force (JIATF)

A JIATF is a potential source for fused interagency information and intelligence analysis. For example, Joint Interagency Task Force-West is United States Indo-Pacific Command's (USINDOPACOM's) lead for DOD support to law enforcement for counterdrug and drug-related activities in the USINDOPACOM AOR. Its assigned mission is to protect national security interests and promote regional stability by providing US and foreign law enforcement with fused interagency information and intelligence analysis and with counterdrug training and infrastructure development support.

VII. Multinational Partner Considerations

Collective security is a strategic objective of the United States which, generally, requires effective integration of diverse multinational partners. This integration effort is often complicated since some of these mission partners have policies, doctrine, procedures, and capabilities that differ from those of the United States. During such operations, joint planning is accomplished within the context of multinational operations. There is no single doctrine for multinational action, and each alliance or coalition develops its own protocols and plans. With regard to information activities and the conduct of OIE, US planning for joint operations accommodates and complements the inherent complexity of multinational partner considerations.

It is essential for the MNF commander to resolve potential conflicts as soon as possible by establishing standard lexicon and procedures, as well as appropriate shared understanding of each other's capabilities. It is also an operational imperative for the MNF commander to integrate multinational partners into joint planning as early as possible. Early integration enables the efficient and effective use of MNF capabilities and resources throughout planning and operations.

Each nation has classified and unclassified capabilities, products, and resources that are useful to the joint force and to the MNF's information activities. For example, NATO's Strategic Communications Division produces an IE assessment that improves joint force understanding by identifying audiences; benchmarking attitudes, perceptions, and behaviors; and identifying communications processes and systems. To maximize the benefits of multinational information activities, each nation must be willing to share appropriate information to accomplish the assigned mission, while excluding the information that each nation is obliged to protect. To enable shared understanding across the MNF, the activities and the structures, systems, and facilities that support them should be classified at the lowest level possible. Information sharing arrangements in formal alliances, such as United States participation in United Nations' missions, are worked out as part of alliance protocols. Conversely, information sharing arrangements in ad hoc multinational operations during which coalitions are working together on a short-notice mission, should be developed during the establishment of the coalition.

Planners and operators consider the capabilities, limitations, and authorities of partners related to the management and application of information by the joint force (e.g., a partner nation with established policies, laws, and means for information dissemination across its country). The policies of each partner regarding the use of information might not align with US/DOD policy, so joint planners, even while collaborating with a partner, always comply with US/DOD policy. From an information joint function perspective, initial requirements for coordinating and synchronizing with and integrating other partners into US planning include:

- Understanding partner agendas, priorities, and objectives.
- Clarifying partner narratives, themes, messages, and activities.
- Establishing deconfliction procedures for narratives, themes, and messages of the MNF that may differ from those of the United States/DOD.
- Identifying threats to, vulnerabilities of, and opportunities for the MNF.
- Developing options to deter or defeat MNF threats and to mitigate MNF vulnerabilities.
- Identifying MNF authorities, capabilities, and capacities.
- Determining appropriate access of partners to US systems, services, and information, to include unclassified and appropriate levels of classification validated as mission-essential.

Refer to JP 3-16, Multinational Operations, for additional information on multinational partners and operations.

IV. (OIE) Operations in the Information Environment

Ref: JP 3-04, Information in Joint Operations (Sept '22), chap. VII.

I. Operations in the Information Environment (OIE)

OIE are military actions involving the integrated employment of multiple information forces to affect drivers of behavior by informing audiences; influencing foreign relevant actors; attacking and exploiting relevant actor information, information networks, and information systems; and by protecting friendly information, information networks, and information systems.

OIE Across the Competition Continuum (See pp. 2-8 to 2-9.)

Military operations vary in scope, purpose, and intensity in cooperation, adversarial competition below armed conflict, and armed conflict. Throughout the competition continuum, the JFC integrates **operations in the information envrionment (OIE)** into joint plans and synchronizes it with other operations to create desired behaviors, reinforce or increase combat power, and gain advantage in the IE. Each joint operation has a unique strategic context, so the nature of OIE and its activities will vary according to the distinct aspects of the mission and OE. While OIE may be conducted as an independent operation, it is never done in isolation. OIE are conducted throughout all campaigns or operations and at any level of conflict.

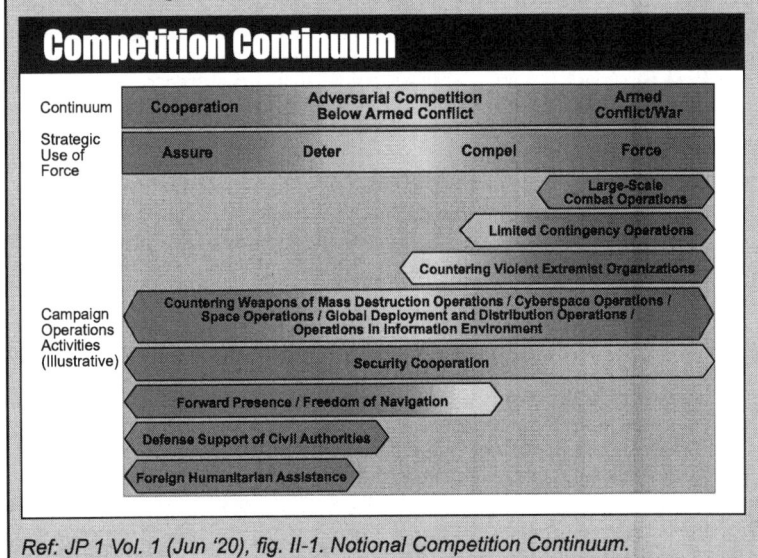

Competition Continuum

Continuum	Cooperation	Adversarial Competition Below Armed Conflict		Armed Conflict/War
Strategic Use of Force	Assure	Deter	Compel	Force
Campaign Operations Activities (Illustrative)				Large-Scale Combat Operations
			Limited Contingency Operations	
			Countering Violent Extremist Organizations	
	Countering Weapons of Mass Destruction Operations / Cyberspace Operations / Space Operations / Global Deployment and Distribution Operations / Operations in Information Environment			
	Security Cooperation			
	Forward Presence / Freedom of Navigation			
	Defense Support of Civil Authorities			
	Foreign Humanitarian Assistance			

Ref: JP 1 Vol. 1 (Jun '20), fig. II-1. Notional Competition Continuum.

OIE leverage information for the purpose of affecting the will, awareness, and understanding of adversaries and other relevant actors and denying them the ability to act in and through the IE to negatively affect the joint force, while protecting joint force will, awareness, understanding, and the ability to take actions in and through the IE.

OIE may provide commanders with a decisive advantage over adversaries by helping to maintain the credibility and legitimacy of joint force actions, preserving the joint force will to fight, maintaining situational understanding, and keeping the joint force free of prohibitive interference due to cyberspace or EMS activity, which cumulatively preserve freedom of action throughout the OE.

OIE are conducted as an integral part of all operations and campaigns and help shape the IE for future operations. As such, joint forces will always be conducting one or more OIE to remain continuously engaged in and through the IE. OIE are conducted in support of all operations and may be a main effort or supporting effort.

Any organization or capability may be tasked to conduct activities to support OIE, whether or not assigned to an OIE unit. For example, a JTF hosting a visit by local journalists, an aviation unit conducting a show of force, or a naval strike group conducting a freedom of navigation mission may all be carrying out these activities to inform or influence relevant actors in support of OIE.

OIE are not a substitute for the joint forces' deliberate leveraging of the inherent informational aspects of military activities. The joint force should still integrate information and informational considerations and capabilities into strategic art and operational design, planning guidance, and planning processes.

See facing page for an overview of OIE limitations and mission considerations.

II. Organizing for Operations in the Information Environment

JFCs may choose to create a task force for the integrated employment of the specialized capabilities required to conduct OIE.

A. Organizations and Personnel *(See pp. 2-21 to 2-36.)*

OIE units consist of a headquarters organization with C2 of assigned and attached information forces. OIE unit personnel include information planners and support personnel (e.g., intelligence, logistics). Information professionals are information force personnel who are specifically trained to inform audiences; influence external relevant actors; attack and exploit relevant actor information, information networks, and information systems; and protect friendly information, information networks, and information systems. Information planners serve in OIE units and as OIE and specialized capability SMEs on JTF and other headquarters planning staffs.

OIE unit headquarters are composed of a commander and a staff of information planners who possess a depth of knowledge and experience in their respective fields, as well as broad experience working alongside planners from other fields.

B. Information Forces *(See pp. 2-32 to 2-33.)*

Information forces, the building blocks of OIE units, are those Active Component and Reserve Component forces specifically organized, trained, and equipped to create, and/or support the creation of effects on the IE. Information forces aggregate military personnel, weapon systems, equipment, and necessary support that provide expertise and specialized capabilities (e.g., CMO, MISO, PA, EMSO, CO) that leverage information and conduct activities central to OIE. See paragraph d., "Information Forces," below, for a discussion of the types of information forces that make up OIE units.

C. Facilitate the JFC's Integration of Information into Joint Force Operations

OIE units have the responsibility of supporting the JFC's integration of information into the planning and execution of all joint force operations and activities. This encompasses maintaining an understanding for the JFC of how information affects their OE; providing advice and assistance on how to best leverage the inherent information aspects of all joint force activities; collaborating with the JFC staff on the

OIE Limitations & Mission Considerations

Ref: JP 3-04, Information in Joint Operations (Sept '22), pp. VI-2.

OIE Limitations

- The ability of the joint force to conduct OIE is limited by the availability of OIE units.
- OIE may be limited by the capabilities and authorities of OIE units or of their higher headquarters.
- OIE may be more successful when integrated with other joint, Service, interorganizational, and other US and foreign mission partners.
- JFCs must obtain and delegate authorities to conduct specific OIE activities to be effective during crisis situations of short duration.
- OIE units and other information forces require prior engagement in an OE to have an understanding of, and experience in dealing with, the aspects of a problem set to be effective during crises situations.
- OIE require intelligence collection support during the conduct of operations to determine the effectiveness of activities.

OIE Mission Considerations

When planning or conducting OIE, commanders and staffs should consider:

- How the mission will support CCPs, the operation, campaign, OPLAN, or contingency response plan.
- The risks of OIE before making employment decisions. OIE may have strategic and transregional impacts beyond the employing JFC's area of operations, and commanders should consider US diplomatic and informational interests in risk calculations.
- Authorities and permissions required for the conduct of activities and the lead times necessary to obtain those authorities and permission.
- The coordination required with other joint, Service, interorganizational, and other US and foreign mission partners to align and synchronize activities and achieve unity of effort.
- Coordination for appropriate SME support (e.g., cultural knowledge and language skills, specialized intelligence support) and capabilities. OIE rely on joint, Service, and other mission partners for SME support.
- The establishment of an assessment framework during initial planning (i.e., baseline, clear MOEs, and MOE indicators).

protection of information, information networks, and information systems; and assessing the effectiveness of joint force activities from an informational perspective.

OIE units accomplish this with assigned, attached, or supporting intelligence capabilities and analysts in conjunction with the joint force intelligence staff. It includes providing analysis of the informational, physical, and human aspects of the environment; identifying threats, vulnerabilities, and opportunities in the IE; and identifying and analyzing relevant actors.

D. OIE Unit Core Activities *(See facing page.)*

OIE units have the responsibility of supporting the JFC's integration of information into the planning and execution of all joint force operations and activities. OIE units conduct this core activity by providing original products to the staff (e.g., analysis of the informational, physical, and human aspects of the environment), input to staff products (e.g., military narrative, information estimate), or participating in JPP with the staff. OIE units do this via planners serving on, or as liaisons to, higher headquarters staffs or through coordination between their staff and higher headquarters staff.

See facing page for an overview of OIE unit core activities.

II. The Joint Functions and OIE

JP 3-0, Joint Campaigns and Operations, describes the seven joint functions common to joint operations: C2, information, intelligence, fires, movement and maneuver, protection, and sustainment. Each joint function is a grouping of tasks and systems that provide a critical capability to help JFCs synchronize, integrate, and direct joint operations. Commanders leverage the capabilities of multiple joint functions during operations to achieve objectives. This section presents an overview of how commanders use joint functions to integrate, synchronize, and direct OIE in support of all DOD missions.

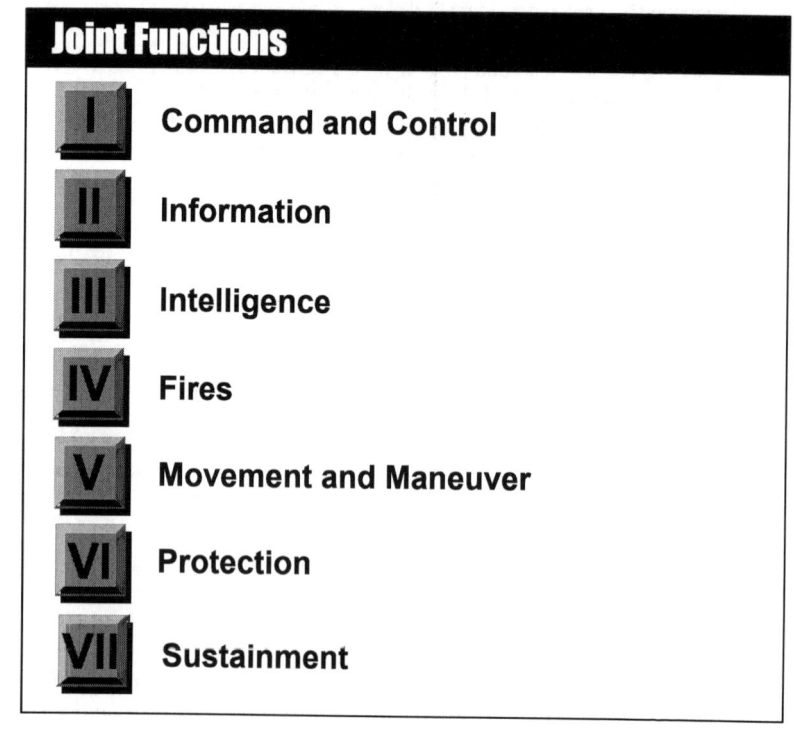

Joint Functions

- I. Command and Control
- II. Information
- III. Intelligence
- IV. Fires
- V. Movement and Maneuver
- VI. Protection
- VII. Sustainment

OIE Unit Core Activities (See also pp. 2-10 to 2-11 and 2-19.)

Ref: JP 3-04, Information in Joint Operations (Sept '22), pp. VII-9 to VII-11.

OIE unit core activities include conducting OIE and facilitating the JFC's integration of information into joint force operations. Other joint force elements conduct some of the information activities associated with these core activities during their operations.

OIE. OIE are the primary focus of OIE units. OIE encompass critical tasks that OIE units must perform to achieve JFC objectives by leveraging information. OIE units accomplish these tasks using military capabilities in a coordinated and synchronized manner to collectively achieve objectives affecting the IE by informing audiences; influencing foreign relevant actors; attacking and exploiting information, information networks, and information systems; and by protecting friendly information, information networks, and information systems. OIE are conducted in support of the JFC's operation or campaign objectives or in support of other components of the joint force. Joint forces continuously conduct OIE to remain engaged with relevant actors.

INFORM. The inform task involves actions taken to accurately communicate with domestic and foreign audiences to build understanding and support for operational and institutional objectives. It seeks to reassure allies and partners and to deter and dissuade competitors, adversaries, and enemies. The inform task uses accurate and timely information and visual media to counter disinformation; correct misinformation; and put operations, activities, and polices in context. It involves communication with domestic and international audiences and with joint force personnel. Planning and executing tasks to inform include public engagement and the acquisition, production, and dissemination of communication and other information products. The inform task facilitates educated perceptions by establishing facts and placing joint force activities in context, correcting inaccuracies and misinformation, and discrediting propaganda with counter-narratives. The primary means used for the inform task is PA; however CA, cyberspace, and psychological operations forces can facilitate the release of truthful information through their respective CMO, CO, and MISO activities.

INFLUENCE. The purpose of the influence task is to affect the perceptions, attitudes, and other drivers of relevant actor behavior. This task is focused on impacting the human aspects of the OE, so planners should consider elements of these aspects as they relate to decision makers (e.g., each decision maker's culture, life experiences, relationships, outside events, ideology, and the influences of those people inside and outside the decision maker's group) during OIE planning, execution, and assessment. Planners integrate influence activities into the existing targeting process. Activities designed to contribute to the influence task include MISO, CMO, CO, OPSEC, and MILDEC operations. Influence may also involve the use of STO. Commanders consider the influence potential of all available capabilities in design, planning, and targeting. OIE units conduct all influence tasks in accordance with approved authorities.

ATTACK & EXPLOIT. The attack and exploit task comprises activities meant to impact or use opponent information, information systems, and information networks in ways that affect decision making and other drivers of behavior to create relative advantages for the joint force. OIE units execute these actions to manipulate or paralyze the adversary or enemy decision-making processes. Attack activities encompass affecting the real or perceived accuracy, integrity, authenticity, or confidentiality of information or the availability of information. OIE units accomplish attack tasks through technical means, such as CO, EMSO, and STO, though maneuver forces and joint fires can also be employed in support of these tasks. Exploit activities include accessing information, information networks, or information systems to gain intelligence and support operational preparation of the environment (OPE) for current or future operations. OPE may subsequently support inform and influence tasks of OIE. OIE units accomplish the exploit task through technical means, such as CO or EMSO.

A. Command & Control (C2) Joint Function

C2 primarily focuses on the exercise of authority and direction by commanders over assigned and attached forces in the accomplishment of their assigned mission. That authority and direction are exercised through a C2 system that consists of the facilities, equipment, communications, staff functions and procedures, and personnel essential for planning, preparing for, monitoring, and assessing operations. The C2 systems enable the JFC to maintain communication with higher, supporting, and subordinate commands to control all aspects of current operations while planning for future operations. The C2 joint function enables the commander to balance the art of command with the science of control and integrate the other joint functions. C2 of the information planners on the staff and OIE units encompasses the exercise of authority and direction by a commander over assigned and attached information forces to accomplish the mission.

B. Information Joint Function *(See pp. 2-7 to 2-20.)*

The three tasks of the information joint function support all the other joint functions and provide commanders with the ability to understand how information impacts the OE, use information to support human and automated decision making, and leverage information through offensive and defensive actions. OIE is closely tied to the tasks of the information joint function. The understand task of the information joint function is used to understand the threats, opportunities, and vulnerabilities required to conduct OIE. It is the preparatory work that sets the stage for OIE. Additionally, the understand task should identify access points and lines of influence that can be exploited through OIE to create effects and ultimately change behavior. It also helps identify the operational signatures that need to be managed or controlled to maintain essential secrecy. OIE uses that understanding to reveal or conceal those signatures to ensure relevant actors see what we want them to see and not see what we do not want them to see. The second task of the information joint function, support to human and automated decision making, is a critical prerequisite of effective OIE. It enables joint forces to preserve and protect our ability (and our trust in that ability) to make sense of the IE. All operations perform the third task of the information joint function, but leveraging information is the primary effort of joint OIE units.

C. Intelligence Joint Function

Understanding the OE, which encompasses aspects of the IE, is fundamental to all operations to include OIE. The intelligence joint function helps to inform the JFC and staff about the opponent's intent, capabilities, vulnerabilities, and future COAs. It also helps them to understand friendly, neutral, and threat information networks and information systems; the ways that information is received, transmitted, and processed; and how information may impact the opponent's own decision making and drivers of their behavior. Using the continuous JIPOE analysis process, properly tailored JIPOE products can enhance understanding of the OE and clarify the impacts of information. This understanding enables the JFC to act inside the opponent's decision cycle. JIPOE provides a socio-cultural analysis of all relevant actors to reveal their decision-making process, norms beliefs, power structures, perceptions, attitudes, and other drivers of behavior. JIPOE also reveals how relevant actors might apply information to exploit vulnerabilities in the joint force's information networks and information systems and how they might leverage information to affect drivers of joint force behavior. Intelligence support to OIE follows the same all-source intelligence process used by all other operations, with unique attributes necessary for support of planning and assessment for OIE. The intelligence necessary to understand the drivers of behavior of enemies, adversaries, or other audiences often requires that units position and employ specific sources and methods (e.g., counterintelligence, human intelligence, targeted social media monitoring) to collect the information and conduct the analyses needed.

D. Fires Joint Function *(See pp. 7-6 to 7-7.)*

Fires is the use of weapon systems or other actions to create specific lethal or nonlethal effects on a target. The nature of the target or threat, the conditions of the mission variables (i.e., mission, enemy, terrain and weather, troops and support available, time available, and civil considerations), and desired outcomes determine how lethal and nonlethal capabilities are employed. OIE may leverage the inherent informational aspects of joint fires. Fires in and through the IE encompass a number of tasks, actions, and processes, including targeting, coordination, deconfliction, and assessment (e.g., BDA).

OIE tasks and capabilities leverage information through fires to create specific effects. To integrate effectively, information planners participate in the joint targeting process by selecting and prioritizing targets for fires or TAs for other actions. OIE units create fires that typically result in nonlethal effects. OIE can also indirectly create effects that result in physical destruction (e.g., manipulating computers that control physical processes). Additionally, OIE can leverage the inherent informational aspects of fires to reinforce the psychological effect of those fires. OIE may rely on joint fires support to transmit information to relevant actors and to deliver nonlethal payloads to affect information, information systems, and information networks (e.g., leveraging CO to deliver computer code designed to deny network access to an adversary, PA releases to inform friendly audiences, or MISO products to influence foreign audiences).

The integration of OIE into the targeting process—a task managed within the fires function—is important to creating effects in and through the IE that will achieve objectives. Even when OIE do not require joint fire support to create effects, they still depend upon the joint targeting process to integrate and deconflict fires effects that may impact strategic- and operational-level objectives. It is important to note that not all forms of information fires dovetail into the targeting process; in some instances, these fires bypass the targeting process to go directly to the effects board. For instance, if the JFC's intent is to influence a relevant actor to participate in peace negotiations during armed conflict, all participants in the targeting process must ensure lethal fires and joint force combat actions do not inhibit or dissuade that participation. Information planners participate in the targeting process as members of the joint targeting coordination board, which plans, coordinates, and deconflicts joint targeting.

Like all forms of fires, fires in support of OIE are included in the joint planning and execution processes to facilitate synchronization and unity of effort. JFCs use coordination and control measures to enable joint action. These measures include strategic and operational mission narratives, PAG, other communication-related guidance, the law of war, and rules of engagement. Additionally, information planners identify control measures for OIE that have the potential to conflict with the OIE of other CCMDs or interorganizational partners. OIE units work with maneuver and fires elements to establish fire control measures to reduce the impact of combat operations on the civilian populace. If multiple USG or allied entities have requirements to create effects or collect intelligence on the same target in the IE, then synchronization and deconfliction across all USG entities are critical to prevent uncoordinated actions from exposing or interfering with each other.

Finally, units conducting OIE contribute to, and benefit from, the joint fires task of assessing the results of employing fires. That task includes assessing the effectiveness and performance of fires, as well as their contribution to the larger operation or objective.

E. Movement and Maneuver Joint Function

Maneuver is the employment of forces in the JOA through movement in combination with the other joint functions to gain a position of advantage in respect to the enemy. The movement and maneuver joint function encompasses the disposition of joint forces to conduct operations by securing advantage and exploiting tactical success to achieve operational and strategic objectives. Movement and maneuver to and within the JOA can signal adversaries, allies, and neutral actors and may have a deterrent or assuring effect in support of JFC objectives. Movement and maneuver involve deploying forces and capabilities into a JOA and positioning them within that area to gain operational advantage in support of mission objectives, including accessing and, as necessary, controlling key terrain. Movement and maneuver of forces have inherent informational aspects that affect the achievement of JFC objectives and should be accounted for during planning and execution.

OIE, the art of maneuvering in the IE, is conducted to enhance the effects of the inherent informational aspects of the movement and maneuver of forces. The pervasive nature of information and the IE provides the joint force with operationally significant access to relevant actors within the JOA, as well as outside the JOA. OIE contribute to the joint force's freedom of action and control of the operational tempo necessary to conduct its activities at a time and place of its choosing to produce the operational reach necessary to create an advantage over the adversary.

F. Protection Joint Function

The protection joint function provides the JFC with the capabilities needed to protect the joint force, its bases, necessary infrastructure, and lines of communication from attack. The protection joint function complements the information joint function by ensuring the use of appropriate physical defensive measures necessary to safeguard information. With respect to OIE, the protection joint function attends to the physical security necessary for the deployment, storage, employment, and redeployment of SAP capabilities necessary for classified OIE. As part of OIE, DODIN and DCO secure and defend the joint force's information, information networks, and information systems that form the backbone of the JFC's C2 joint function. Due to their global and commercial connectivity, protection of these assets is complicated. OIE reinforces the protection function by degrading the opponent's ability to target the joint force by attacking its information, information networks, information systems, and human and automated decision making. For example, OIE that include OCO and JEMSO (e.g., jamming communication frequencies) can protect the joint force by disrupting the opponent's targeting and C2 systems. OPSEC also supports the protection joint function by protecting critical information.

G. Sustainment Joint Function

The sustainment joint function provides the JFC with the capabilities necessary to provide the logistics and personnel services required to maintain and prolong joint operations until mission objectives are achieved. Successful execution of OIE requires that information be regarded as a mission-essential resource that must be sustained (e.g., assuring its integrity, accuracy, confidentiality, accessibility, nonrepudiation, and flow). Joint operations, especially globally integrated operations in the IE, may require those portions of the joint force that conduct OIE to be geographically dispersed and virtually connected, which will require special considerations for sustainment. Sustainment support for the OIE unit is essential and requires coordination with the joint force's logistics staff. Sustainment contributes to the joint force's ability to generate effects and operate in the IE. From an operational perspective, OIE can help protect sustainment efforts by manipulating or masking the inherent informational aspects of joint force sustainment activities in ways that impair an opponent's ability to sense and target these efforts.

Chap 2
V. OIE: Planning, Coordination, Execution, & Assessment

Ref: JP 3-04, Information in Joint Operations (Sept '22), chap. IV - VII.

*See following pages (pp. 2-46 to 2-49) for an overview of the **integration of information** during the planning, execution, and assessment of joint operations.*

I. Planning (See pp. 2-55 to 2-66. See also chap. 4.)

Commanders integrate OIE into their operations at all levels. Plans should address how OIE affect the will, awareness, and understanding of adversaries and other relevant actors; deny competitors the ability to act in and through the IE to undermine the joint force; and protect joint force will, awareness, understanding, and the joint force ability to take actions in and through the IE.

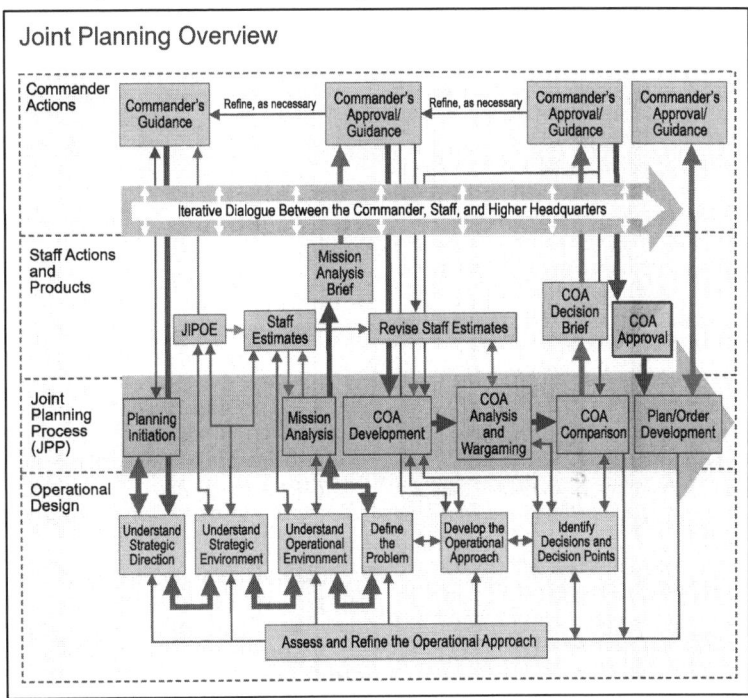

Joint Planning Overview

Commander Actions: Commander's Guidance → Refine, as necessary → Commander's Approval/Guidance → Refine, as necessary → Commander's Approval/Guidance → Commander's Approval/Guidance

Iterative Dialogue Between the Commander, Staff, and Higher Headquarters

Staff Actions and Products: JIPOE, Staff Estimates, Mission Analysis Brief, Revise Staff Estimates, COA Decision Brief, COA Approval

Joint Planning Process (JPP): Planning Initiation → Mission Analysis → COA Development → COA Analysis and Wargaming → COA Comparison → Plan/Order Development

Operational Design: Understand Strategic Direction, Understand Strategic Environment, Understand Operational Environment, Define the Problem, Develop the Operational Approach, Identify Decisions and Decision Points

Assess and Refine the Operational Approach

Ref: JP 3-04 (Sept '22), fig. V-1. Joint Planning Overview.

Refer to JFODS6: The Joint Forces Operations & Doctrine SMARTbook, 6th Ed. (Guide to Joint Warfighting, Operations & Planning). JFODS6 is updated for 2023 with new/updated material from the latest editions of JP 3-0 Joint Campaigns and Operations (Jun '22), JP 5-0 Joint Planning (Dec '20), JP 3-33 Joint Force Headquarters (Jun '22), and JP 1 Volumes I and II Joint Warfighting and the Joint Force (Jun '20), Additional topics and references include Joint Air, Land, Maritime and Special Operations (JPs 3-30, 3-31, 3-32 & 3-05).

Guide for the Integration of Information in Joint Operations

Ref: JP 3-04, Information in Joint Operations (Sept '22), app. C. Fig. C-1 is a reference guide for the integration of information during planning, execution, and assessment of joint operations.

A. Operational Environment Awareness & Understanding

(Continually Ongoing) Develop and maintain an integrated understanding of the OE spanning geographic, functional, domain, classification, and organizational boundaries.

1. Characterize Overall IE
 a. Understand why and how information moves through the OE, how it is received, processed, and employed, by whom, and for what purposes.
 b. Establish IE baseline to create a reference point of relevant actor perceptions, beliefs, and attitudes. Assess changes over time.
 c. Distinguish relevant information and characterize its sources and methods of movement or transmission.
 d. Identify misinformation and disinformation and credible from non-credible sources of information.
 e. Understand the information networks and systems used by relevant actors.
 f. Understand social/cultural norms needed for effective influence.

2. Identify and Understand Relevant Actors
 a. Identify humans and automated systems that are potential relevant actors.
 b. Describe what drivers of behavior are most likely to affect relevant actors.
 c. Understand how relevant human actors sense and process information to trigger a behavior that can positively or negatively impact joint operations.
 d. Understand how relevant automated systems sense and process information.
 e. Describe how relevant actors communicate and make decisions.
 f. Identify relevant actors that are decision makers, key influencers, or both.
 g. Identify key influencers for relevant actors both inside and outside the operational area.

B. Strategy and Course of Action Development

Establish operational approach and develop COA options for attaining and maintaining conditions that enable achievement of JFC intent and advancement of campaign objectives.

1. Initiation: Receive and Refine Planning Guidance
 a. Review overall approach to integrating efforts with overall joint force, allies/partners, and Interagency.
 b. Describe relevant actor desired behaviors (e.g., specifics in terms of assure, deter, induce, compel)
 c. Articulate current authorities for information activities at JFC and subordinate levels.
  • CCMD-approved MISO program.
  • CCMD-approved CO/MILDEC/Space activity.
 d. Identify forces available to conduct or support OIE (via OPCON, TACON, direct support, or general support relationships).
 e. Identify risks that can or cannot be accepted related to activities in the IE.
 f. Update the information estimate.
 g. Provide updates on changes in the IE, status of information forces, and results of information activities.

2. Mission Analysis
 a. Analyze planning directives and strategic guidance from HHQ.
  • Determine national and HHQ objectives.
  • Determine desired relevant actor behaviors.
 b. Develop mission narrative.
  • Develop JFC operational narrative and supporting themes and messages based on CCMD guidance.
 c. Determine facts and planning assumptions.
  • Identify relevant actors within context of JFC's objectives.

- Identify relevant actors outside the JOA who may present opportunities for secondary/tertiary influence or affect joint force activities.
- Describe relevant actor current known/relevant behaviors and key informational capabilities.
- Describe perceptions, attitudes, beliefs, and other drivers of relevant actor behaviors.
- Characterize relevant actor physical and informational escalation thresholds.
- Characterize relevant actor equities, interests, and decision-making processes.
- Identify authoritative sources used by relevant actors.
- Identify information conduits used by relevant actors.
- Identify partner nation sensitivities, constraints, and restraints

d. Identify operational limitations.
- Review current OIE-related authorities and permissions, and identify requirements for obtaining additional authorities and permissions as early as possible.
- Determine constraints and restraints related to engaging relevant actors, use of certain capabilities, and use of specific themes or messages, given that information and its effects may not be geographically constrained or limited to intended audiences.

e. Determine specified, implied, and essential tasks and develop the mission statement.
- Identify specified and implied tasks for leveraging information and supporting decisionmaking.

f. Conduct initial force and resource analysis.
- Identify the lead time to deploy information forces and specialized capabilities into theater or provide direct support to the joint force from outside the JOA.
- Request for forces or personnel with unique skills (e.g., linguists, sociocultural experts, social media experts, and automated-intelligence and machine-learning experts).
- Identify mission partners with information forces and specialized capabilities and/or capacities to fill joint force resource gaps.

- Evaluate requirements against existing or potential contracts or task orders.

g. Develop military objectives.
- Determine and articulate attainable behavioral goals.
- Develop MOEs and MOE indicators to assess how well the joint force leverages information.
- Identify indicators of trending success or failure into the monitoring and assessment plan.

h. Develop COA evaluation criteria.
- Information activities that produce desired behaviors in prioritized relevant actors.
- Information activities that protect the joint force from opponent attempts to undermine the joint force narrative or the legitimacy of the joint force mission and actions.
- Information activities that prevent, counter, and mitigate attempts to undermine joint force decision making and C2.
- Identify potential external information activities that negatively impact the achievement of the JFC's objectives.

i. Develop risk assessment.
- Identify strategic risks to JFC's narrative.
- Identify risk to joint force from malign influence.
- Refine analysis of risks to strategy, force, and mission. Develop mitigation approaches.

j. Develop CCIRs.
- Relevant actor behaving in a way that was not anticipated.
- Demonstration by an opponent of a new information capability against the joint force not foreseen during planning.
- Loss of access to communications channels used to inform or influence relevant actors.
- Emergent events that present a challenge to or opportunity for the JFC narrative.
- Relevant actor desired behaviors and perceptions.

k. Develop Information Estimate *(Refer to JP 3-04, Appendix B)*.

l. Prepare and deliver mission analysis brief.

(Joint Ops) V. OIE: Planning, Coordination, Execution, & Assessment 2-47

3. COA Development
a. Develop initial COA.
- Establish approach to informing domestic, international, internal audiences.
- Establish approach to influencing primary relevant actors.
- Establish approach to attacking and protecting information, information networks, and information systems.
- Validate operational objectives by developing preliminary MOE indicators for relevant actor behaviors and evaluating ability to assess them.
- Identify forces desired to support informational power actions (including mission partners).
- Develop COA narrative, key themes, and messages.

b. Refine COA (Based on COA analysis/wargaming).
- Select relevant actors inside or outside JOA that may be opportunities for secondary or tertiary influence.
- Depict primary, secondary, and tertiary regions of influence inside and outside the JOA.
- Determine relative timing, tempo, intensity, scope, and linkage of physical force and informational power activities to create reinforcing effects in such a way to maximize their potential value.
- Identify opportunities to leverage/exploit PAI through MILDEC or other means.
- Determine inherent informational aspects of activities and develop OIE approach for leveraging them to shape relevant actor behavior.
- Integrate OIE approaches into main and supporting lines of effort.
- Anticipate how joint force actions will resonate from physical domains to the IE, and how to respond to reactions by any potential relevant actors.
- Anticipate the opponent's OIE approaches and develop flexible options for countering them.
- Establish weights of effort for informational power and physical force.

c. Establish timelines for executing proposed COA under available and accessible authorities.

d. Identify OIE tasking mechanisms and associated timelines for employment to synchronize effects as required.

e. Develop layered assessment plan.
- Develop MOEs and MOE indicators for relevant actor behavior changes and overall strategic gain.
- Identify anticipated timeframes for gathering useful MOE indicators – initial reactions followed by long-term sentiment analysis.
- Develop collection requirements to observe MOE indicators.

4. COA Analysis/Wargaming
a. Assess how relevant actors react to changes that each COA causes in the IE/OE.
b. Anticipate how relevant actors might exploit PAI.
c. Identify new relevant actors that may emerge as a result of the COA.
d. Identify high priority relevant actors for influence.
e. Evaluate ability to gather MOE indicators of relevant actor behavior changes within operationally relevant timeframes.

C. Detailed Planning
Develop detailed plans that affect relevant actor behavior through the integration of informational power w/capabilities & activities using assigned, attached, & supporting forces.

1. Develop Detailed Plan
a. Incorporate behaviorally-focused objectives into existing targeting processes and practices.
b. Conduct ROE/JA review of proposed informational power effects.
c. Draft collection plan to observe informational power-related MOE indicators.
d. Develop integrated force package options to create desired effects.
- Identify and select specialized capabilities that can best enable/support other capabilities and activities (C2, fires, intelligence, movement and maneuver, sustainment, or protection).
- Identify and select specialized capabilities that can best leverage inherent informational aspects of activities.
- Identify and select specialized capabilities that can best directly affect relevant actor attitudes, perceptions, and other drivers of behaviors.

e. Ensure access or ability to use specialized capabilities when required.
f. Identify capability and capacity shortfalls related to the management and application of information and develop potential solutions.

g. Develop synchronization matrix to align informational power and physical force (fires, movement and maneuver, sustainment, protection, intelligence, and intelligence) activities.

D. Execution

Synchronize the creation of integrated effects. Adapt approach as evolving circumstances require.

1. Check Executive Conditions
a. Red-team informational power approaches prior to execution, leveraging up to date understanding of operational environment.

2. Monitor Execution
a. Collect MOE indicators and maintain understanding of the IE.
b. Monitor how joint force activities are resonating through the IE.
c. Update Information Estimate.

3. Manage and Adapt Execution
a. Synchronize execution of OIE activities with other joint force activities.
   - Maintain synchronization matrix.
   - Align OIE activities within overall targeting cycle.
   - Maintain updated narrative.
b. Ensure operating within limits of applicable OIE authorities throughout execution.
c. Identify and resolve conflicting OIE approaches with mission partners.
d. Anticipate/adapt OIE approach to evolving situation in accordance with JFC objectives.
e. Counter and compete with the opponent's emergent narratives and other OIE of concern.
f. Integrate OIE into approach to handling escalation.

E. Assessment

Evaluate effects created against relevant actor perceptions, behavior, and capabilities. Identify new opportunities created to advance JFC objectives.

1. Evaluate Ability to Execute OIE and Synchronize with Other Activities (MOPs)
a. Determine if OIE and other activities were sequenced and executed as intended.
b. Identify capability shortfalls and resource issues impeding effectiveness.
c. Identify gaps in authorities/permissions impeding effectiveness.
d. Identify communications (technical or human) issues that impeded effectiveness.

2. Evaluate MOE Indicators and MOEs
a. Evaluate MOE indicators to characterize changes in relevant actor perceptions, attitudes, beliefs, and other drivers of behaviors.
   - Ascertain if inherent informational aspects of activities were interpreted as intended.
   - Track echoing/re-communication of JFC and DOD messaging (accurately or inaccurately).
b. Evaluate OIE gain/loss.
   - Establish extent to which non-overt actions were attributed to the joint force.
   - Determine what was revealed about joint force capabilities through OIE activities, and if that met gain/loss expectations for those capabilities.
   - Evaluate effects of component-level OIE in contributing to overall campaign-level joint force strategic gain.

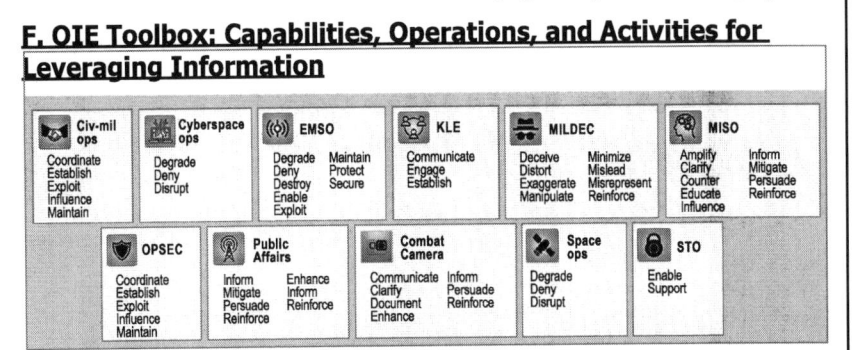

F. OIE Toolbox: Capabilities, Operations, and Activities for Leveraging Information

Capability	Actions
Civ-mil ops	Coordinate, Establish, Exploit, Influence, Maintain
Cyberspace ops	Degrade, Deny, Disrupt
EMSO	Degrade, Deny, Destroy, Maintain, Protect, Secure, Enable, Exploit
KLE	Communicate, Engage, Establish
MILDEC	Deceive, Distort, Exaggerate, Manipulate, Minimize, Mislead, Misrepresent, Reinforce
MISO	Amplify, Clarify, Counter, Educate, Influence, Inform, Mitigate, Persuade, Reinforce
OPSEC	Coordinate, Establish, Exploit, Influence, Maintain
Public Affairs	Inform, Mitigate, Persuade, Reinforce, Enhance, Inform, Reinforce
Combat Camera	Communicate, Clarify, Document, Enhance, Inform, Persuade, Reinforce
Space ops	Degrade, Deny, Disrupt
STO	Enable, Support

Operational Design

Operational design is the analytical framework that underpins planning. Operational design supports commanders and planners in understanding the JFC's OE as a complex interactive system. As commanders and staffs apply operational design methodology to develop the operational approach, they account for how information impacts the OE and the potential inherent informational aspects of their activities. In doing so, joint force planners gain an understanding of relevant actors and consider how information is used by, and affects the behavior of, those actors.

Refer to JP 3-04, pp. IV-2 to IV-14.

Joint Planning Process (JPP) *(See pp. 2-55 to 2-66.)*

OIE are planned using the joint planning process (JPP). The Joint Planning Process (JPP) is an orderly, analytical process that consists of a logical set of steps to analyze a mission, select the best course of action and produce a campaign or joint OPLAN or order. Throughout the JPP steps, information planners assist other joint planners in incorporating their understanding of how information impacts the OE to identify how to best support human and automated system decision making and how to best leverage information to achieve the JFC's objectives during operations.

JFCs integrate OIE into operations, as main or supporting efforts, or conduct an OIE as a stand-alone effort. During plan development, JFC provides planning guidance that describes the desired conditions that must exist in the IE to support mission accomplishment, how the joint force will leverage the inherent informational aspects of its activities to support the JFC's objectives, and the types and level risk that the JFC will accept in the IE. Specifically for OIE units, the JFC provides guidance on how OIE will support the JFC's scheme of maneuver. The JFC ensures supporting OIE plans and concepts describe the role and scope of OIE in the JFC's effort and address how OIE support the execution of the JTF plan.

See OIE planning considerations, facing page.

Information Planners

All members of the JFC's staff are responsible for accomplishing or contributing to tasks of the information joint function, to include understanding how information affects joint force operations, understanding how their respective activities impact and are impacted by the IE, and integrating that understanding into their respective portions of joint plans.

Information planners assigned to the staff enhance the JFC staff's ability to carry out information joint function tasks. Information planners are trained professionals from across specialized capabilities (e.g., MISO, CMO, CO, EMSO, PA). Those planners have subject matter expertise with specialized capabilities, experience working with and in OIE units, and an understanding of the inherent informational aspects of capabilities and activities of other units (e.g., a bomber task force or a carrier strike group executing a show of force, an armored task force conducting a feint). Information planners collaborate with the rest of the staff to develop and plan activities in a manner that most effectively leverages the informational aspects of joint force operations, as well as planning OIE, to support achieving the JFC's objectives. They ensure the joint force remains aware of interagency activities that may either support or potentially conflict with achieving objectives and, when possible, collaborate with external organizations to coordinate and synchronize information activities that support achieving shared objectives.

Information Cross-Functional Team (CFT) *(See pp. 2-30 to 2-31.)*

Information planners comprise the information planning cell and are the core of the information CFT with responsibility for incorporating input from the information CFT into the operational design and planning of joint operations and maintaining the information estimate. Some information planners are assigned to serve in the JS J-5, or as liaisons to external organizations, particularly with OIE units and information forces.

OIE Planning Considerations

Ref: JP 3-04, Information in Joint Operations (Sept '22), pp. VII-12 to VII-13.

Information planners have the same operational design considerations and challenges as planners for operations in the physical domains but also have some unique considerations for planning OIE. While OIE plans are developed to inform and influence, and to affect or protect information, information networks, and information systems, but there are factors outside the control of OIE that will have impacts in and through the IE that undermine those plans. These factors range from unanticipated adversary or mission partner actions inside the JOA to natural disasters or unforeseen domestic social or political developments that occur outside of the JOA that, nonetheless, affect the JFC's OE.

Different Planning Considerations for Contributing Information Forces

Each of the information forces that contribute to OIE (e.g., psychological operations forces, cyberspace forces) has their own unique planning considerations that increase the complexity of planning OIE. For each capability or activity employed, OIE planners will need to understand the different authorities and permissions, coordination requirements, intelligence requirements, and account for the lead time necessary to satisfy these requirements prior to the execution of activities.

Planning and Execution Timelines

Related to the above, the applicable authorities will vary depending upon when and where the activities occur and what or whom they will affect (e.g., if effects are likely to impact other relevant actors outside of a JFC's operational area). This includes accounting for the lead time required to obtain the necessary intelligence for target development and target access; confirm the appropriate authorities; and complete necessary coordination, including interagency coordination and/or synchronization. Additionally, planners will need to understand the length of time it will take for certain actions to have the desired effects and the duration of those effects. This may require OIE to begin prior to other joint force activities or even continue after some of those activities cease.

Language, Regional, Cultural, or Technical Expertise

Leveraging information for the purpose of affecting the behavior of relevant actors requires an understanding of the drivers of human or automated systems behavior. These drivers include language, regional, cultural, and often technical aspects. Planning teams will need to obtain support from various SMEs with an understanding of these aspects to understand relevant actors and develop feasible plans.

Plan for Monitoring Effects and Adjusting Activities

The dynamic nature of the IE often makes it challenging to determine whether OIE are effective. Planners of OIE should determine MOEs and MOE indicators during initial planning, incorporate monitoring tasks as essential elements of all OIE plans, and obtain adequate support to fulfill information and intelligence requirements.

Unintended Effects in and through the IE

The inherent informational aspects of activities and the lack of boundaries in the IE guarantee that military activities will often have impacts in and through the IE beyond the intended area or relevant actor. This makes the evaluation of potential effects particularly important when conducting OIE. Information activities can cause effects in and through the IE in ways that are not evident to planners. Some of these effects may affect other commanders' areas of operations and objectives or have strategic impacts. Coordinating plans and activities with joint, USG, and other mission partners will help identify potential effects beyond those intended and allow planners to avoid or mitigate effects that jeopardize their own or mission partner objectives.

II. Execution *(See chap. 6.)*

Joint operations span the competition continuum from recurring cooperative activities to sustained combat operations in armed conflict. The information joint function enables the application of informational power by expanding commanders' range of options for action across the competition continuum. Employing the information joint function may be the primary option available to a JFC during long-duration cooperation and competition short of armed conflict, where the use of physical force is inappropriate or restricted.

At its most basic level, execution involves synchronizing activities to maximize their combined effects during the conduct of operations, monitoring those activities and the effects they have on the OE, and adjusting activities based upon threats, vulnerabilities, and opportunities in the OE. The dynamic nature of the IE makes it vital that the JFC have the organizations, processes, and tools in place to rapidly recognize the informational aspects of activities and adapt joint force activities in response to failures or to exploit successes in and through the IE. The following are **essential elements** that facilitate that rapid adaptation.

Organization

The JFC modifies current or establishes new command and staff structures, as necessary, to facilitate joint force unity of effort to use and leverage information. This includes the option of establishing an OIE unit with the personnel, authorities, and other resources to conduct OIE. This requires an overarching strategy to gain operational advantage through the use of information and synchronizing the execution of that strategy between the command, subordinate units, and supporting or related operations of component commanders.

Monitoring and Analyzing for Effects In and Through the IE

Monitoring and analysis comprise the observation and evaluation of how information impacts the JFC's OE, how joint force activities affect relevant actors, and how those activities resonate in and through the IE to affect other JFC's OEs They include observing and evaluating the informational, physical, and human aspects of the OE for potential threats, vulnerabilities, and opportunities (to include a nuanced view of relevant actors) that could impact the JFC's decisions concerning mission requirements. The JFC may establish and resource an information CFT with the means to monitor and recommend adjustments to joint force operations to align them with objectives and the strategic and operational narratives.

See pp. 2-30 to 2-31 for an overview and discussion of an Information CFT.

The Synchronization Matrix

An information synchronization matrix, built around the CONOPS, contains the phasing of the operation and enables planners to graphically display the activities, linked to the scheme of maneuver, that leverage information to affect behavior and impact the OE. The matrix displays the mechanics of physical movement but, more importantly, reveals how the informational aspects of the operations are knitted together with other functional elements of joint maneuver to deliberately show or hide joint force activities or intentions. The body of the matrix contains critical tasks, arrayed in time and linked to responsible elements for execution. Joint force planners may use a matrix to display progress against actual execution and recommend adjustments as needed.

Commander's Critical Information Requirements (CCIRs) *(p. 2-61.)*

Information planners update the critical information requirements to provide the JFC with an understanding of how information impacts the OE; an awareness of the threats, vulnerabilities, and opportunities in the IE; and the status of organizations critical to supporting human and automated decision making and leveraging information.

Additional OIE Considerations

Ref: JP 3-04, Information in Joint Operations (Sept '22), pp. VII-12 to VII-15.

Intelligence Support to OIE

The complexity of OIE requires dedicated intelligence support. Intelligence professionals will need to work closely with OIE planners throughout the planning, execution, and assessment of operations to ensure they understand and meet the unique OIE information and intelligence requirements.

- **Intelligence Requirements.** During OIE mission analysis, the planners identify significant information gaps about the adversary and other relevant aspects of the OE. Information requirements related to the IE will include questions about the informational, physical, and human aspects of the environment; the questions about the characteristics of relevant actors; and the impact of the aspects of the environment on relevant actor behavior.
- **Requests for Information (RFIs).** Planners can submit RFIs to obtain intelligence products that support their activities or trigger collection efforts in any part of the OE. RFIs are specific, time-sensitive, ad hoc requirements for intelligence information to support an ongoing crisis or operation and not necessarily related to standing requirements or scheduled intelligence production.

Targeting

Commanders may choose to engage relevant actors through lethal and/or nonlethal fires as part of OIE. Relevant actors selected for engagement through joint fires are developed, vetted, and validated within the established targeting process. Planning and targeting staffs develop and select relevant actors for targeting in and through the IE based on the commander's objectives rather than on the capabilities available to achieve them. The focus is on creating effects that accomplish targeting-related tasks and achieve objectives, not on using a particular capability simply because it is available.

C2 of OIE Units

The complex and dynamic nature of the IE, where all joint force activities cause effects in and through that environment, make unity of effort crucial for all effective operations, including OIE. The JFC promotes unity of effort through the integration of information considerations into the planning and execution of all joint force operations. The JFC may reinforce this unity of effort with unity of command by establishing a subordinate task force so the preponderance of information forces' capabilities and activities are the responsibility of one commander under the JFC. The JFC assigns OIE units missions to create effects in and through the IE to set conditions that support the JFC's objectives and enduring outcomes.

Synchronization of OIE Activities

By its nature, OIE involves the synchronization of multiple capabilities and activities to aggregate their effects and achieve operational objectives. Synchronization comprises the coordination, tracking, and direction of all OIE activities to ensure they are aligned with the JFC's overarching narrative and objectives, and synchronized and deconflicted with activities external to the command.

Due to the interconnectedness of the IE, the effects of activities in and through the IE may cross geographical boundaries and, if not carefully planned and synchronized, may have unanticipated effects on tactical up through strategic-level objectives. OIE should be coordinated with other DOD entities, the interagency, and multinational partners so objectives and activities are deconflicted and, to the greatest extent possible, synchronized to create greater effects. Coordination of OIE with external organizations is through information planners or other personnel serving on higher headquarters staff or at adjacent joint and mission partner units.

The Narrative

The narrative can be thought of as a unifying story that acts as an information control measure to avoid conflicting messages and promote unity of effort. This is analogous to control and coordination measures used for maneuver and movement control, airspace coordination, and fire support coordination. The JFC and staff monitor the effects in and through the IE of the activities of the joint force to ensure those activities support the narrative.

Refer to JP 3-04, App. A.

Information and KM

IM and KM ensure users are aware of and can access critical information for decision making, and enables shared understanding. During execution, IM and KM facilitate synchronization, monitoring, and direction of activities. Effective IM and KM are essential for staying inside the enemy's decision-making cycle. Combined with effective planning, IM and KM help commanders anticipate enemy actions and develop branches, sequels, or adjustments.

The Information Staff Estimate

The information planning cell is responsible for the information staff estimate. The information staff estimate is a continual evaluation of how factors related to the IE impact the planning and execution of operations. The purpose of the information staff estimate is to inform the commander, staff, and subordinate commands on how information can be used to support mission accomplishment. The estimate helps feed the commander's estimate and contributes to the JFC's common operational picture of the OE for planning, mission coordination, and assessment of all operations. The information planning cell on the JFC's staff produces this consolidated estimate as an overview of all capabilities and activities available to perform tasks related to the information joint function. It includes the analysis of the informational, physical, and human aspects of the environment; the status of friendly OIE units and information forces and their activities; and an assessment of adversarial capabilities and intent.

Refer to JP 3-04, App. B.

III. Assessment *(See chap. 8.)*

Assessment helps the commander determine progress toward achieving joint force objectives and mission accomplishment. This requires identifying current (baseline) conditions of the OE and determining those desired conditions that define achievement of objectives, then monitoring for change from the current to desired conditions. Measuring this progress toward the mission objectives and delivering feedback into the planning process to adjust operations during execution involves deliberately comparing the planned effects of OIE with actual outcomes to determine the overall effectiveness of OIE unit activities.

The assessment process for OIE begins during planning and includes developing MOEs and MOPs of OIE activities, as well as their contribution to the larger operation or objective. This includes identifying MOE indicators and incorporating monitoring tasks as essential elements of those OIE plans. Historically, combat assessment has emphasized the BDA component of measuring physical and functional damage, but this approach does not always represent the most complete effect, particularly with respect to OIE. OIE often seeks to have effects outside the scope of battle and often do not create physical damage. While assessing the effects of OIE may require typical BDA analysis and assessment of physical, functional, and target system components, the higher-order effects of actions in and through the IE are often subtle. Assessment of second- and third-order effects of OIE activities can be difficult and may require significant intelligence collection and analysis efforts. Clearly articulating the desired effects and creating and resourcing an assessment plan for OIE during the planning processes increases the likelihood that all objectives are met. Planners should emphasize JIPOE, COG analysis, target systems analysis, and collection management activities to inform assessment. *Refer to JP 3-04, chap. VI.*

Chap 2

VI. Information & the Joint Planning Process

Ref: JP 3-04, Information in Joint Operations (Sept '22), chap. 4. (See chap. 4.)

The joint planning process (JPP) is an orderly, analytical process that consists of a logical set of steps to analyze a mission, select the best COA and produce a campaign or joint OPLAN or order. Like operational design, it is a logical process to approach a problem and determine a solution. It is a tool to be used by planners but is not prescriptive. Throughout the JPP steps (see Figure IV-3), information planners assist other joint planners in incorporating their understanding of how information impacts the OE to identify how to best support human and automated system decision making and how to best leverage information to achieve the JFC's objectives during operations. The result of the JPP is a plan or order that clearly specifies how the joint force will use and leverage information as part of the overall operation.

See pp. 2-45 to 2-51 for discussion of OIE information planning considerations.

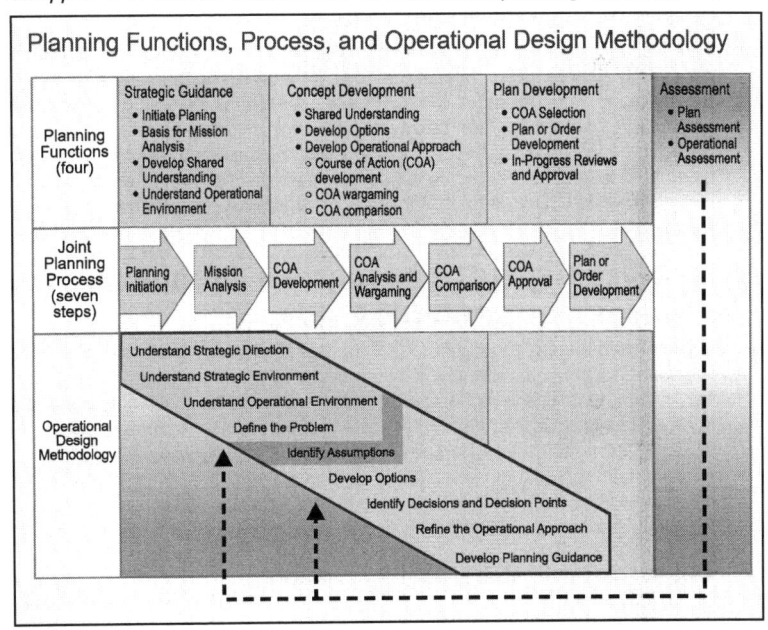

Planning Functions, Process, and Operational Design Methodology

Planning Functions (four)	Strategic Guidance	Concept Development	Plan Development	Assessment

	• Initiate Planning
	• Basis for Mission Analysis
	• Develop Shared Understanding
	• Understand Operational Environment
		• Shared Understanding
		• Develop Options
		• Develop Operational Approach
		  ○ Course of Action (COA) development
		  ○ COA wargaming
		  ○ COA comparison
			• COA Selection
			• Plan or Order Development
			• In-Progress Reviews and Approval
				• Plan Assessment
				• Operational Assessment

Joint Planning Process (seven steps): Planning Initiation → Mission Analysis → COA Development → COA Analysis and Wargaming → COA Comparison → COA Approval → Plan or Order Development

Operational Design Methodology:
- Understand Strategic Direction
- Understand Strategic Environment
- Understand Operational Environment
- Define the Problem
- Identify Assumptions
- Develop Options
- Identify Decisions and Decision Points
- Refine the Operational Approach
- Develop Planning Guidance

Ref: JP 3-04 (Sept '22), fig. IV-3. Planning Functions, Process, and Operational Design Methodology.

Refer to JFODS6: The Joint Forces Operations & Doctrine SMARTbook, 6th Ed. (Guide to Joint Warfighting, Operations & Planning). JFODS6 is updated for 2023 with new/updated material from the latest editions of JP 3-0 Joint Campaigns and Operations (Jun '22), JP 5-0 Joint Planning (Dec '20), JP 3-33 Joint Force Headquarters (Jun '22), and JP 1 Volumes I and II Joint Warfighting and the Joint Force (Jun '20), Additional topics and references include Joint Air, Land, Maritime and Special Operations (JPs 3-30, 3-31, 3-32 & 3-05).

Step 1—Planning Initiation

During planning initiation, information planners use their specific expertise to assist the JPG in:

- Reviewing commander's planning guidance for information activities and explicit and implied tasks that will impact planning.
- Identifying external stakeholders that the joint force should collaborate with for planning and executing information activities (e.g., DOS Global Engagement Center, country teams, JIATF or JIACG). See Chapter III, "Unity of Effort," for organizations to consider.
- Determining initial information planning support requirements to augment the staff (e.g., information professionals to serve as information planners, language/regional/cultural expertise).
- Gathering and analyzing the information required to plan operations that affect relevant actor behavior and identified networks.
- Updating the information estimate, providing updates on changes in the IE, updating the status of information forces, and providing the results of any ongoing information activities.

Step 2—Mission Analysis

The JFC and staff develop a restated mission statement that allows subordinate and supporting commanders to begin their own estimates and planning efforts for higher headquarters' concurrence. The joint force's mission is the task or set of tasks, together with the purpose, that clearly indicates the action to be taken and the reason for doing so. Mission analysis is used to study the assigned tasks and to identify all other tasks necessary to accomplish the mission. Mission analysis focuses the commander and the staff on the problem at hand and lays a foundation for effective planning.

A. Analyze Higher Headquarters' Planning Directives and Strategic Guidance

Information planners contribute to the analysis of strategic guidance and higher headquarters' planning directives by understanding and advising the JFC on how national leadership and higher headquarters intend for the military to support the informational instrument of national power. In particular, information planners determine higher headquarters' perspective of how the military will leverage information to achieve national strategic and military objectives, what behaviors that higher leadership wants from relevant actors to support those objectives, and what role the joint force has in leveraging information to obtain those desired behaviors.

The Operational Mission Narrative

During this step of mission analysis, CCMD and operational-level headquarters staffs use strategic guidance to begin developing the operational mission narrative. The operational mission narrative will include themes and messages that nest under the strategic mission narrative. The development of the operational mission narrative is a collaborative effort that should include planners with regional and cultural expertise. Operational mission narratives focus on the theater/region and seek to advance the legitimacy of the mission while countering adversary narratives. A compelling narrative at this level guides planning, targeting, and execution. Likewise, the joint force should make every effort to ensure operations, activities, words, and images are perceived as being consistent with the narrative, thereby preventing audiences from perceiving a conflict between the joint force's actions and its words.

See facing page.

Operational Mission Narrative

Ref: JP 3-04, Information in Joint Operations (Sept '22), pp. IV-16 to IV-18.

When developing the operational mission narrative, planners should recognize that narratives are not created in a vacuum. There are pre-existing narratives in the OE and others may emerge. These narratives may be from adversaries, friendly forces, or relevant neutral groups. These other narratives may reinforce or run counter to the joint force narrative. Awareness of these narratives leads to greater understanding of how to leverage operations and messaging activities to achieve friendly objectives.

Analyzing existing narratives provides insight into the messages that relevant actors are conveying, how they are disseminated and propagated, how the intended audiences and relevant actors react to the themes and messages in those narratives, and potential avenues for influence. In addition to informing mission analysis and the development of the operational mission narrative, the results from narrative analysis should be incorporated into JIPOE and operational assessment processes. Figure IV-4 shows some sample questions that an analysis of existing narratives can answer.

Questions for Narrative Analysis
- How do the relevant actors frame and explain their ideology?
- How do relevant actors make their ideology appear enduring and natural to the local culture?
- Do joint force activities challenge their assumptions, beliefs, and meanings?
- What are the local culture/society goals that the joint force can leverage?
- Are there inconsistencies in a relevant actors' narrative? If so, how does the relevant actor deal with those inconsistencies? Do those inconsistencies present a vulnerability that can be exploited?
- What is the structure of the existing narratives?
- How do existing narratives resonate with relevant actors?

Ref: JP 3-04, (Sept '22), fig. IV-4. Questions for Narrative Analysis.

Additionally, information planners identify operations worldwide in execution and ongoing activities, to include information activities, which will limit the JFC's range of possible COAs, as well as impact plans and operations. This awareness of other ongoing operations and activities includes those of multinational partners.

Finally, as part of mission analysis, information planners identify existing authorities and permissions and what additional authorities and permissions that the JFC will require for the conduct of information activities. This is done as early as possible in the JPP because of the time required to obtain those additional authorities and permissions. Use of some capabilities or activities that leverage information to affect behavior may require unique authorities and permissions. Joint force planners should also review the authorities for the use of capabilities and conduct of activities in their own AOR that could affect the OEs of other JFCs through the IE. Achieving a shared understanding of authorities vertically across echelons of command and horizontally across mission partners is key to successful execution. Information planners can advise the planning team on which authorities for leveraging information may require additional time, legal review, or subject matter expertise to request.

Refer to JP 3-04, App. A, Narrative Development.

B. Review Commander's Initial Planning Guidance

Information planners use the commander's initial planning guidance as the basis for continuing the analysis of the OE begun during operational design, which focused on describing the relationship between the informational, physical, and human aspects of the environment, and on identifying and describing relevant actors and their range of potential behaviors.

C. Determine Known Facts and Develop Planning Assumptions

Information planners provide facts and assumptions related to the joint force understanding of how information impacts the OE, the joint force's ability to manage and share information to support decision making, and the joint force's ability to leverage information. Potential facts and assumptions include but are not limited to:

1. The identity of relevant actors and why they are relevant to the JFC's mission.

2. The degree to which the joint force understands the perceptions, attitudes, beliefs, and other drivers of relevant actor behaviors (see Chapter I, "Fundamentals of Information," paragraph 3.d., "Information can affect behavior," for examples of drivers of relevant actor behaviors).

3. The access that the joint force will have to humans and automated systems to affect the behavior of relevant actors.

4. The impact that joint force operations will have upon the OE and relevant actors. This includes the range of potential and likely behaviors of relevant actors in response to joint force or others' activities.

5. The availability and capacity of specialized capabilities for the joint force to conduct OIE and information activities, to include those of mission partners.

6. The ability of the joint force to affect relevant actor behavior within the parameters of the mission. In other words, will the joint force be able to affect relevant actor behavior to the degree necessary and in sufficient time to support the achievement of the JFC's objectives?

7. The authorities and permissions available to the joint force to use specialized capabilities, to target specific relevant actors, and to undertake information activities.

8. The ability of relevant actors to attack or exploit the joint force's information, information networks, and information systems.

9. The ability of the joint force to protect its information, information networks, and information systems from relevant actor action and the resilience of those information networks and systems.

10. The ability of the joint force to manage and share friendly information to support effective decision making and C2 during operations, especially during multinational operations.

D. Determine and Analyze Operational Limitations

Some operational limitations may arise due to the inherent informational aspect of military activities, the effects of which are not geographically constrained or limited to a joint force's intended audiences. The joint force cannot control the spread of information or its impact on audiences, within or beyond their specified JOA. This may restrict a commander's freedom of action if the informational aspect of a COA undermines higher-priority national objectives or negatively impacts the operations of other JFCs. Based upon their understanding of how information impacts the OE, information planners work with the other joint planners to develop a list of limitations related to relevant actors, the employment of specialized capabilities or conduct of information activities, and the use of specific themes and messages. Many of these limitations will be specified in authorities and permissions from higher headquarters.

E. Determine Specified, Implied, and Essential Tasks and Develop the Mission Statement

The commander and staff review the planning directive's specified tasks and discuss implied tasks during planning initiation, then confirm the tasks during mission analysis. Information planners identify specified and implied tasks to understand how information impacts the OE, leverage information, and support decision making. Information planners identify other implied tasks based upon their analysis of the informational, physical, and human aspects of the OE and on an understanding of the relevant actors and how to affect their drivers of behavior. From the lists of specified and implied tasks, the commander and staff determine the essential tasks and use them to develop the mission statement.

F. Conduct Initial Force and Resource Analysis

During mission analysis, the commander and staff team begin to develop a list of required forces and capabilities necessary to accomplish the specified and implied tasks. Information planners contribute to this list by identifying those forces and capabilities required to understand how information impacts the OE, support human and automated decision making, and leverage information. In resource-constrained environments, military forces or capabilities may be unavailable or not readily available to meet all requirements. As part of their initial force and resource analysis, information planners should consider:

1. The lead time to deploy information forces and specialized capabilities into theater or direct support to the joint force from a home location.

2. The lead time to coordinate approval of information authorities and activities.

3. RFF or personnel with unique skills such as linguists, sociocultural experts, social media experts, experts in analyzing publicly available information, as well as experts on artificial intelligence and machine-learning.

4. Collaborating with mission partners who have information forces and specialized capabilities and the capacity to fill joint force resource gaps.

5. Planners evaluate appropriate requirements against existing or potential contracts or task orders to determine if a contracted support solution can meet the requirement.

G. Develop Military Objectives

Each military objective establishes a clear goal toward which all the actions and effects of a LOO or LOE are directed. While military objectives commonly describe the condition and/or the relative position of the joint or enemy forces, the JFC may also express objectives as a particular behavior that the military operation will bring about. Information planners work with the rest of the staff to determine attainable behavioral goals that are based upon the analysis of the OE, including the previously identified potential behaviors in response to joint force or others' activities. Information planners use these objectives to develop measures of effectiveness (MOEs) and MOE indicators to assess how well the joint force leverages information. These include identifying and incorporating indicators of trending success or failure into the monitoring and assessment plan before finalization of the overall plan. Planners should keep in mind that it may take a considerable amount of time to observe the effects of information activities, and cause and effect relationship may be difficult to assess.

H. Develop COA Evaluation Criteria

Information planners help develop evaluation criteria that measure the relative effectiveness and efficiency of a COA to address threats and avoid or mitigate hazards in or through the IE. Potential evaluation criteria may include whether and how well the COA:

1. Aligns planned actions with strategic and operational mission narratives to establish the legitimacy of the joint force mission and actions with relevant actors.

2. Includes information activities focused on producing the desired behaviors in prioritized relevant actors.

3. Includes information activities that protect the joint force from adversary attempts to undermine the joint force narrative or the legitimacy of the joint force mission and actions (e.g., coordinated PA efforts to engage foreign and domestic publics, leadership outreach to mission partners, and HN military and civilian leadership).

4. Includes information activities that prevent, counter, and mitigate adversary or enemy attempts to undermine the joint force decision making and C2 (e.g., hardening of information systems against known enemy capabilities, building resiliency into C2 systems)

5. Identifies and accounts for the potential second- and third-order effects and potential risks to enduring strategic objectives (e.g., hardening of information systems against known enemy capabilities, building resiliency into C2 systems).

6. Accounts for the potential impacts on the joint force from the activities that resonate in and through the IE.

I. Develop Risk Assessment

Information planners characterize the risk of obstacles or actions having effects in and through the IE that could preclude mission accomplishment. This includes actions that counter the narrative and indicate a "say-do gap" in joint operations. Information planners are responsible for carefully articulating this risk characterization so that commanders have a clear understanding of the potential benefits and dangers associated with information activities. Many of these impediments can be derived from an examination of friendly strategic and operational COGs and include, but are not limited to, the following:

1. Likelihood and impact of an adversary denying friendly C2 through technical means (e.g., EMSO or CO).

2. Likelihood and impact of allied or partner nation withdrawing support from a multinational operation.

3. Likelihood and impact of the collateral effects of joint force actions (e.g., civilian casualties, economic hardship, cultural offense) undermining the strategic or operational narrative and/or legitimacy of the joint force operation.

4. Likelihood and impact of adversary propaganda efforts undermining joint force strategic or operational narrative and/or legitimacy of the joint force operations.

5. Likelihood and impact of friendly force casualties undermining domestic support for joint force operations.

6. Likelihood and impact of international pressure causing cessation of joint force operations prior to strategic objectives being achieved.

See Chairman of the Joint Chiefs of Staff Manual (CJCSM) 3105.01, Joint Risk Analysis, for additional information and guidance on risk determination.

J. Determine Initial Commander's Critical Information Requirements (CCIRs)

Ref: JP 3-04, Information in Joint Operations (Sept '22), pp. IV-22 to IV-23.

Commander's Critical Information Requirement (CCIRs) identify key elements of information the commander identifies as being critical to timely decision making.

Priority Intelligence Requirements (PIRs)

Information planners should consider the following as potential priority intelligence requirements (PIRs):

- Intelligence required to resolve any remaining assumptions related to adversary actions or capabilities or unresolved assumptions regarding the IE.
- Intelligence required to detect the existence of any obstacles or any adversary actions that were characterized during risk assessment as moderate or higher risk.
- Intelligence about the pending or actual conduct of activities by opponents or other actors that will create effects in and through the IE that will likely impact the JFC's or strategic objectives (e.g., an opponent's announcement that they will withdraw forces from a contested area, a political announcement that would cause partner nations to doubt US or joint force resolve to continue operations, corruption in a supported government that would cause locals to oppose that government and the joint force).

Friendly Force Information Requirement (FFIR)

Information planners focus on the following as potential friendly force information requirement (FFIR):

- Information required to resolve any remaining assumptions related to the availability and capabilities of friendly information forces and OIE units or of authorities and permissions to employ those capabilities or conduct information activities.
- Any change in status of OIE units or specialized information forces' capabilities, to include MNF partners conducting information activities.
- Any change in the authorities or permissions to employ specialized capabilities or conduct information activities.
- Information on the planned or actual conduct of activities by other commands that will create effects in and through the IE that will likely impact the JFC's or strategic objectives.
- Planned or actual activities by or related to mission partners that would undermine the composition or cohesiveness of the MNF (e.g., political developments in a partner nation that could jeopardize continued support by forces from that nation, operations by the forces of one mission partner that are publicly opposed by another).
- Degradation or loss of any communication capability resulting in the JFC's inability to C2 the joint force.
- Loss of access to social media or other outlets the joint force is using to understand, inform, and influence relevant actors.
- Loss of critical access point or other conduit the joint force is using to attack, exploit, or deny information, information, networks, and information systems of relevant actors.

K. Prepare Staff Estimates

The information planners produce the information staff estimate in conjunction with OIE units and Service component information planners. That estimate includes the status and capabilities of OIE units or other forces tasked with leveraging information or elements that are critical to joint force protection of the joint force's information, information networks, or information systems. The information staff estimate includes an analysis of how information impacts the OE, as well as an assessment of how the inherent informational aspects of activities planned by each of the functional areas might impact the IE in ways to support or to undermine achieving the JFC's objectives. The information staff estimate will also identify additional capabilities to augment organic assets.

Refer to JP 3-04, App. B, "Information Staff Estimate Format."

The intelligence estimate includes an information section. This section should include relevant aspects of the IE, such as:

- Inputs from capabilities, operations, and activities that gather operational information. These include, but are not limited to, CA, KLE, PA, MISO, OPSEC, JEMSO, COMCAM, space operations, and CO.
- Those likely and dangerous transitions of enemy, adversary, or competitor behavior that challenge US objectives. This section enables planners to estimate the interests, intent, capability, capacity, and likely disruptive actions of relevant actors to support or counter USG interests.

For additional information on the intelligence estimate, refer to CJCSM 3130.03, Planning and Execution Formats and Guidance.

L. Prepare and Deliver Mission Analysis Brief

Upon conclusion of the mission analysis, the staff, including information planners, will present a mission analysis brief to the commander.

A key portion of the information planners' input to the mission analysis briefing is the development of the operational mission narrative. The operational mission narrative will immediately follow the commander's intent in the final plan or order. The commander's intent describes the desired outcome and the operational mission narrative communicates the "why," "how," and "by whom" of an operation. A well-crafted mission narrative and commander's intent provides coherence to military actions and activities and facilitates synchronization of communications and actions. Tactical units use the commander's intent and the operational mission narrative to develop a tactical or local narrative that lends continuity to operations and communications.

Step 3—COA Development

A COA is a potential way to accomplish the assigned mission. After the mission analysis briefing, the staff begins developing COAs for analysis and comparison based on the commander's intent, operational mission narrative, restated mission, and planning guidance. A good COA accomplishes the mission within the commander's guidance, advances the narrative, provides flexibility to meet unforeseen events during execution, and positions the joint force for future operations.

> **Information Planners use their Specific Expertise to:**
>
> 1. Identify ways land, maritime, air, space, cyberspace, and special operations forces contribute to each of the tasks of the information joint function.
>
> 2. Advise on how the joint force can leverage the inherent informational aspects of activities to create or shape the desired perceptions to achieve the commander's objectives.
>
> 3. Advise on how to integrate actions in the physical domains, IE (including cyberspace), and EMS to align with the operational mission narrative.
>
> 4. Identify threats, vulnerabilities, and opportunities in the IE.
>
> 5. Determine how to task organize and employ OIE units and other information forces in support of objectives. This includes identifying how OIE will amplify or conceal physical actions in a manner that increases or decreases ambiguity.
>
> 6. Identify critical capabilities required to inform domestic, international, and internal audiences; influence relevant actors; and attack and exploit information, information networks, and information systems.
>
> 7. Identify any friendly information systems or segments of friendly information networks that need to be prioritized for defensive actions based on each COA.
>
> 8. Identify critical information that the joint force needs to protect for each COA and recommend appropriate protection measures.
>
> 9. Determine communication channels that are most credible to and are most effective for reaching the selected audiences.
>
> 10. Identify how to integrate lethal and nonlethal actions required to create specific effects in and through the IE (e.g., destruction of a radio tower) into existing targeting and fires planning processes.

A. Review Objectives and Tasks and Develop Ways to Accomplish Tasks

During COA development, planners review and refine objectives from the initial work done during the development of the operational approach. Information planners determine the tasks required to effectively leverage information to achieve the refined objectives. These objectives and tasks are assigned in plans or orders to joint force units, including OIE units. COAs should include tasks to inform domestic and international, and internal audiences; influence relevant actors; and attack and exploit information, information networks, and information systems.

See pp. 2-10 to 2-11 for a discussion of these tasks.

B. Select and Prioritize Audiences, TAs, and Targets

Ref: JP 3-04, Information in Joint Operations (Sept '22), pp. IV-25 to IV-26.

Information planners participate in the joint targeting process during COA development to identify and prioritize relevant actors with whom the joint force will interact.

Relevant actors are categorized as audiences, TAs, or targets depending upon their relationship to a threat and the means with which the joint force will interact with them (i.e., whether through lethal or nonlethal engagement).

Audiences

Audiences are a broadly defined group that contains stakeholders and/or publics relevant to military operations. Audiences are not the enemy and do not directly perform a function for the enemy. KLE, PA, and CMO are examples of activities that use the term audiences to characterize the relevant actors selected for engagement. Information planners aid in selecting and prioritizing audiences to ensure activities are synchronized and deconflicted and to prevent or mitigate any negative effects caused by fires or other information activities.

Target Audiences (TAs)

A TA is an individual or group selected for influence. Individuals or groups are designated as such when a change in their behavior is necessary to achieve the commander's objectives. TAs will sometimes also meet the criteria of a target if they perform a function for a threat, whether they do so knowingly or not, willingly or unwillingly. In those cases, TAs are included on one of the joint target lists to be prioritized, vetted, and approved in accordance with JFC priorities; legal, political, and operational constraints; rules of engagement; collateral damage restrictions; political considerations; and operational requirements.

Targets

A target is an entity or object that performs a function for the threat considered for possible engagement or other action. A TA may be a target if it is the adversary or performs a function for the adversary. A target's importance derives from its potential contribution to achieving a commander's objective(s) or otherwise accomplishing assigned tasks. Offensive military activities (e.g., electromagnetic attack, cyberspace attack, MISO) should be coordinated and deconflicted within the joint targeting process. Information planners participate in the targeting working groups and boards to nominate targets, identify targets for inclusion in the joint restricted fires list, and evaluate targets for their psychological impact on relevant actors. The traditional methodology of identifying target systems, sets, components, and their critical elements remains valid for OIE. Some capabilities used for OIE may require long lead time for development of the JIPOE and release authority and should be identified as early in the target process as possible.

C. Identify the Sequencing (Simultaneous, Sequential, or Combination) of Actions for each COA

Understand which resources become available and when during the operation or campaign. Resource availability will significantly affect sequencing operations and activities. Sequencing of inform, influence, and attack tasks rely on the relevance of the information in relation to the timing of an event. Any gap between publicized information and performance of an activity has the potential to undermine the intent of the activity and negatively affect the achievement of objectives. Information planners carefully consider how the sequencing of activities will impact the inherent informational aspects of each COA. This includes consideration of how information activities will be synchronized with other activities to enhance the effectiveness of the COA (e.g., synchronizing jamming against an IADS in support of air interdiction). The timing and synchronization of activities of each COA should consider how it can pre-empt, undermine, or counter adversary and enemy use of narratives, especially those that convey misinformation or disinformation. This is critical due to the extreme difficulty to change minds or beliefs, even when presented with facts and evidence once an audience has been influenced by misinformation or disinformation. Therefore, the goal is to provide accurate and useful information to relevant actors in a timely manner to increase its credibility and relevance. Information planners can advise on how each COA can communicate information in a timely fashion, multiple times, and from multiple sources to create the desired effects.

Step 4—COA Analysis and Wargaming

COA analysis is the process of closely examining potential COAs to reveal details that enable the commander and staff to tentatively evaluate COA validity and identify the advantages and disadvantages of each proposed friendly COA. Wargaming is a primary means for COA analysis. Wargames are representations of conflict or competition in a synthetic environment, in which people make decisions and respond to the consequences of those decisions.

During COA analysis and wargaming, information planners examine how well each COA leverages information to achieve objectives. Wargaming helps the staff to visualize the flow of the operation and, in doing so, facilitates understanding the effects of the joint force's leveraging of information. During wargaming planners also examine the extent to which joint force activities align with and support JFC's operational mission narrative. Information planners help examine friendly and adversary information activities (i.e., those activities that inform audiences; influence foreign relevant actors; and attack and exploit relevant actor information, information networks, and information systems) to determine their potential effects in relation to the objectives. To the extent possible, those personnel or organizations tasked to conduct such activities participate in the wargaming process. Wargaming might identify activities that were previously not identified. During COA analysis and wargaming, information planners help the staff:

1. Determine the likelihood that joint force activities will affect relevant actor behavior. This includes consideration of how relevant actors are likely to react to information activities and the inherent informational aspect of physical activities.

2. Determine the relative importance of relevant actors and identify the potential emergence of new relevant actors.

3. Identify high-value targets related to inform, influence, attack, or exploit activities.

4. Identify decision points related to the joint force's leveraging of information to change or maintain perceptions, attitudes, and other drivers of relevant actor behaviors.

5. Identify how the joint force reacts to threats, vulnerabilities, and opportunities in the environment.

6. Identify and recommend adjustments to information tasks conducted by information forces, including OIE units.

7. Recommend adjustments to task organization of joint force elements to better support leveraging information and the inherent informational aspects of activities.

8. Identify and provide time, space, and purpose input for synchronization matrices or other decision-making tools.

9. Identify tasks that leverage information for branches and sequels.

10. Identify PIRs and FFIRs.

11. Refine information concept of support (the description of how information will support the CONOPS).

12. Refine sequencing and timing of information activities.

13. Refine risks associated with joint force use and leveraging of information.

14. Review and update the information estimate (see Appendix B, "Information Staff Estimate Format") from applicable OPLANs and concept plans.

COA analysis and wargaming benefits from the participation of red teams, green cells, and white cells. Because they bring a different perspective into COA analysis and wargaming, these elements help joint planners reduce mirror-imaging and better understand and evaluate the potential actions and reactions of relevant actors. SMEs for red teams, and for green and white cells may include multinational partners, behavioral scientists, and cultural anthropologists. If not resident to the core planning staff, these experts may be available through reachback support.

Step 5—COA Comparison

COA comparison is both a subjective and objective process, whereby COAs are considered independently and evaluated/compared against a set of criteria that are established by the staff and commander. COA comparison starts with all staff elements analyzing and evaluating the advantages and disadvantages of each COA from their respective viewpoints. Each of the COA evaluation criteria should contain information considerations. During mission analysis, information planners helped develop the evaluation criteria used in COA to measure the relative effectiveness and efficiency of a COA to address threats and avoid or mitigate hazards in or through the IE. How well the joint force uses information and leverages information may indirectly affect the rating of that COA evaluation criteria.

Step 6—COA Approval

In this JPP step, the staff briefs the commander on the COA comparison and the analysis and wargaming results and provides the commander with a recommended COA. The commander combines personal analysis with the staff recommendation, resulting in a selected COA. It gives the staff a concise statement of how the commander intends to accomplish the mission and provides the necessary focus for planning and plan development. The information planner helps the staff refine the commander's COA selection into a clear decision statement, then completes the commander's estimate. The commander's estimate provides a concise statement of how the commander intends to accomplish the mission and provides the necessary focus for campaign planning and contingency plan development. The commander's estimate will include the refined commander's intent along with the commander's operational mission narrative.

Step 7—Plan or Order Development

This final JPP step includes development of the CONOPS and publication of a plan or order. During plan or order development, the staff further develops and refines component missions and tasks that specify how the joint force will use information and leverage information to achieve objectives. The final plan or order will assign those missions and tasks to OIE units and other information forces.

Chap 3
(Information) CAPABILITIES

* In accordance with the changes in joint and Army doctrine (editor's note, p. 1-2), Army forces will no longer use the terms information operations, information-related capabilities, or information superiority. Neither JP 3-04 (Sept '22) nor ADP 3-13 (Nov '23) provide a specific alternative term in lieu of "information-related capabilities", instead generically referring to "information capabilities." Below outlines how these capabilities are addressed.

I. Joint Force <u>Capabilities, Operations, and Activities</u> for Leveraging Information (JP 3-0, Jan '22)

In addition to planning all operations to benefit from the inherent informational aspects of physical power and influence relevant actors, the JFC also has <u>additional means with which to leverage information in support of objectives</u>. Leveraging information involves the generation and use of information through tasks to inform relevant actors; influence relevant actors; and/or attack information, information systems, and information networks:

- Key Leader Engagement (KLE)
- Public Affairs (PA)
- Civil-Military Operations (CMO)
- Military Deception (MILDEC)
- Military Information Support Operations (MISO)
- Operations Security (OPSEC)
- Signature Management
- Electronic Warfare (EW)
- Combat Camera (COMCAM)
- Historians
- Space Operations
- Special Technical Operations (STO)
- Cyberspace Operations (CO)

II. Army Doctrine (ADP 3-13, Nov '23) *(See p. 3-4.)*

Although there are multiple mentions of "information capabilities" throughout ADP 3-13, they are neither identified specifically nor listed. Examples, however, are provided under a section on "TECHNICAL TRAINING AND EDUCATION" *(ADP 3-13, p. 8-19).*

...the Army trains and educates technical specialists who possess the ability to counter and defeat threat activities. <u>Examples of technical informational training for select information specialists include</u>—

- Civil affairs operations.
- Counterintelligence (CI).
- Cyberspace operations.
- Electromagentic warfare (EW).
- Information management (IM).
- Intelligence and the various intelligence disciplines
- Knowledge management.
- Military deception (MILDEC).
- Military Information in Support of Operations (MISO).
- Network operations
- Public affairs operations.
- Space operations.
- Operational law.

See following pages for an overview of how these capabilities (and operations) were previously described in FM 3-13 (Dec '16).

Information Operations & the IRCs
(* as previously defined in FM 3-13, Dec '16)

Ref: FM 3-13, Information Operations (Dec '16), pp. 1-2 to 1-6.

> **Information Operations (IO)** is the integrated employment, during military operations, of information-related capabilities in concert with other lines of operation to influence, disrupt, corrupt, or usurp the decision-making of adversaries and potential adversaries while protecting our own (JP 3-13).

Breaking down the definition into constituent parts helps to understand its meaning and implications for land forces:

Information Operations (IO)* is the...

Integrated Employment of Information-Related Capabilities (IRCs)...
IO brings together IRCs at a specific time and in a coherent fashion to create effects in and through the information environment that advance the ability to deliver operational advantage to the commander. While IRCs create individual effects, IO stresses aggregate and synchronized effects as essential to achieving operational objectives.

During Military Operations...
Army forces, as part of a joint force, conduct operations across the conflict continuum and range of military operations. Whether participating in security cooperation efforts or conducting major combat operations, IO is essential during all phases (0 through V) of a military operation.

In Concert with Other Lines of Operation...
Commanders use lines of operations and lines of effort to visualize and describe operations. A line of operations is a line that defines the directional orientation of a force in time and space in relation to the enemy and that links the force with its base of operations and objectives (ADRP 3-0). Lines of operations connect a series of decisive points that lead to control of a geographic or force-oriented objective. A line of effort is a line that links multiple tasks using the logic of purpose rather than geographical reference to focus efforts toward establishing operational and strategic conditions (ADRP 3-0). Lines of effort are essential to long-term planning when positional references to an enemy or adversary have little relevance. Commanders may describe an operation along lines of operations, lines of effort, or a combination of both. Commanders, supported by their staff, ensure information operations are integrated into the concept of operation to support each line of operation and effort. Based on the situation, commanders may designate IO as a line of effort to synchronize actions and focus the force on creating desired effects in the information environment. Depending on the type of operation or the phase, commanders may designate an IO-focused line of effort as decisive.

To Influence, Disrupt, Corrupt, or Usurp...
IO seeks to create specific effects at a specific time and place. Predominantly, these effects occur in and through the information environment. Immediate effects (disrupt, corrupt, usurp) are possible in the information environment's physical and informational dimensions through the denial, degradation, or destruction of adversarial or enemy information-related capabilities. However, effects in the cognitive dimension (influence) take longer to manifest. It is these cognitive effects—as witnessed through changed behavior—that matter most to achieving decisive outcomes.

The Decision Making of Enemies and Adversaries...
While there are differences among the terms adversaries, threats, and enemies, all three refer to those individuals, organizations, or entities that oppose U.S. efforts. They

therefore must be influenced in some fashion to acquiesce or surrender to or otherwise support U.S. national objectives by aligning their actions in concert with commanders' intent. [The joint phrasing "adversaries and potential adversaries" is revised to "enemies and adversaries" to better align with Army terminology.]

While Protecting Our Own... Friendly commanders, like enemy and adversary leaders, depend on an array of systems, capabilities, information, networks, and decision aids to assist in their decision making. Gaining operational advantage in the information environment is equally about exploiting and protecting the systems, information, and people that speed and enhance friendly decision making, as it is about denying the same to the threat.

Information-Related Capabilities (IRCs)*

An **information-related capability (IRC)** is a tool, technique, or activity employed within a dimension of the information environment that can be used to create effects and operationally desirable conditions (JP 1-02). The formal definition of IRCs encourages commanders and staffs to employ all available resources when seeking to affect the information environment to operational advantage. For example, if artillery fires are employed to destroy communications infrastructure that enables enemy decision making, then artillery is an IRC in this instance. In daily practice, however, the term IRC tends to refer to those tools, techniques, or activities that are inherently information-based or primarily focused on affecting the information environment.

The information-related capabilities (IRCs) include—

- Military deception
- Military information support operations (MISO)
- Soldier and leader engagement (SLE), to include police engagement
- Civil affairs operations
- Combat camera
- Operations security (OPSEC)
- Public affairs
- Cyberspace electromagnetic activities
- Electromagentic warfare
- Cyberspace operations
- Space operations
- Special technical operations

All unit operations, activities, and actions affect the information environment. For this reason, whether or not they are routinely considered an IRC, a wide variety of unit functions and activities can be adapted for the purposes of conducting information operations or serve as enablers, to include:

- Commander's communications strategy or communication synchronization
- Presence, profile, and posture
- Foreign disclosure
- Physical security
- Physical maneuver
- Special access programs
- Civil military operations
- Intelligence
- Destruction and lethal actions

III. INFO Capabilities (INFO2 SMARTbook)

The INFO2 SMARTbook discusses the following capabilities in greater detail:

Public Affairs See pp. 3-5 to 3-16.
Army public affairs is communication activities with external and internal audiences (JP 3-61). Public affairs operations help to establish conditions that lead to confidence in the Army and its readiness to conduct unified land operations.

Civil Affairs & Civil-Military Operations See pp. 3-17 to 3-26.
Civil affairs operations encompass actions planned, executed, and assessed by civil affairs forces. Civil-military operations are activities of a commander performed by designated civil affairs or other military forces that establish, maintain, influence, or exploit relations between military forces, indigenous populations, and institutions.

Military Deception (MILDEC) See pp. 3-27 to 3-32.
Military deception (MILDEC) involves actions executed to deliberately mislead adversary military, paramilitary, or violent extremist organization decision makers. The intent of MILDEC is to feed information that deliberately misleads the enemy decision makers as to friendly military capabilities, intentions, and operations and lead the enemy to take actions (or inactions) that contribute to accomplishment of the friendly mission.

Military Information Support Operations (MISO) See p. 3-33.
Military information support operations are planned operations to convey selected information and indicators to foreign audiences to influence their emotions, motives, objective reasoning, and ultimately the behavior of foreign governments, organizations, groups, and individuals in a manner favorable to the originator's objectives (JP 3-13.2).

Operations Security (OPSEC) See pp. 3-39 to 3-44.
Operations security is a capability that identifies and controls critical information, indicators of friendly force actions attendant to military operations, and incorporates countermeasures to reduce the risk of an adversary exploiting vulnerabilities (JP 3-13.3).

Cyberspace Electromagnetic Activities (CEMA) See p. 3-45.
Cyberspace electromagnetic activities is the process of planning, integrating, and synchronizing cyberspace and electronic warfare operations in support of unified land operations (ADRP 3-0).

Cyberspace Operations (CO) See pp. 3-47 to 3-54.
Cyberspace operations are the employment of cyberspace capabilities where the primary purpose is to achieve objectives in or through cyberspace (JP 3-0).

Electromagentic Warfare (EW) See pp. 3-55 to 3-60.
Electromagnetic Warfare (EW) is military action involving the use of electromagnetic and directed energy to control the electromagnetic spectrum or to attack the enemy.

Space Operations See pp. 3-61 to 3-68.
Space operations are operations that occur in the space domain and seek to gain superiority over enemies and adversaries in the space domain and its corresponding environment.

Additional Capabilities See pp. 3-69 to 3-72.
Additional capabilities discussed include integrated joint special technical operations (IJSTO); special access programs (SAP); personnel recovery (PR); physical attack; physical security; presence, profile, and posture (PPP); soldier and leader engagement (SLE); police engagement; and social media.

Chap 3
I. Public Affairs (PA)

Ref: JP 3-61 (w/Chg 1), Public Affairs (Aug '16) and FM 3-61, Communication Strategy and Public Affairs Operations (Feb '22).

See pp. 1-35 to 1-44 for discussion of the INFORM activity as related to information advantage (ADP 3-13) and p. 2-19 for leveraging information by "informing domestic, international and internal audiences" (JP 3-04).

Public affairs (PA) doctrine and principles apply across the range of military operations. PA is a command responsibility and should not be delegated or subordinated to any other staff function below the command group. The public should perceive information communicated by PA as accurate.

Public Affairs Guidance (PAG)
Public affairs guidance (PAG) supports the public discussion of defense issues and operations and serves as a source document when responding to media representatives and the public. PAG also outlines planning guidance for related public affairs responsibilities, functions, activities, and resources. The development and timely dissemination of PAG ensures that all information is in consonance with policy when responding to the information demands of joint operations. PAG also conforms to operations security and the privacy requirements of the members of the joint forces.

The US military has an obligation to communicate with its members and the US public, and it is in the national interest to communicate with international publics. The proactive release of accurate information to domestic and international audiences puts joint operations in context, facilitates informed perceptions about military operations, undermines adversarial propaganda, and helps achieve national, strategic, and operational objectives.

Over the past two decades, there have been dramatic changes in the information environment. Notably, traditional media is no longer the only voice influencing key publics. The abundance of information sources, coupled with technology such as smart phones, digital cameras, video chat, and social media enterprises, allows information to move instantaneously around the globe. As such, it is imperative for PA personnel to rapidly develop themes and messages to ensure that facts, data, events, and utterances are put in context. Coordination and synchronization of themes and messages take place to ensure unity of effort throughout the information environment.

These tools provide the US military the ability to reach various audiences without mass media, as well as create the opportunity to join the conversation (as opposed to simply delivering a message) with an audience. Two-way conversation permits greater transparency and clarity. Joint operations will be supported by tailored communication that addresses friendly, neutral, and adversarial audiences. Often, these audiences want to both listen to and be heard by US forces. PA personnel will focus their communication efforts to a given public or publics. The speed of modern communications and the disparity of multiple audiences increase the importance of quickly and agilely synchronizing communication.

The First Amendment guarantees freedom of the press, but within the Department of Defense (DOD) this right must be balanced against the military mission that requires operations security (OPSEC) at all levels of command to protect the lives of US or multinational forces and the security of ongoing or future operations. These competing goals sometimes lead to friction between the media and the military. The Privacy Act of 1974 prevents the release of certain personal information to the media, but does not forbid individuals from releasing information about themselves in social media. In addition, stringent restrictions exist for protecting personally identifiable information, and there are strict reporting requirements if personally identifiable information is released, even inadvertently.

The tempo of military operations, OPSEC concerns, and the number and variety of other information sources competing for the attention of the populace complicate the joint force commanders' (JFCs') ability to provide information to diverse publics at the same pace as the media and other sources. The ability of anyone with Internet access to share information and provide graphic visuals without validating facts as an event unfolds further complicates the military's effort to accurately inform the media and populace. JFCs and public affairs officers (PAOs) should evaluate missions to identify public information and visual information (VI) requirements, as well as the means to acquire and move those products in a timely manner. PA planning should include considerations to reduce the time lag between an event and when information about it, if any, can be shared.

The public can get information about the military and its operations from official DOD and unofficial sources (e.g., information disseminated by Service members, distributed by the public, the media, or by groups hostile to US interests). Regardless of the source, intention, or method of distribution, information in the public domain either contributes to or undermines the achievement of operational objectives. Official information can help create, strengthen, or preserve conditions favorable for the advancement of national interests and policies and mitigate any adverse effects from unofficial, misinformed, or hostile sources.

PA is a command responsibility. Official communication with US and international audiences will have a significant impact on the operational environment (OE). Effective PA is a key enabler for the commander to build and maintain essential relationships.

Public support for the US military's presence or operations is likely to vary. The PAO, in conjunction with others on the staff, must be able to quickly and accurately assess the information environment to provide valuable guidance and courses of action (COAs) to the commander. Such assessments enable the commander to better inform relevant audiences about ongoing operations and engender their support.

I. Public Affairs and the Operational Environment (OE)

Information in the public domain affects the OE and influences operations. Commanders should carefully evaluate how various friendly, enemy, adversary, and neutral actions, images, and words impact planned and ongoing operations. PA understands that various audiences have differing information needs and works closely with other information providers to ensure consistency of messaging and accuracy of content. By conveying the facts about joint force activities in a proactive manner, PA helps the JFC to impact the information environment, particularly as it relates to public support. The joint force must coordinate all of its messages; further, it must integrate those messages with its partner nations' message as part of the ongoing alignment to maintain unity of effort and stand out in a saturated information environment. The information environment is the aggregate of individuals, organizations, and systems that collect, process, disseminate, or act on information.

For additional discussion of the OE, see pp. 0-6 to 0-10. See facing page for a discussion of public perception.

Public Perception

Ref: JP 3-61 (w/Chg 1), Public Affairs (Aug '16), pp. I-3 to I-4,

Perceptions Can Become Reality. A first impression on the perceptions and attitudes of decision makers, leaders, and other individuals cannot be underestimated. First impressions influence perceptions and judgments, which bias how individuals process subsequent information. Additionally, information that contradicts first impressions may be dismissed altogether. Enemies take advantage of this and often communicate lies or misleading information before we can verify details and communicate the truth. The first side that presents information often sets the context and frames the public debate. It is extremely important in maintaining legitimacy and public trust to get accurate information and VI out first, even information that may portray DOD in a negative manner. Maintaining legitimacy through disseminating rapid and accurate information helps disarm the enemy's propaganda and defeats attempts by the adversary to use negative information against friendly forces. JFCs should be prepared to assume some risk to ensure that public communication activities can be executed in time to ensure the most accurate and contextual information is publicly available.

Timeliness and Repetition. Timeliness is a key component of newsworthy information. Providing accurate and useful information in a timely manner increases credibility and relevancy. For information to have an impact, the audience must receive the information in a timely fashion, multiple times, and from multiple sources.

Cultural Considerations. The JFC staff and PA must understand who they are communicating with to enhance reception and understanding of the message. News is produced by people who adhere to the values and cultural system of the society they serve. News media coverage does not always reflect reality, but frames reality by choosing what events to cover and how to present them.

Impact of Propaganda. Propaganda is any form of communication misleading in nature designed to influence the opinions, emotions, attitudes, or behavior of any group to benefit the sponsor. It should not be assumed that all propaganda is misleading or outright lies. While the term propaganda generally implies lies and deceit, enemy propaganda may in fact be honest and straightforward. Propaganda is compelling as it often uses elements that make information newsworthy. Many people are drawn to conflict or violence. Adversaries use conflict and violence reports to influence public opinion to further their objectives and minimize our effectiveness. Anticipating events that adversaries may exploit with propaganda can allow us to mitigate the value of that propaganda through the preemptive release of information.

Media Landscape Complexity. The type and diversity of media by which a specific audience receives information impacts the effectiveness of communication with that audience. Prior to the advent of internet communication and social media, governments or interest groups could more easily affect public perception and shape a commonly held narrative about specific events, due to the limited number of media outlets to which their publics had access. Today, the ability of these groups to influence audiences has decreased dramatically with the proliferation of media platforms tailored to specific points of view. This more fragmented media environment has allowed the coexistence of multiple, conflicting narratives, making the "defeat" of unfavorable narratives or "memes" difficult, if not impossible. The ability of an organization to influence audiences in this environment has become complicated, and traditional assumptions that have in the past shaped engagement strategies are often no longer valid.

PA Across the Range of Military Operations. PA supports military activities spanning the range of military operations. It helps manage and deliver public information and is synchronized with other communication disciplines as well as other PA assets of interagency and mission partners to facilitate unity of effort.

Effective PA contributes to:

Enhanced Morale and Readiness

PA activities enable military personnel, DOD civilians, and their family members to better understand their roles by explaining the legitimacy of policies, programs, and operations affecting them. PA activities can help alleviate uncertainty and concern regarding Service member participation in crisis and contingency operations, living conditions in the operational area and at home, the duration of separation, the lack of daily communications between family members, and many other factors impact morale and readiness at home and within the unit. Additionally, PA assists Service members and their families in preparing for media events, to include providing relevant, legitimate, and responsive information on topical issues. As global media interest expands to include human interest stories, military personnel and family members can expect the foray of media to cover the impact of military operations on their lives and livelihood, to include their daily activities. Family members, including spouses and children, may be approached for interviews. This will have a direct and indirect impact on morale. This support requires planning and resources and should be incorporated into the command's planning efforts.

Public Trust and Support

A builds public trust and understanding for the military's contribution to national security. PA provides US citizens information concerning the legitimacy of military roles and missions. This information helps sustain support for military operations.

Enhanced Global Understanding

JFCs should employ PA in concert with other information-related capabilities (IRCs) to develop and implement communication that inform global, and specifically regional, publics about US military operations. This provides opportunity to explain the US narrative as well as counter potential adversary information campaigns regarding US forces in the area.

Deterrence

The credible threat of US military action can be an effective deterrent to adversary action. PA teams assist combatant commanders (CCDRs) to plan deterrence efforts and convey possible responses to the adversary, potentially avoiding the need to use force. PA clearly communicates the legitimacy of US military goals and objectives, what the adversary is illegitimately doing, why international concern is important, and what the United States Government (USG) intentions are for its armed forces if the adversary refuses to comply. Additionally, adversary propaganda frequently targets a known center of gravity, the resolve of the US public. PA's efforts to counter adversary propaganda are focused on informing the US public of the illegitimacy of the threat while simultaneously legitimizing the efforts of the Armed Forces. PA activities may involve highlighting the military's deployment preparations, activities, and force projections to show the domestic, multinational, and adversary public what the commander is actually doing to prepare for conflict. When adversaries are not deterred from conflict, information about US military capabilities and resolve may still shape the adversary's planning and actions in a manner beneficial to the US.

Institutional Credibility

PA activities are essential to preserving the credibility of DOD before, during, and after a specific mission, crisis, or other activity. By adhering to the principle of "maximum disclosure, minimum delay," PA is a critical component for defending, maintaining, and when necessary, repairing the reputation of DOD. The application of transparency, especially during crises, is essential to maintaining public trust.

II. (Army) Public Affairs Core Tasks

Ref: FM 3-61, Communication Strategy & Public Affairs Operations (Feb '22), pp. 1-4 to 1-8.

PA activities support the commander's communication strategy (CCS). There are five PA activities: **public information, command information, CCS, community engagement, and visual information.**

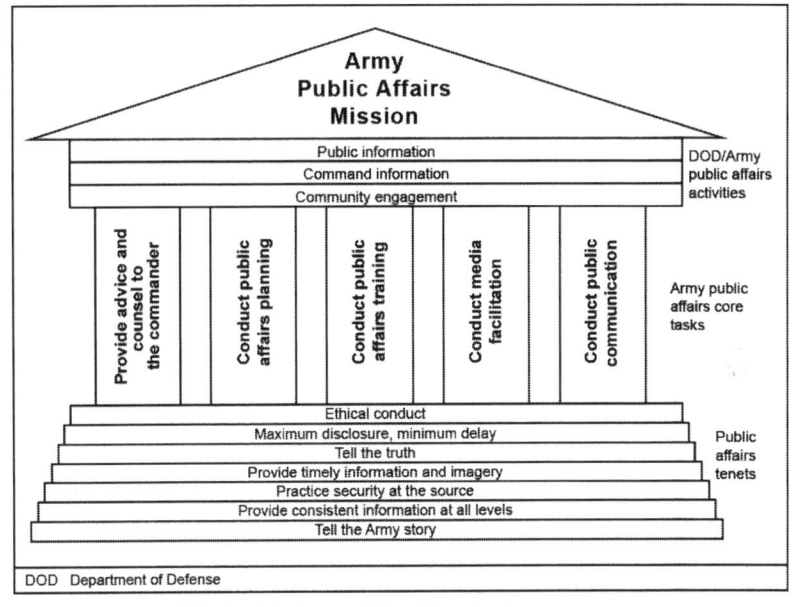

```
                        Army
                   Public Affairs
                      Mission

   Public information                          DOD/Army
   Command information                         public affairs
   Community engagement                        activities

   |  Provide  | Conduct  | Conduct  | Conduct | Conduct  |
   |  advice   | public   | public   | media   | public   |
   |  and      | affairs  | affairs  | facili- | communi- |    Army public
   |  counsel  | planning | training | tation  | cation   |    affairs core
   |  to the   |          |          |         |          |    tasks
   |  commander|          |          |         |          |

              Ethical conduct
         Maximum disclosure, minimum delay
              Tell the truth                              Public
       Provide timely information and imagery             affairs
         Practice security at the source                  tenets
      Provide consistent information at all levels
              Tell the Army story

   DOD  Department of Defense
```

Ref: FM 3-61 (Feb '22), fig. 1-1. Army public affairs structure.

The Army Universal Task List (known as AUTL) outlines the Army PA **core tasks**. Core tasks are key activities PA personnel perform to ensure mission success. PA core tasks make the PA section or unit essential for commanders, staffs, media representatives, and publics. The core tasks, as outlined in the Army Universal Task list, are:

Provide Advice and Counsel to the Commander

The PAO is the commander's senior advisor on communication strategy and PA activities. The PAO establishes and sustains commander and staff relationships and maintains direct and timely access to the commander. The more the PAO understands the operational environment, the more valuable the advice and counsel and the more developed the communication strategy.

PAOs ensure commanders understand implications of their decisions as well as the strength of public perception. Commanders understand the implications of their actions and decisions on PA. With the evolution of the global information environment, PA activities have become an increasingly critical element in determining the success of support to strategic end states. Commanders recognize the strength and influence of public opinion and perception on the morale, confidence, and effectiveness of Soldiers' abilities to achieve mission success.

PA professionals assist the commander in providing complete, accurate, and timely information to the public while developing the commander's communication strategy and the plans that achieve that strategy. All information must abide by the constraints

II. (Army) Public Affairs Core Tasks (Cont.)

Ref: FM 3-61, Communication Strategy & Public Affairs Operations (Feb '22), pp. 1-4 to 1-8.

Continued from previous page

of operations security (OPSEC). Providing timely information within such constraints enables the commander to achieve a balanced, fair, and credible information exchange and relationship with the public while deterring competitors and defeating adversaries.

PA professionals assist the commander in understanding the information needs and expectations of Soldiers, Family members, the home station community, and all other affected publics. Commanders consider these expectations when developing their communication strategies. PA professionals also tailor communication plans to meet the information needs and expectations of the affected publics both foreign and domestic.

Conduct Public Affairs and Visual Information Planning

PA section or unit conducts PA and VI planning. Communication strategy takes continuous, collaborative planning. Developing a synchronized, cohesive, and comprehensive PA and VI plan is vital in meeting the commander's communication objective requirements. The PA and VI section or unit must articulate and synchronize PA and VI planning within the military decision-making process (MDMP), to include incorporating COMCAM assets where necessary. Commanders incorporate the communication strategy and objectives in the initial plan. The PA section or unit provides detailed analysis of PA activities beyond article counts and positive, neutral, and negative evaluations. PA professionals also plan for and incorporate traditional, nontraditional, and digital media platforms into the PA plan, analysis, and assessment.

Conduct Public Affairs Training

PA professionals conduct PA training. PA and VI qualification training occurs through the Defense Information School as a joint Service program. PA leaders and Soldiers are trained to follow the operations process of plan, prepare, execute, and assess in unit training and leader development by using unit training management as discussed in ADP 7-0. PA professionals participate in and use the MDMP to plan PA training.

The training may be group media familiarization, one-on-one interview techniques with subject matter experts, and appropriate use of digital media that can impact the information environment and strategic level actions. Such training applies to all Soldiers within the command. PA professionals must be prepared to train and assist unified action partners. Effective training replicates operational realities and teaches the fundamentals of media and military interactions. Such training emphasizes that the media is a communication channel to internal and external audiences and not an adversary. OPSEC must be a consideration for all PA training.

Continued from previous page

Commanders are responsible for establishing a unit public affairs representatives (UPARs) training program in their commands. (Refer to AR 360-1 for UPAR.) Commanders provide Soldiers who have been officially designated and placed on additional duty orders the requisite training in PA.

Conduct Media Facilitation

PA staff conduct media facilitation. Media facilitation involves planning, preparing, executing, and assessing a media engagement. A media engagement is a specified instance of media interaction between a spokesperson and a member or members of the media. Media interactions occur when the media interacts with Soldiers, often on the battlefield, without PA presence.

The meaning of the word media has evolved from radio, televised, and print mass communication and now includes information technology and social media. PA staff also evolve media facilitation to support digital media platforms needs and increased nontraditional media interest in Army activities. Facilitation must include traditional, nontraditional, and digital media. PA personnel require access to information and operations centers, along with adequate media facilitation facilities, to assist the media in reporting the Army story properly.

Media facilitation includes the support of embedded media. Media embeds can be included into a command over an extended period. Commanders ensure that media embeds are credentialed and that they are provided a set of ground rules.

Conduct Public Communication

PA professionals conduct public communication in all military activities. Through public communication, PA personnel manage and deliver public information. Public communication is communication through coordinated programs, plans, themes, and messages among the Army and international, national, and local publics, as well as competitors and adversaries. It involves the receipt and exchange of ideas and opinions that contribute to shaping public understanding of, and discourse with, the Army.

Counter Misinformation and Disinformation

PA professionals counter misinformation and disinformation. Misinformation is a subset of information that includes all incorrect information. Disinformation is the deliberate use of incorrect or false information with the intention to deceive or mislead. Army PA enables commanders to preempt, identify, and counter adversary attempts at malign narrative. PA distribute legitimate, timely, and truthful information regarding Army operations, equipment, and personnel across multiple platforms within OPSEC constraints.

Commanders play a critical and strategic role in countering misinformation and disinformation. Misinformation, disinformation, or a combination of both is not an adversary but a tactic used by adversaries. Adversaries use information as propaganda to minimize the effect of military operations and programs. To combat this use, commanders must understand that the release of timely and accurate information is paramount. While conducting this task along with conducting communication assessments, PA personnel can identify the adversarial narrative. With an identified narrative, the commander can shape the communication objectives to counter the narrative through timely and accurate information.

Conduct Communications Assessments

Commanders and staff conduct communications assessments. Communications assessments follow the general assessments process discussed broadly in ADP 5-0 and in detail in FM 6-0. ADP 5-0 provides overarching guidance on assessment. However, commanders have unique considerations for PA assessment. Proper PA assessments may require additional resources not readily available within the command. Additionally, commanders must examine legal authorities when considering aspects of assessing an operational environment as it relates to use of capabilities not typically available within the PA staff.

Commanders and staff conduct communications assessments throughout an operation and measure whether the unit achieved communication objectives as planned. These assessments inform the commander's decision on whether or not to change course. Commanders and staff continuously assess an operational environment and the progress of the commander's objectives within it. The PAO and other staff representatives monitor the operational environment, which influences the outcome of operations and then provide the commander timely information needed for decisions. Planning for the assessment identifies key aspects of the operation in which the commander is interested in closely monitoring and where the commander wants to make decisions. The assessment identifies and evaluates the information environment relevant to the commander's intent, mission, area of operations, and echelon of unit. It includes an examination of the physical and social infrastructure from a PA perspective.

III. Public Affairs Fundamentals (See also pp. 1-36 to 1-37.)

Ref: JP 3-61 (w/Chg 1), Public Affairs (Aug '16), pp. I-7 to I-9.

> **Principles of Information** (See p. 1-37.)
> DOD is responsible for making timely and accurate information available so that the public, Congress, and the news media may assess and understand facts about national security and defense strategy. Requests for information from organizations and private citizens should be answered quickly.
>
> For more information, see Department of Defense Directive (DODD) 5122.05, Assistant Secretary of Defense for Public Affairs (ASD[PA]).

Tenets of Public Affairs

The PA tenets described below normally result in more effective relationships and help JFCs conduct efficient PA operations and activities and build and maintain relationships with the media. They complement the DOD principles of information and describe best practices. The tenets should be reviewed and applied during all stages of joint operation planning and execution.

Tell the Truth

PA personnel will release only accurate, fact-based information. The long-term success of PA activities depends on the integrity and credibility of officially released information. Deceiving the public undermines the perception of legitimacy and trust in the Armed Forces. Accurate, credible presentation of information leads to confidence in the Armed Forces and the legitimacy of military operations. Denying unfavorable information or failing to acknowledge it can lead to media speculation, the perception of cover-up, and degradation of public trust. These issues should be openly and honestly addressed as soon as possible. Once an individual or unit loses the public perception of integrity, it is nearly impossible to recover.

Provide Timely Information

Commanders should be prepared to release timely, factual, coordinated, and approved information and VI about military operations. Information and VI introduced into the public realm have a powerful effect on friendly, neutral, and adversary decision-making cycles and perceptions. The PAO who releases timely and accurate information and VI often becomes the media's preferred source of information. PAOs need to establish expeditious processes for release of information. VI enhances communication by adding imagery (graphics, still photos, and video) to text, sounds and words that inform the public about the joint operation. VI supports other PA functions, and stands on its own via displays and video productions. VI efforts, particularly imagery collection, should be synchronized and integrated with operational planning. Social media is an integral part of DOD operations, requiring that PAOs and staffs assist commanders in making the best use of appropriate platforms. Social media is a dynamic, rapidly changing environment, so it is important to learn and adapt as appropriate.

Refer to the DOD Social Media Hub (http://www.defense.gov/socialmedia/) for guidance on Web and Internet-based capabilities policies.

Practice Security at the Source

All DOD personnel and DOD contractors are responsible for safeguarding sensitive information. DOD members should not disclose critical information identified by the OPSEC process, whether through media interviews, social media, or community engagement. Official Information should be approved for release prior to dissemination to the public.

Likewise, it is important for interview participants to understand how what they say will be used. There are four categories of attribution for interviews: on the record, background, deep background, and off the record.

- **On the Record Interviews.** Information provided in the interview is attributable to the source by name. This is the preferred type of media engagement.
- **Background and Deep Background Interviews.** For background interviews, information is attributable to a military official, but not by name. For deep background interviews, both the person and the source are not attributable but the information can be used.
- **Off the Record Interviews.** Information provided in the interview cannot be used for direct reporting with any kind of attribution. Off the record interviews are used when there is a need to give reporters a larger context for a subject or event than can be given with any level of attribution. Off the record interviews are not preferred by reporters as they cannot directly report from the conversation and present increased risk for PAOs in that the reporter could break the confidence of the interview.

Provide Consistent Information at all Levels

The public often receives information from a variety of official DOD sources at various levels simultaneously. When this information conflicts, DOD's credibility is put in jeopardy. Before information is released to the public, it must be in compliance with all applicable guidance.

Tell the DOD Story

Although commanders designate specific military personnel or DOD civilian employees as official spokespersons, they should educate and encourage all military and civilian employees to tell the DOD story by providing them with information that is appropriate to share. By projecting confidence and commitment during interviews or in talking to family and friends, DOD personnel can help promote public understanding of military operations and activities. Social media has become a popular means for Service members to tell their story, which can be an important means of validating official releases of information. Social media use should conform to all relevant DOD and Service guidance and take into account OPSEC, operational risk, and privacy. JFCs also use social media as another means to communicate with the various publics. Official and personal blogging in local and regional languages may be helpful in reaching the local population, but also poses OPSEC risks and must be carefully monitored.

Public Affairs and Commander's Communication Synchronization (CCS) *(See p. 1-39.)*

JFCs can use the commander's communication synchronization (CCS) process to coordinate and synchronize themes, messages, images, and actions (i.e., planning, deployments, operations). The CCS process aligns communication concerning the joint force's mission with the broader national strategic narrative. The JFC should determine who will lead the CCS process for the command, but normally it is the PA office.

CCS focuses USG efforts to understand and communicate with key audiences to create, strengthen, or preserve conditions favorable for the advancement of USG interests, policies, and objectives through coordinated programs, plans, themes, messages, and products synchronized with the actions of all instruments of national power. As the primary coordinator of public information within the military, PA plays a key role in the CCS process. Official information released in a timely manner can help create, strengthen, or preserve conditions favorable for the advancement of national interests and policies, and help mitigate unofficial information, misinformation, and propaganda.

IV. Audiences, Stakeholders, and Publics

Ref: JP 3-61 (w/Chg 1), Public Affairs (Aug '16), pp. I-7 to I-9.

Scholars in public relations, marketing, and social sciences have a variety of sometimes conflicting definitions for the terms "audience," "stakeholder," and "public." The joint force communicates with audiences, but must also be able to identify and communicate with stakeholders and publics that can affect mission success. Continual assessment and evaluation of stakeholders, publics, and the information environment is critical to effective joint force decision making.

See p. 2-64 for related discussion of audiences, targets, and target audiences.

Audiences

An audience is a broad, roughly defined group based on common characteristics. It defines a population that contains relevant stakeholders. Military communications to audiences are generally one-way and are often indirect and without feedback. Audience examples include the American people; DOD military, civilians, contractors, and family members; international, host nation (HN), and local communities; and adversaries. For joint force planners, audiences are not groups on which to formulate a communication approach, but simply the beginning of the planning process for determining stakeholders and assessing publics. Stakeholders are part of the same system or environment as the joint force. More specifically, what stakeholders know, feel, or do has a potential to impact the joint force and vice versa.

Stakeholders

Individuals or groups of people are stakeholders when they are affected by–or are in a position to affect–joint force efforts. Stakeholders could be key individuals in government, nongovernmental organizations (NGOs), individuals that live outside a military base, etc. For joint force planners, identifying stakeholders means early assessment of the information environment and understanding joint force plans and their potential impacts. PA planners assess the need to communicate with stakeholders based on the extent to which they may be affected by–or might affect–joint force operations, actions, or outcomes.

Publics

A public is a stakeholder individual or group that has become more active in its communication efforts. Publics often develop during joint force efforts as opposed to stakeholders, but may also exist prior to the start of any mission. Publics warrant special attention, because they may be attempting to affect joint force operations. Examples may include lobbying groups, adversaries, or other stakeholders that are now actively seeking to communicate about issues or take actions. For joint force planners this means continually assessing the information environment to identify the development of publics, as they require more resources and a greater share of the communication effort. Planners prioritize the need to communicate with publics based on their level of activity. Furthermore, it is important to note that the joint force may need to communicate with one public or stakeholder because of the actions of another public. For example, adversaries will attempt to spread disinformation concerning joint force efforts. The joint force may have identified both another public (key local leaders in the operational area) and a stakeholder group (reporters currently embedded with deployed forces). The joint force may decide that the best use of time and resources is to communicate with both the local leaders and the reporters concerning the misinformation.

In the joint and international environment, a clear understanding between the public affairs professional and the commander of the terms 'audience,' 'public,' and 'stakeholder' is critical for the overall communication effort. The terms 'audience' and 'public' are used synonymously with the term 'stakeholder' frequently used when planning communication efforts for 'target audiences.'

V. Narrative, Themes, and Messages
Ref: JP 3-61 (w/Chg 1), Public Affairs (Aug '16), pp. I-11 to 1-14

Narrative
A narrative is a short story used to underpin operations and to provide greater understanding and context to an operation or situation.

- **Narrative in National Security Strategy.** The national security narrative is formed primarily by broad national policies, as articulated in strategic documents like the National Security Strategy and National Military Strategy. More specific national strategy is developed in National Security Council (NSC) meetings and executed by the relevant departments. For every military operation, the President or NSC staff may create the national/strategic narrative to explain events in terms consistent with national policy.

- **Conflicting Narratives.** Across areas of responsibility (AORs) and during operations within a specified operational area, there can be a struggle to define the prevailing narrative at all levels (internationally, nationally, and within the operational area) on favorable terms. To gain superiority over the adversary's narrative, diminish its appeal and followership, and supplant it or make it irrelevant, the USG needs to establish the reasons for and desired outcomes of the conflict, in terms understandable and acceptable to all relevant publics.

Refer to JP 3-04, App. A for joint discussion of narrative development.

Supporting Themes and Messages
Themes are developed by the NSC staff, Department of State (DOS), DOD, and other USG departments and agencies. JFCs support strategic themes by developing themes appropriate to their mission and authority. Figure I-3 depicts how United States Forces Korea established a theater-strategic narrative linked to a long-term campaign plan. Themes at each level of command should support the themes of the next higher level, while also supporting USG strategic themes.

Operational-level themes are often created for each phase of an operation. Operational themes are nested with strategic themes and enduring national narratives to mitigate the risk that phase-by-phase themes appear to give conflicting messages.

Messages support themes by delivering tailored information to a specific public and can also be tailored for delivery at a specific time, place, and communication method. While messages are more dynamic, they must always support the more enduring themes up and down the chain of command. The more dynamic nature and leeway inherent in messages provide joint force communicators and planners more agility in reaching publics.

Theater and operational themes should nest within the CCDR's and USG's strategic themes. Theater and operational-level messages must also support themes at their level. This enables consistent communications to local and international audiences, which supports strategic objectives.

Sources of information for the national narrative include Presidential speeches and White House communications (www.whitehouse.gov), Secretary of State speeches and DOS communications (www.state.gov and rapid response unit products), Secretary of Defense speeches and DOD communications (www.defense.gov), Chairman of the Joint Chiefs of Staff (CJCS) speeches and communications (www.jcs.mil), and CCDR speeches and combatant command (CCMD) communications. Sources of information for the joint force themes should include the mission, commander's intent, and any other guidance contained within the warning order, planning order, operation order (OPORD), and execute order (EXORD). This is not an exhaustive list; other official sources providing national strategic narratives can contribute to a joint force's narrative. The Defense Press Office (DPO) can help joint force communications with strategic guidance. The DPO routinely coordinates DOD communications with the NSC staff and participating USG departments and agencies.

VI. PA Actions in the Joint Planning Process

Ref: JP 3-61 (w/Chg 1), Public Affairs (Aug '16), pp. III-6 to III-10.

PA planners participate throughout the planning process. Examples of specific PA activities conducted or steps taken during the joint planning process (JPP) are depicted in Figure III-2.

See pp. 2-55 to 2-66 for discussion of information planning within the seven steps of JPP.

Joint Operation Planning Process and Public Affairs Actions

JOPP Step	Public Affairs Actions

1. Initiation
2. Mission Analysis
	Begin analysis of the operational environment.
	Participate in JIPOE.
	Review the following for PA implications:
	    National strategic guidance
	    Higher headquarters planning directive
	    Initial JFC intent
	Provide PA perspective during mission analysis.
	Identify intelligence requirements for PA support to planning.
	Identify specified, implied, and essential PA tasks.
	Develop PA input to the mission statement.
	Conduct initial PA force structure analysis including the need for VI support, AFRTS, and the DOD National Media Pool.
	Develop PA facts and assumptions.
	Develop PA estimates.
	Participate in all cross functional staff organizations related to planning.

3. Course of Action Development 4. COA Analysis and Wargaming 5. COA Comparison 6. COA Approval	Participate in COA development; identify needed PA capabilities and forces required as well as shortfalls. Participate in COA analysis and wargaming; identify advantages and disadvantages of each COA from a PA perspective. Revise the PA staff estimate as needed based on wargaming. Provide PA input on COA recommendation. Continued participation in all cross functional staff organizations.
7. Plan or Order Development	Refine PA requirements (capabilities, force structure, equipment/logistics, and other resources) to support the COA. Provide PA personnel requirements of the request for forces. Participate in the time-phased force and deployment data build/validation as applicable. Continued participation in all cross functional staff organizations related to planning. Provide input to the operational planning process for all applicable annexes including B, C, D, G, O, V, and draft annex F. Coordinate any administrative or contracting requirements. Develop and submit proposed PAG to higher headquarters for review/approval. Coordinate with subordinate PA staffs to ensure plan synchronization and a smooth transition to deployed operations.

Legend

AFRTS	American Forces Radio and Television Service	JOPP	joint operation planning process
COA	course of action	PA	public affairs
DOD	Department of Defense	PAG	public affairs guidance
JFC	joint force commander	VI	visual information
JIPOE	joint intelligence preparation of the operational environment		

JP 3-16, fig. III-2. Joint Operation Planning Process and Public Affairs Actions.

PA activities should be synchronized across the joint force, and with other agencies, early in JPP. Authorities to plan, integrate, approve, and disseminate information and imagery should be clearly established. Legal considerations regarding release of information on investigations in the operational area, including those regarding alleged law of war violations, should be addressed as early as possible in the PAG. Coordination of themes, as well as support to media coverage and PAG, should be approved prior to deployment.

II. Civil Affairs and Civil-Military Operations (CMO)

Ref: JP 3-57, Civil-Military Operations (Jul '18) and FM 3-57, Civil Affairs (Jul '21).

I. Civil Affairs and Civil-Military Operations

Civil-military operations (CMO) are the activities performed by military forces to establish, maintain, influence, or exploit relationships between military forces and indigenous populations and institutions (IPI). CMO support US objectives for host nation (HN) and regional stability. CMO may include activities and functions normally performed by the local, regional, or national government. These activities may occur prior to, during, or subsequent to other military actions or operations. They may also occur in the absence of other military operations. CMO are conducted across the conflict continuum.

Civil-Military Operations

Unified Action
- The synchronization, coordination, and integration of the activities of governmental and nongovernmental entities with military operations to achieve unity of effort
- Takes place within unified commands, subordinate unified commands, and joint task forces under the direction of these commanders

Civil-Military Operations
- The responsibility of a commander
- Normally planned by civil affairs personnel, but implemented by all elements of the joint force

Civil Affairs
- Conducted by civil affairs forces
- Provides specialized support of civil-military operations
- Applies functional skills normally provided by civil government

Ref: JP 3-57 (Jul '18) fig. I-3. Unified Action, Civil-Military Operations, and Civil Affairs Operations.

Refer to TAA2: Military Engagement, Security Cooperation & Stability SMARTbook (Foreign Train, Advise, & Assist) for further discussion. Topics include the Range of Military Operations (JP 3-0), Security Cooperation & Security Assistance (Train, Advise, & Assist), Stability Operations (ADRP 3-07), Peace Operations (JP 3-07.3), Counterinsurgency Operations (JP & FM 3-24), Civil-Military Operations (JP 3-57), Multinational Operations (JP 3-16), Interorganizational Cooperation (JP 3-08), and more.

CMO planners identify, evaluate, and incorporate civil considerations into courses of action (COAs) that support the commander's mission by synchronizing and building synergy between multiple military and civil entities, focusing on the stabilization of the operational environment (OE).

CMO planners incorporate relevant messages and themes through the commander's communication synchronization (CCS) process while CMO enablers and other forces disseminate these messages and themes to local leaders and the HN population. Both are done to support the commander's lines of effort (LOEs) and achievement of objectives.

Feedback generated during CMO provides data to specific information-related activities as the assessments reveal sentiments of targeted HN populations or organizations.

CMO Operational Categories

JFCs conduct CMO across the range of military operations in three primary categories:

- Military engagement, security cooperation (SC), and deterrence.
- Crisis response and limited contingency operations.
- Large-scale combat operations.

Joint forces conduct CMO and civil affairs operations (CAO) to enable unified action. Unified action is the synchronization, coordination, and integration of the activities of governmental and nongovernmental entities with military operations to achieve unity of effort. In addition, the JFC should integrate CA with other military forces to work alongside HN agencies, military, and security forces (e.g., national, border, and local police) and to support unified action by interacting and consulting with other government agencies, IPI, international organizations, nongovernmental organizations (NGOs), HN, foreign nations (FNs), and the private sector to provide the capabilities needed for successful CMO.

II. Civil-Military Operations Range of Activities

The joint force conducts CMO to facilitate military operations by establishing, maintaining, influencing, or exploiting relationships between military forces and the civilian populace. CMO usually include governmental organizations and NGOs and HN authorities in an OA. The environment may be permissive, uncertain, or hostile. All CMO should support USG objectives and end states.

CMO include a range of activities that integrate civil and military actions. These activities are particularly suited to support the achievement of objectives that promote stability. CMO are necessitated by the requirement to achieve unified action and the inclusion of the civil component into the planning and execution to attain the desired stable end state.

GCCs employ forces in their area of responsibility (AOR) to accomplish their missions. Promoting stability in support of national objectives is often a key responsibility. Stability activities include military missions conducted by GCCs and subordinate commanders, in coordination with other USG departments and agencies and their implementing partners. JFCs, in conjunction with the chief of mission (COM), may use components of CMO as a tool to mitigate suffering.

A. Military Government

Military government is established during military operations when replacement or sustainment of civil authority is required to maintain stability and governance in an area that has been legally occupied. Military government supports the US instruments of national power abroad through executing governance tasks mandated by US policy and international law. The objective of a military government is to establish law and order for USG stabilization and reconstruction efforts, the end state of which is a reconstructed indigenous government that employs governing policies consistent with US interests.

Strategic Aspects of Civil-Military Operations

Ref: JP 3-57, Civil-Military Operations (Jul '18), pp. I-2 to I-3.

Frequently, the threat will use common culture, religion, generosity, and coercion to destabilize or depose legitimate governments and then exploit their success to advance political goals and objectives. These tactics can sometimes inspire support from local civilian populations. Effective efforts to separate the populace from the threat requires a joint force skilled in building relationships through cooperation to counter the influence of the threat. Otherwise, the joint force risks winning tactical and even operational victories but losing the war.

CMO are an inherent command responsibility. They are the activities joint force commanders (JFCs) engage in to establish and maintain relations with civil authorities, the general population, and other organizations. Ultimately, the JFC should understand transregional, multi-domain, and multifunctional threats to the joint force and how CMO impacts a global perspective and responsiveness. As a part of unified action, JFCs are responsible for the organization of CMO in their operational area (OA) designed to reduce and/or prevent detrimental effects to the HN population to facilitate the seamless execution of military operations in support of strategic and military objectives. US forces conduct CMO to coordinate civil and military activities, minimize civil-military friction, reduce civil component threats, and maximize support for operations. CMO also meet the commander's legal obligations and moral responsibilities to the civilian populations within the OA. Civil affairs (CA) conducts operations that nest within the overall mission and commander's intent.

The JFC synchronizes capabilities of the joint force and coordinates with the other instruments of national power to achieve the national strategic objectives of the mission.

Contending with the civil component of any society presents challenges. Foreign governments face challenges to security, stability, and peace, which may include ethnic and religious conflict, cultural and socioeconomic friction, and terrorism and insurgencies. Other civil vulnerabilities include the proliferation of weapons of mass destruction (WMD), international organized crime, incidental and deliberate population migration, environmental security, infectious diseases, and increasing competition for or exploitation of dwindling natural resources. The joint force may also encounter these challenges when participating in foreign internal defense (FID) or support to governance while engaging in stability activities. US forces may be required to establish a military government in the absence of governing entities by providing support to civil administration or asserting transitional military authority.

United States Government (USG) policy initiatives, national security directives, joint strategies, and military doctrine reflect a growing appreciation of the need to leverage more nonmilitary tools and representative elements of the instruments of national power, such as interagency partners (e.g., Department of State [DOS]) and private sector, in order to build a more effective and balanced strategy.

At the strategic, operational, and tactical levels of warfare, during all military operations, CMO facilitate unified action between military forces and nonmilitary entities within the OA. CMO facilitate this unified action, particularly in support of shaping, stability, counterinsurgency (COIN), and other activities that counter asymmetric and irregular threats. CMO may permeate other aspects of national security and military strategy for an operation or campaign.

III. Civil-Military Operations and the Levels of War

Ref: JP 3-57, Civil-Military Operations (Jul '18), pp. I-6 to I-8.

The three levels of warfare—strategic, operational, and tactical—link tactical actions to achievement of national objectives. There are no finite limits or boundaries between these levels, but they help commanders design and synchronize operations, allocate resources, and assign tasks to the appropriate command. CMO may be applied at the strategic, operational, and tactical levels of warfare. Specific actions at one level of warfare may affect all three levels simultaneously with different effects at each level. CMO guidance should include higher headquarters objectives and end states synchronized with USG policy and guidance. Individuals and units conducting CMO must understand how tactical CMO actions may have strategic implications.

Civilian and military organizations often have differing perspectives. Some civilian leaders may object to specific CMO because the civilian populations might confuse the military with independent NGO or international organization efforts. Some international organizations or NGOs have filed objections with senior HN or USG officials when they feel CMO have compromised their neutrality. When these differing perspectives cause tactical or operational friction, they may escalate to strategic, time-consuming issues for the JFC. Prevention of such friction may entail military leaders, ensuring non-military entities have the lead in nonmilitary related efforts as opposed to being seen as subordinated to military actions.

Recognizing that military and nonmilitary organizations use different decision-making processes and philosophies can help reduce friction among all stakeholders and set conditions for common understanding. Most civilian agencies do not organize themselves or make decisions based on the tactical, operational, and strategic levels in regards to organizing or decision making. For example, civilian organizations may also organize activities around sectors of activity, such as health services or education, rather than geographically, such as by district or province. Most civilian agencies do not recognize the tactical, operational, and strategic levels in regards to organizing or decision making.

A. Strategic
At the strategic level, CMO focus on larger and long-term issues that may be part of USG shaping, stabilization, reconstruction, and economic development initiatives in failing, defeated, or recovering nations. CMO are an essential tool used to improve the HN in improving the capacity, capability, and willingness required to regain governance. Strategic CMO are part of a geographic combatant commander's (GCC's) SC guidance in the theater campaign plan (TCP). During certain contingency operations, the Secretary of Defense (SecDef) and the Secretary of State will integrate stabilization and reconstruction contingency plans with military contingency plans and develop a general framework to coordinate stabilization and reconstruction activities and military operations.

B. Operational
At the operational level, CMO synchronize stability activities with other activities and operations (offense and defense) within each phase of any joint operation. CMO also integrate the stabilization and reconstruction efforts of USG interagency, international organization, and NGO activities with joint force operations.

Joint force planners and interagency partners should identify civil-military objectives early in the planning process. CMO are integrated into plans and operations through interagency coordination, multinational partnerships, and coordination with international organizations and NGOs. Coordination of CMO for current and future operations is conducted at the operational level. Information management (IM) enables CMO and facilitates interorganizational cooperation to efficiently distribute resources and measure success using nontraditional operational indicators.

C. Tactical

A civil-military team or civil-military operations center (CMOC) may facilitate tactical CMO among the military, the local populace, NGOs, and international organizations. Commanders can coordinate, integrate, and synchronize with the civil component through military engagement, civil reconnaissance (CR), a civil-military support element, or through an established CMOC. Tactical CMO are normally focused on specific areas or groups of people and have more immediate effects.

Annex G (Civil-Military Operations) describes CAO and larger CMO in a plan or operation order (OPORD). CMO require coordination among CA, logistical support, maneuver, health service support, military police (MP), engineer, transportation, and special operations forces. CMO involve cross-cutting activities across staff sections and subordinate units. Annex G identifies, consolidates, and deconflicts the activities of the various sections and units. Planning and coordination at lower echelons require significantly more details than discussed in annex G.

Changes in the OE, such as changes in the military or strategic situation, natural or man-made disasters, or changes in the other operational variables, can divert the joint force's main effort from CMO. By continually analyzing the OE, the JFC can identify warnings of changes in the OE and allocate resources to monitor these changes in order to anticipate changes in force requirements and planning. Branch and sequel planning and preventive action may mitigate disruption of CMO. For example, a branch may call for the use of a show of force to deter aggressive action by a group while CMO are being conducted. The JFC can task a unit with a "be prepared to" mission in order to facilitate execution of branches or sequels. This can occur in the context of the commander on the ground tasking a subordinate element or in a larger context of a GCC tasking a Service component like the Navy with the be-prepared-to mission while the Army conducts the civil-military operation.

Possible escalation Indicators include:

- Political activities and movements
- Food or water shortages
- Outbreaks of disease
- Military setbacks
- Natural disasters
- Crop failures
- Fuel shortages
- Onset of seasonal changes (winter may exacerbate fuel and food shortages, for example)
- Police force and corrections system deterioration
- Judicial system shortcomings
- Insurgent attacks
- Sharp rise in crime
- Terrorist attack
- Disruption of public utilities, e.g., water, power, sewage, and economic strife due to socioeconomic imbalance
- Increases in dislocated civilians

IV. CMO in Joint Operations
Ref: JP 3-57, Civil-Military Operations (Jul '18), pp. I-13 to I-16.

CMO Across the Conflict Continuum

CMO tasks support a variety of military operations across the conflict continuum. The precise arrangement of CMO will depend on the objectives of the plan. These activities may be performed by CA, other military forces, or a combination and are described in the following discussion.

Shape
During implementation of the CCDR's SC planning objectives, CMO can mitigate the need for other military operations in response to a crisis. CA support FID and contribute to planning. Before a crisis, CA working with HNs, regional partners, and IPI can shape the OE. Shaping operations can include regional conferences to bring together multiple stakeholders with competing concerns and goals, economic agreements designed to build interdependency, or regional aid packages and other capability/capacity building activities to enhance stability.

Deter
CMO should be integrated with flexible deterrent options to generate maximum strategic or operational effect. CMO, in conjunction with deterrence activity, builds on shaping activities and can provide a stabilizing effect on the OE, reduce uncertainty, and influence the perception of joint force intentions.

Seize the Initiative
In conjunction with other joint force activities to seize the initiative, CMO are conducted to gain access to theater infrastructure and to expand friendly freedom of action in support of JFC operations. CMO are planned to minimize civil-military friction and support friendly political-military objectives. CMO conducted before the outbreak of conflict may also be used to create opportunities to aid in seizing the initiative, such as development of dual-use infrastructure and relationship building with IPI.

Dominate
CMO also help minimize HN civilian interface with joint operations so that collateral damage to IPI from offensive, defensive, or stability activities is limited. Limiting collateral damage may reduce the duration and intensity of combat and stability activities. Stability activities are conducted as needed to ensure a smooth transition to stabilization activities, relieve suffering, and set conditions for civil-military transition.

Stabilize
Stabilize actions are required when there is no fully functional, legitimate civil governing authority. This condition can be caused by a natural or man-made disaster, major combat operation, or regime collapse. The joint force may be required to occupy territory, perform limited local governance, or take on full governing responsibilities through a transitional military authority. It must then integrate the efforts of other supporting or contributing multinational, international organization, NGO, or USG department and agency participants until legitimate local entities are functioning. CMO facilitate humanitarian relief, civil order, and restoration of public services as the security environment stabilizes. Throughout these activities, the JFC continuously assesses whether current operations enable transfer of overall regional authority to a legitimate civil entity.

Enable Civil Authority
These activities are predominantly characterized by joint force support to legitimate civil governance in the OA. This includes coordination of CMO with interagency, multina-

tional, IPI, international organization, and NGO participants; establishing and assessing measures of effectiveness (MOEs) and measures of performance (MOPs); and favorably influencing the attitude of the HN population regarding both the US and the local civil authority's objectives.

CMO in Support of Joint Operations include:

Displaced civilian (DC) operations are planned to minimize civil-military friction, reduce civilian casualties, alleviate human suffering, and control DC movements.

Forces executing CMO coordinate with civilian agencies to implement measures to locate and identify population centers. These forces also coordinate with civilian agencies to create, restore, and maintain public order; coordinate resources (e.g., labor, supplies, and facilities); coordinate immediate life sustaining services to civilians in the OA(s); and assist with planning for disease control measures to protect joint forces.

CMO assets may designate routes and facilities for DCs to minimize their contact with forces engaged in combat. CMO may help contribute to logistics operations. CMO planners can help logistic planners identify available goods and services by using their contacts within the civilian sector.

Civil-military implications for most operations depend on the mission scope, national strategic end state, and the characteristics of the civil sector in the OA CMO in support of joint operations include:

Counterinsurgency Operations (COIN). COIN operations are comprehensive civilian and military efforts taken to simultaneously defeat and contain insurgency and address its root causes. CMO support to COIN operations includes using military capabilities to perform traditionally civilian activities to help the HN or FN deprive insurgents of popular support. CMO combine military-joint operations with diplomatic, political, economic, and informational initiatives of the HN and participating interorganizational partners to foster stability. The principle purpose of CMO in COIN operations or FID is to isolate the insurgents from the populace, thus depriving them of recruits, resources, intelligence, and credibility.

Peace Operations (PO). PO are normally multiagency and multinational contingencies involving all instruments of national power and may include international humanitarian and reconstruction efforts. CMO foster a cooperative relationship between the military forces, civilian organizations, and the governments and populations in the OA.

Noncombatant Evacuation Operations (NEO). CMO support to a NEO should limit local national interference with evacuation operations; maintain close liaison with embassy officials to facilitate effective interagency coordination; obtain civil or indigenous support for the NEO; help DOS identify US citizens and others to be evacuated; and help embassy personnel receive, screen, process, and debrief evacuees.

Countering Threat Networks (CTN). To effectively counter threat networks, the joint force must apply a comprehensive network engagement approach by partnering with friendly networks and engaging neutral networks through the building of mutual trust and cooperation. Network engagement is the interactions with friendly, neutral, and threat networks, conducted continuously and simultaneously at the tactical, operational, and strategic levels, to help achieve the commander's objectives. CMO may support CTN operations through obtaining information on and influencing of friendly and neutral networks.

Stabilization. Stabilization activities are necessary to consolidate military gains into lasting strategic success and may include efforts to establish civil security, provide access to dispute resolution mechanisms, deliver targeted basic services, and establish a foundation for the voluntary return of displaced people. The joint force's core responsibilities during stabilization are security, basic public order, and the immediate needs of the population. These activities are separate from, but complementary to, humanitarian assistance.

V. Civil-Military Operations Center (CMOC)

Ref: JP 3-57, Civil-Military Operations (Jul '18), pp. II-19 to II-22. (See also p. 2-35.)

A CMOC (see Figure II-2) is an organization, normally comprising CA, established to plan and facilitate coordination of activities of the Armed Forces of the United States within IPI, the private sector, international organizations, NGOs, MNFs, and other governmental agencies in support of the JFC. The CMOC is employed whenever CMO planning, coordination, synchronization, and integration is necessitated to a higher degree than can be achieved by the organically assigned CMO staff. CMOCs can be established at every echelon tailored to specific operations and are often utilized from operational to tactical level.

CMOCs are tailored to the mission and augmented by engineer, medical, and transportation assets to the supported commander. The CMOC is the primary coordination interface for US forces and IPI, humanitarian organizations, international organizations, NGOs, MNFs, HN government agencies, and other USG departments and agencies. The CMOC facilitates coordination among the key participants.

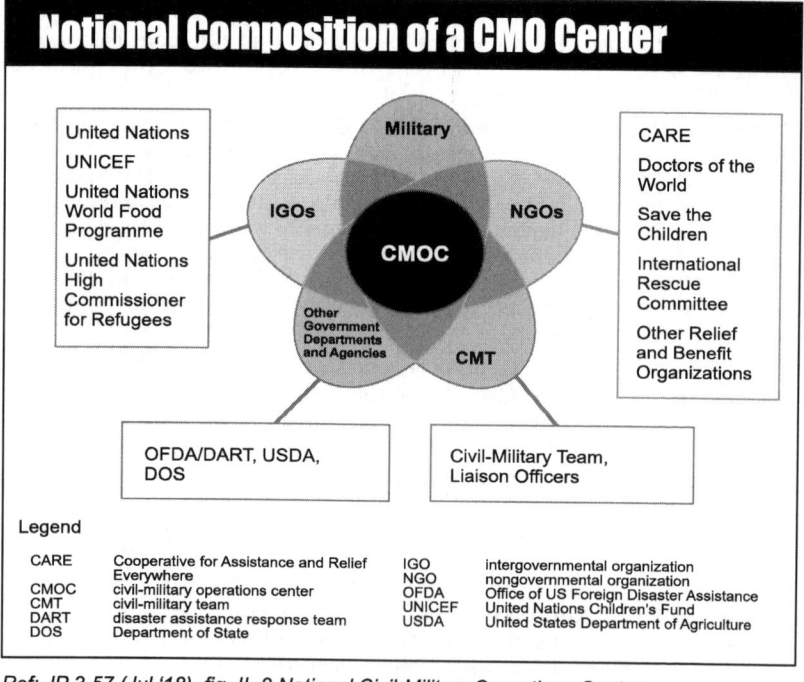

Notional Composition of a CMO Center

IGOs:
- United Nations
- UNICEF
- United Nations World Food Programme
- United Nations High Commissioner for Refugees

NGOs:
- CARE
- Doctors of the World
- Save the Children
- International Rescue Committee
- Other Relief and Benefit Organizations

Other Government Departments and Agencies: OFDA/DART, USDA, DOS

CMT: Civil-Military Team, Liaison Officers

Legend
- CARE — Cooperative for Assistance and Relief Everywhere
- CMOC — civil-military operations center
- CMT — civil-military team
- DART — disaster assistance response team
- DOS — Department of State
- IGO — intergovernmental organization
- NGO — nongovernmental organization
- OFDA — Office of US Foreign Disaster Assistance
- UNICEF — United Nations Children's Fund
- USDA — United States Department of Agriculture

Ref: JP 3-57 (Jul '18), fig. II-2 Notional Civil-Military Operations Center.

When joint operations are tailored specifically to FHA or foreign disaster relief, the HN may establish a humanitarian operations center (HOC) or the CCDR may establish a humanitarian assistance coordination center (HACC). The interagency can also provide representation to these centers through liaison or civil-military teams.

A joint force public affairs officer (PAO) or PA representative and information operations (IO) representative should attend recurring CMOC meetings as the PAO is the only official spokesperson for the JFC, other than the JFC. The USG may establish a crisis reaction center (CRC) as part of a response package prior to the arrival of a JTF. CMOC personnel may integrate with the established USG CRC to mitigate duplication of effort. USG departments and agencies, IPI, international organizations, NGOs, and MNFs coordinating with an established CRC may not interface with a CMOC. However, the JTF should be prepared to create a CMOC in the event the CRC is overwhelmed by the situation or in another location based on the JTF's mission. See Figure II-3 for a comparison between a HOC, HACC, and CMOC.

Operations Center Comparisons

	Establishing Authority	Function	Composition
Humanitarian Operations Center (HOC)	Designated individual of affected country, United Nations (UN), or United States Government (USG) department or agency	Coordinates overall relief strategy at the national (country) level.	Representatives from: • affected country • UN • US embassy or consulate • joint task force • other nonmilitary agencies • concerned parties (private sector)
Humanitarian Assistance Coordination Center	Combatant commander	Assists with interagency coordination and planning at the strategic level. Normally is disestablished once a HOC or CMOC is established.	Representatives from: • combatant command • nongovernmental organizations • international organizations • regional organizations • concerned parties (private sector)
Civil-Military Operations Center (CMOC)	Joint task force or component commander	Assists in coordination of activities at the operational level and tactical level with military forces, USG departments and agencies, nongovernmental and international organizations, and regional organizations.	Representatives from: • joint task force • nongovernmental organizations • international organizations • regional organizations • USG departments and agencies • local government (host country) • multinational forces • other concerned parties (private sector)

The authority of all three centers is coordination only.

HOC. The HOC is a senior-level international and interagency coordinating body designed to achieve unity of effort in a large FHA operation. All members are responsible to their own organizations or countries. The HOC normally is established under the direction of the government of the affected country or the UN or possibly the USAID Office of United States Foreign Disaster Assistance (OFDA) during a US unilateral operation.

HACC. During FHA operations, CCDRs may organize an HACC to assist with interagency partners, international organization, and NGO coordination and planning. Normally, the HACC is a temporary body that operates in early planning and coordination stages. Once a CMOC or HOC is established, the role of the HACC diminishes, and its functions are accomplished through normal CCDR's staff and crisis action organization.

JCMOTF. JFCs may establish a JCMOTF when the scope of CMO requirements and activities are beyond the JFC's organic capability.

Refer to TAA2: Military Engagement, Security Cooperation & Stability SMARTbook (Foreign Train, Advise, & Assist) for further discussion. Topics include the Range of Military Operations (JP 3-0), Security Cooperation & Security Assistance (Train, Advise, & Assist), Stability Operations (ADRP 3-07), Peace Operations (JP 3-07.3), Counterinsurgency Operations (JP & FM 3-24), Civil-Military Operations (JP 3-57), Multinational Operations (JP 3-16), Interorganizational Cooperation (JP 3-08), and more.

B. Support to Civil Administration (SCA)

Support to civil administration (SCA) is assistance given to a governing body or civil structure of a foreign country, whether by assisting an established government or interim civilian authority or supporting a reconstructed government. SCA occurs when military forces support DOS in the implementation of interim civil authority or US foreign policy in support of HN internal defense and development. SCA supports the US instruments of national power abroad through executing tasks affiliated with cooperative security, SC, and FID as a function of stability activities.

C. Populace and Resources Control (PRC)

PRC consists of two distinct, yet linked, components:

Populace control provides security to people, mobilizes human resources, denies personnel to the enemy, and detects and reduces the effectiveness of enemy agents. Populace control measures include curfews, movement restrictions, travel permits, identification and registration cards, and voluntary resettlement. Dislocated civilian (DC) operations involve populace control that requires extensive planning and coordination among various military and civilian organizations.

Resources control regulates the movement or consumption of materiel resources, mobilizes materiel resources, and denies materiel to the enemy. Resources control measures include licensing, regulations or guidelines, checkpoints (e.g., roadblocks), ration controls, amnesty programs, and inspection of facilities.

D. Foreign Humanitarian Assistance (FHA)

FHA consists of DOD activities conducted outside the US and its territories to directly relieve or reduce human suffering, disease, hunger, or privation. These activities are governed by various statutes and policies and range from steady-state engagements to limited contingency operations. FHA includes foreign disaster relief operations and other activities that directly address a humanitarian need and may also be conducted concurrently with other DOD support missions and activities such as DC support; security operations; and international chemical, biological, radiological, and nuclear response (ICBRN-R).

E. Foreign Assistance

Foreign assistance is civil or military assistance rendered to a nation by the USG within that nation's territory based on agreements mutually concluded between the US and that nation. Foreign assistance supports the HN by promoting sustainable development and growth of responsive institutions. The goal is to promote long-term regional stability. It is a strategic, economic, and moral imperative for the US and vital to US national security. These activities often include security assistance (SA) programs; FID; and other Title 10, USC, programs and activities performed on a reimbursable basis by USG departments and agencies. Common HCA projects include rudimentary engineering, medical, dental, or veterinary activity in the HN.

F. Overseas Humanitarian, Disaster, and Civic Aid (OHDACA) Appropriation

OHDACA funding is for DOD humanitarian programs to support achievement of a GCC's TCP objectives in support of larger US national security objectives. OHDACA is a common funding source for FHA activities. Each combatant command (CCMD) normally submits an annual FHA strategy and OHDACA budget request via the Overseas Humanitarian Assistance Shared Information System.

III. Military Deception (MILDEC)

Ref: FM 3-13.4, Army Support to Military Deception (Feb '19), chap. 1.

Military deception is actions executed to deliberately mislead adversary military, paramilitary, or violent extremist organization decision makers, thereby causing the adversary to take specific actions (or inactions) that will contribute to the accomplishment of the friendly mission (JP 3-13.4). Deception applies to all levels of warfare, across the range of military operations, and is conducted during all phases of military operations. When properly integrated with operations security (OPSEC) and other information-related capabilities (IRCs), deception can be a decisive tool in altering how the enemy views, analyzes, decides, and acts in response to friendly military operations.

Deception is a commander-driven activity that seeks to establish conditions favorable for the commander to achieve objectives. It is both a process and a capability. As a process, deception employs an analytic method to systematically, deliberately, and cognitively target individual decision makers. The objective is to elicit specific action (or inaction) from the enemy. As a capability, deception is useful to a commander when integrated early in the planning process as a component of an operation focused on causing an enemy to act or react in a desired manner. Deception greatly enhances the element of surprise. Deception aligns with surprise and the displacement of critical threat capabilities away from the friendly point of action. Due to the potentially sensitive nature of deception activities and selected means, planners must implement appropriate security and classification measures to properly safeguard deception tactics, techniques, and procedures.

I. Functions of Military Deception

Planners must have a thorough understanding of the functions and the scope of what deception can and cannot accomplish. A deception plan serves as a part of the overall mission. Every deception plan must clearly indicate how it supports the commander's objectives. The functions of deception include, but are not limited to—

- Causing delay and surprise through ambiguity, confusion, or misunderstanding.
- Causing the enemy to misallocate personnel, fiscal, and materiel resources.
- Causing the enemy to reveal strengths, weaknesses, dispositions, and intentions.
- Causing the enemy to waste combat power and resources with inappropriate or delayed actions.

II. Categories of Deception

Deception activities support objectives detailed in concept plans, operation plans (OPLANs), and operation orders (OPORDs) associated with approved military operations or activities. Deception applies during any phase of military operations to establish conditions to accomplish the commander's intent. The Army echelon that plans a deception activity often determines its type. The levels of war define and clarify the relationship between strategic and tactical actions. The levels have no finite limits or boundaries. They correlate to specific authorities, levels of responsibility, and planning. The levels help organize thought and approaches to a problem. Decisions at one level always affect other levels. Table 1-1 shows the three types of deception.

A. Military Deception (MILDEC)

Military deception (MILDEC) is planned, trained, and conducted to support military campaigns and major operations. MILDEC activities are planned and executed to cause adversaries to take actions or inactions that are favorable to the commander's objectives. The majority of MILDEC planned for and executed by the combatant command (CCMD) to create operational-level effects. MILDEC is normally planned before, and conducted during, combat operations. CCMD instructions add guidelines, policies, and processes that must be adhered to in their respective commands. MILDEC is a joint activity to which the Army, as the primary joint land component, contributes. Army forces do not unilaterally conduct MILDEC. MILDEC must adhere to the regulatory requirements found in Army policy and regulations, CJCSI 3211.01 series, and applicable CCMD instructions.

B. Tactical Deception (TAC-D)

Tactical deception is an activity planned and executed by, and in support of, tactical-level commanders to cause enemy decision makers to take actions or inactions prejudicial to themselves and favorable to the achievement of tactical commanders' objectives. Commanders conduct tactical deception (TAC-D) to influence military operations to gain a relative, tactical advantage over the enemy, obscure vulnerabilities in friendly forces, and enhance the defensive capabilities of friendly forces. In general, TAC-D is a related subset of deception that is not subject to the full set of MILDEC program requirements and authorities. In most circumstances, Army commanders can employ TAC-D unilaterally if certain criteria are met. In description, TAC-D differs from MILDEC in four key ways:

- MILDEC is centrally planned and controlled through CCMD-derived authorities, but TAC-D is not. TAC-D can be employed unilaterally by tactical commanders with an approved plan.

- TAC-D actions are tailored to tactical requirements of the local commander and not always linked or subordinate to a greater MILDEC plan.

- The TAC-D approval process differs from the MILDEC approval process in that it is only required to be approved at two echelons higher, provided that it adheres to the joint policy for MILDEC addressed in CJCSI 3211.01. CCMD instructions add guidelines, policies, and processes that must be adhered to in their respective commands.

- Planning for TAC-D is usually more abbreviated, but still focuses on influencing the action or inaction of enemy decision makers, to gain a tactical advantage over an enemy. TAC-D gains this relative advantage using deception activities that affect the enemy's perceptions of friendly activities and possibly targeting lower-echelon enemy combatants to affect their operations.

C. Deception in Support of Operations Security (DISO)

Deception in support of operations security (DISO) is a deception activity that conveys or denies selected information or signatures to a foreign intelligence entity (FIE) and limits the FIE's overall ability to collect or accurately analyze critical information about friendly operations, personnel, programs, equipment, and other assets. The intent of DISO is to create multiple false, confusing, or misleading indicators to make friendly force intentions harder to interpret by FIE. DISO makes it difficult for FIEs to identify or accurately derive the critical information and indicators protected by OPSEC. Deception and OPSEC are mutually supporting activities. DISO prevents potential enemies from accurately profiling friendly activities that would provide an indication of a specific course of action (COA) or operational activity. DISO differs from joint MILDEC and TAC-D plans in that it only targets FIEs and is not focused on generating a specific enemy action or inaction.

III. Deception Means

Ref: FM 3-13.4, Army Support to Military Deception (Feb '19), pp. 1-11 to 1-13.

Deception means are methods, resources, and techniques that can be used to convey information to the deception target (JP 3-13.4).

Physical Means. Physical means are resources, methods, and techniques used to convey or deny information or signatures normally derivable from direct observation or active sensors by the deception target. Most physical means also have technical signatures visible to sensors that collect scientifically or electronically. Planners typically evaluate physical means using characteristics such as shape, size, function, quantity, movement pattern, location, activity, and association with the surroundings. Examples might include—
- Movement of forces.
- Exercises and training activities.
- Decoy equipment and devices.
- Tactical actions.
- Visible test and evaluation activities.
- Reconnaissance and surveillance activities.

Technical Means. Technical means are resources, methods, and techniques used to convey or deny selected information or signatures to or from the deception target. Examples of technical means might include—
- The establishment of communications networks and interactive transmissions that replicate a specific unit type, size, or activity.
- The emission or suppression of chemical or biological odors associated with a specific capability or activity.
- Multispectral simulators that replicate or mimic the known electronic profile of a specific capability or force.
- Selected capabilities that disrupt an enemy sensor or affect data transmission.

Administrative Means. Administrative means are resources, methods, and techniques to convey or deny selected written, oral, pictorial, or other documentary information or signatures to or from the deception target. They normally portray information and indicators associated with coordination for ongoing or planned military activity to the deception target. Examples of administrative means normally visible to an enemy at some level might include—
- Movement, transit, or overflight requests including flight planning, port call, or traffic control coordination.
- Basing inquiries or construction requests.
- Other preparatory coordination associated with a military operation normally done through unclassified channels.

Camouflage, Concealment, And Decoys. Camouflage and concealment are OPSEC measures and survivability operations tasks used to protect friendly forces and activities from enemy detection and attribution. Camouflage makes friendly capabilities or activities blend in with the surroundings. Concealment makes friendly capabilities or activities unobservable or unrecognizable to the enemy. Concealing the location, movement, and actions of friendly forces can delay hostile attack and assist commanders in retaining the tactical advantage. Both use physical, technical, and administrative means to deceive the enemy and protect the deception story. Deception measures use the same signatures for simulating friendly forces and activities.

IV. Types of Military Deception

Ref: FM 3-13.4, Army Support to Military Deception (Feb '19), pp. 1-6 to 1-8.

Any deception aims to either increase or decrease the level of uncertainty, or ambiguity, in the mind of the deception target. This ambiguity has the potential to compel the target to mistakenly perceive friendly motives, intentions, capabilities, and vulnerabilities thereby altering the target's assessment. Two generally recognized types of MILDEC exist:

Ambiguity-Increasing Deception

Ambiguity-increasing deception provides the enemy with multiple plausible friendly COAs. Ambiguity-increasing deception is designed to generate confusion and cause mental conflict in the enemy decision maker. Anticipated effects of ambiguity-increasing deception can include a delay to making a specific decision, operational paralysis, or the distribution of enemy forces to locations far away from the intended location of the friendly efforts. Ambiguity-increasing deception is often directed against decision makers known to be indecisive or risk-adverse.

These deceptions draw attention from one set of activities to another. They can create the illusion of strength where weakness exists, or create the illusion of weakness where strength exists. They can also acclimate the enemy to particular patterns of activity that are exploitable later. For example, ambiguity-increasing deceptions can cause the target to delay a decision until it is too late to prevent friendly mission success. They can place the target in a dilemma for which no acceptable solution exists. They may even prevent the target from taking any action at all. This type of deception is typically successful with an indecisive decision maker who is known to avoid risk.

Ambiguity-Decreasing Deception

Ambiguity-decreasing deceptions manipulate and exploit an enemy decision maker's pre-existing beliefs and bias through the intentional display of observables that reinforce and convince that decision maker that such pre-held beliefs are true. Ambiguity-decreasing deceptions cause the enemy decision maker to be especially certain and very wrong. Ambiguity-decreasing deceptions aim to direct the enemy to be at the wrong place, at the wrong time, with the wrong equipment, and with fewer capabilities. Ambiguity-decreasing deceptions are more challenging to plan because they require comprehensive information on the enemy's processes and intelligence systems. Planners often have success using these deceptions with strong-minded decision makers who are willing to accept a higher level of risk.

Tactics

Deception tactics can be characterized as operational-level constructs that encompass a broad range of deceptive activity and information integrated as a component of the overall plan. Deception plans apply five basic tactics: diversions, feints, demonstrations, ruses, and displays. These tactics are often best employed in TAC-D to support the commander's objectives. The selection of tactics and their use depends on planners' understanding the current situation as well as the desired deception goal and objective.

Diversion

A diversion is the act of drawing the attention and forces of an enemy from the point of the principal operation; an attack, alarm, or feint that diverts attention (JP 3-03). The goal of diversion is to induce the enemy to concentrate resources at a time and place that is advantageous to friendly objectives.

Feint
In military deception, a feint is an offensive action involving contact with the adversary conducted for the purpose of deceiving the adversary as to the location and/or time of the actual main offensive action (JP 3-13.4). A feint is designed to lead the enemy into erroneous conclusions about friendly dispositions and concentrations. A series of feints can condition the enemy to react ineffectively to a future main attack in the same area.

Demonstration
In military deception, a demonstration is a show of force similar to a feint without actual contact with the adversary, in an area where a decision is not sought that is made to deceive an adversary (JP 3-13.4). A demonstration's intent is to cause the enemy to select a COA favorable to friendly goals.

Ruse
In military deception, a ruse is an action designed to deceive the adversary, usually involving the deliberate exposure of false information to the adversary's intelligence collection system (JP 3-13.4). A ruse deceives the enemy to obtain friendly advantage. A ruse in deception is normally an execution based on guile or trickery that contributes to the larger deception plan.

Display
In military deception, a display is a static portrayal of an activity, force, or equipment intended to deceive the adversary's visual observation (JP 3-13.4). Displays include the simulation, disguise, or portrayal of friendly objects, units, or capabilities in the projection of the deception story. Such objects, units, or capabilities may not exist but are made to appear that they exist.

Techniques
The application of techniques varies with each operation depending on time, assets, and objectives. Planners assess which techniques to apply based on feasibility, availability, and effectiveness. Table 1-2 provides sample deception techniques.

Technique	Deception created
Amplifying signatures	To make a force appear larger and more capable or to simulate the deployment of critical capabilities.
Suppressing signatures	To make a force appear smaller and less capable or to conceal the deployment of critical capabilities.
Overloading enemy sensors	To confuse or corrupt their collection assets by providing multiple false indicators and displays.
Repackaging known organizational or capability signatures	To generate new or deceptive profiles that increase or decrease the ambiguity of friendly activity or intent.
Conditioning the enemy	To desensitize to particular patterns of friendly behavior and to induce enemy perceptions that are exploitable at the time of friendly choosing.
Reinforcing the impression	To mislead by portraying one course of action when actually taking a different course of action.
Conditioning the target by repetition	To believe that an apparently standard routine will be pursued, whilst in fact preparing a quite different course of action.
Leading the enemy by substitution	To believe that nothing has changed by covertly substituting the false for the real, and vice versa.
Leading the enemy by mistake	To believe that valuable information has come into their possession through a breach of security, negligence, or inefficiency.

Ref: FM 3-13.4 (Feb '19), Table 1-2. Sample deception techniques.

V. Army Tactical Deception Planning

Ref: FM 3-13.4, Army Support to Military Deception (Feb '19), chap. 2.

The Army tactical deception planning process nests in the steps of the Army's military decisionmaking process (known as MDMP). The deception plan supports the OPLAN. Planners nest and integrate the deception plan with the OPLAN to achieve the deception's desired effect. A successful deception plan unfolds logically and realistically. Deception planning is an iterative process that requires continual reexamination of its goals, objectives, targets, and means. The early integration of deception in the planning cycle ensures optimum application of resources and maximizes the potential for overall success. Table 2-1 shows the Army tactical deception planning process nesting in the military decisionmaking process.

Army tactical deception planning process		Military decisionmaking process
Deception preplanning.	→	Step 1: Receipt of mission
Step 1: Determine the deception goal and the deception objective.	→	Step 2: Mission analysis
Step 2: Identify and analyze the deception target. Step 3: Identify desired perceptions of the deception target. Step 4: Develop deception observables and means. Step 5: Develop the deception story. Step 6: Develop the deception event schedule. Step 7: Develop OPSEC and other protection measures. Step 8: Develop feedback criteria. Step 9: Develop a termination plan.	→	Step 3: Course of action development
		Step 4: Course of action analysis and war gaming
		Step 5: Course of action comparison
		Step 6: Course of action approval
Step 10: Produce Appendix 14 (Military Deception) to Annex C (Operations)	→	Step 7: Orders production, dissemination, and transition
OPSEC operations security		

Ref: FM 3-13.4 (Feb '19), Table 2-1. The Army tactical deception planning process in the military decisionmaking process.

The complexity and sensitivity of deception requires detailed planning that begins with preplanning. MDOs have three preplanning considerations: capability development, planning guidance, and mission analysis. A successful deception plan incorporates preplanning considerations as well as flexibility to lessen the risk of failure. When preplanning, MDOs create a baseline analysis, prepare deception planning guidance, and complete mission analysis.

MDOs need to participate in and have their efforts informed by conventional planning efforts. Conventional and deception planning horizons occur simultaneously in parallel. During mission analysis, the MDO begins with analyzing and assessing an operational environment and information environment. Deception may be a feasible option, if appropriate to the mission, and if there is a possibility of success.

Issues that planners consider when determining if deception is a viable COA include—

- Availability of assets.
- Understanding any potential deception targets.
- Suitability.
- Time.
- Risk.

IV. Military Information Support Operations (MISO)

Ref: JP 3-13.2 (w/Chg 1), Military Information Support Operations (Dec '11).

Today's global information environment is complex, rapidly changing, and requires integrated and synchronized application of the instruments of national power to ensure responsiveness to national goals and objectives. In the current operational environment, effective influence is gained by unity of effort in what we say and do, and how well we understand the conditions, target audiences (TAs), and operational environment. Within the military and informational instruments of national power, the Department of Defense (DOD) is a key component of a broader United States Government (USG) communications strategy. To be effective, all DOD communications efforts must inherently support the credibility, veracity, and legitimacy of USG activities.

Military information support operations (MISO) play an important role in DOD communications efforts through the planned use of directed programs specifically designed to support USG and DOD activities and policies. MISO are planned operations to convey selected information and indicators to foreign audiences to influence their emotions, motives, objective reasoning, and ultimately the behavior of foreign governments, organizations, groups, and individuals in a manner favorable to the originator's objectives. Military information support (MIS) professionals follow a deliberate process that aligns commander's objectives with an analysis of the environment; select relevant TAs; develop focused, culturally, and environmentally attuned messages and actions; employ sophisticated media delivery means; and produce observable, measurable behavioral responses.

The employment of MIS units is governed by explicit legal authorities that direct and determine how their capability is utilized. This legal foundation establishes MISO as a communications means and allows their integration with those strategies that apply the instruments of national power. Leaders and planners interpret relevant laws and policies to conduct MISO in any situation or environment, internationally and domestically.

Joint MISO support policy and commanders' objectives from strategic to tactical levels. Although military leadership and local key communicators are examples of TA engaged at the operational and tactical levels that are capable of affecting the accomplishment of a strategic objective.

MISO are used to establish and reinforce foreign perceptions of US military, political, and economic power and resolve. In conflict, MISO as a force multiplier can degrade the enemy's relative combat power, reduce civilian interference, minimize collateral damage, and maximize the local populace's support for operations.

MISO contribute to the success of both peacetime engagements and major operations. The combatant commander (CCDR) receives functional and theater strategic planning guidance from the Joint Strategic Capabilities Plan (JSCP), Unified Command Plan (UCP), and Guidance for Employment of the Force (GEF). These documents are derived from the Secretary of Defense (SecDef) National Defense Strategy, which interprets the President's national security policy and strategy, and the Joint Chiefs of Staff National Military Strategy.

I. MISO Purpose

Every activity of the force has potential psychological implications that may be leveraged to influence foreign targets. MISO contribute to the success of wartime strategies and are well-matched for implementation in stable and pre-conflict environments. MISO are applied across the range of military operations and, as a communication capability, constitute a systematic process of conveying messages to selected foreign groups to promote particular themes that result in desired foreign attitudes and behaviors. MISO are used to establish and reinforce foreign perceptions of US military, political, and economic power and resolve. In conflict, MISO as a force multiplier can degrade the enemy's relative combat power, reduce civilian interference, minimize collateral damage, and maximize the local populace's support for operations.

MISO are integrated with US ambassador and GCC's theater-wide priorities and objectives to shape the security environment to promote bilateral cooperation, ease tension, and deter aggression. MISO convey the intent of the GCC by supporting public diplomacy efforts, whether to foster relations with other nations or to ensure their collaboration to address shared security concerns.

II. MISO Missions

The purpose of joint MIS forces is further clarified by the application of their activities across the range of military operations.

Missions performed by joint MIS forces include:

MISO in Support of Combat Operations

MISO are planned to influence the perceptions, attitudes, objective reasoning, and ultimately, the behavior of adversary, friendly, and neutral audiences and key population groups in support of US combat operations and objectives. Operations supported by joint MIS forces support include the following:

- Offense.
- Defense.
- Stability operations.

Military Information Support to DOD Information Capabilities in Peacetime

This support can shape and influence foreign attitudes and behavior in support of US regional objectives, policies, interests, theater military plans, or contingencies. Operations or activities supported by MIS forces may include, but are not limited to:

- FHA/disaster relief.
- Noncombatant evacuation operations.
- Maritime interception operations.
- Support to USG country team or host nation (HN) civil programs (e.g., counterdrug, demining, human immunodeficiency virus awareness, security institution building, ethnic tolerance, and reconciliation).

Defense Support to Civil Authorities

MIS forces provide support to public information efforts when authorized by SecDef or the President in accordance with Title 10, United States Code (USC). This support is provided during natural disaster relief following domestic incidents.

Support to Special Operations (SO)

SO are relevant across the range of military operations and the eleven core activities, including MISO should be integral parts of a theater strategy, OPLAN, or campaign plan.

III. Information Roles & Relationships
Ref: JP 3-13.2 (w/Chg 1), Military Information Support Operations (Dec '11), fig. II-1, p. II-9.

There are a variety of functions and capabilities that help a JFC formulate the command's message and communicate with local, international, and US domestic audiences as part of broader policy and in support of operational objectives. DOD information activities include IO, MISO, PA (to include visual information), and DSPD.

Dept of Defense Information Activities

INFORMATION ACTIVITY	PRIMARY TASK	FOCUS OF ACTIVITY	PURPOSE	DESIRED OUTCOME
US Government (USG) Strategic Communication (Department of State Lead)	Coordinate information, themes, plans, programs, and actions that are synchronized with other elements of national power	Understand and engage key audiences	Better enable the USG to engage foreign audiences holistically and with unity of effort	Create, strengthen, or preserve conditions favorable to advance national interests and objectives
Department of Defense (DOD) support to Strategic Communication	Use DOD operational and informational activities and strategic communication processes in support of Department of State's broader public diplomacy efforts	Key audiences	Improve the alignment of DOD actions and information with policy objectives	The conduct of military activities and operations in a shaped environment
Information Operations	Integrate information operations core, supporting, and related capabilities as part of a military plan	Adversary audiences	Influence, disrupt, corrupt, or usurp adversarial human and automated decision making while protecting our own.	Optimum application of capability to desired military outcome
Military Information Support Operations	Influence target audience perceptions, attitudes, and subsequent behavior	Approved foreign audiences	Shape, deter, motivate, persuade to act	Perceptions, attitudes, and behavior conducive to US/multinational partner objectives
Public Affairs	Provide truthful, timely, accurate information about DOD activities (inform)	US, allied, national, international, and internal audiences	Keep the public informed, counter adversary information activities, deter adversary actions, and maintain trust and confidence of US population, and friends and allies	Maintain credibility and legitimacy of US/multinational partner military operations with audience

PA and MISO are separate and unique activities that are governed by policy and practice in terms of audiences, focus, and scope. SC integrates various instruments of national power with other activities across the USG to synchronize crucial themes, messages, images, and actions. SC is policy driven and generally conducted under DOS lead. DOD SC activities are designed to support the continuity of DOD strategic- and operational-level messages and activities with overall USG policy and SC themes.

Information Roles & Responsibilities (Cont.)
Ref: JP 3-13.2 (w/Chg 1), Military Information Support Operations (Dec '11), pp. II-8 to II-12.

Although each of these activities is distinct, commanders must ensure that there is a general compatibility of messages within the broader communications strategy. This must be accomplished without blurring traditional lines of separation between PA and MISO and their respective audiences. To this end, it is critical that all DOD military information activities are conducted in a manner that reinforces the credibility, veracity, and legitimacy of DOD and USG activities. In very narrow circumstances, MISO may support military deception (MILDEC) operations designed to preserve operational surprise and the safety of friendly forces, but this is done only after the commander and staff carefully weigh the likely benefits of a deception operation against a potential short-and long-term loss of credibility with the media or local audiences.

MISO and Strategic Communication
SC consists of a focused USG effort to understand and engage key audiences to create, strengthen, or preserve conditions favorable for the advancement of USG interests, policies, and objectives through the use of coordinated programs, plans, themes, messages, and products synchronized with the actions of all instruments of national power.

MISO are a key capability that supports SC by influencing foreign audiences in support of US objectives. Given its focus on foreign TAs, MISO personnel should possess a good understanding of the language and culture of the TA and ensure this knowledge is effectively used in the preparation of MISO products and related activities.

MISO and Information Operations
MISO play a central role in the achievement of the JFC's information objectives through their ability to induce or reinforce adversary attitudes and behavior favorable to these objectives. MISO can be particularly useful during pre-and post-combat operations, when other means of influence are restrained or not authorized. Because of its wide ranging impact, it is essential MISO be fully coordinated and synchronized with relevant activities and operations. This is normally facilitated through the combatant command IO cell.

IO is the integrated employment during military operations of information-related capabilities in concert with other lines of operation to influence, disrupt, corrupt, or usurp the decision making of adversaries and potential adversaries while protecting our own. MISO must be coordinated with other information-related capabilities, such as computer network operations (CNO), electronic warfare (EW), operations security (OPSEC), and MILDEC, to ensure deconfliction control measures are in place, and that all capabilities within IO are coordinated and synchronized in time, space, and purpose to achieve the objectives established in planning.

In order to ensure all aspects of IO are properly integrated and synchronized into the combatant command planning process, an IO cell chief is chosen (in accordance with Chairman of the Joint Chiefs of Staff Manual [CJCSM] 1630.01, Joint Information Operations Force). This cell chief convenes meetings of the IO cell periodically in order to facilitate the integration of information-related capabilities. Within the IO cell, the MISO representative integrates, coordinates, deconflicts, and synchronizes the use of MISO with other IO capabilities. Specific examples of this kind of interaction between MISO and the other information-related capabilities follow:

MISO and Computer Network Operations
CNO support MIS forces with dissemination assets (including interactive Internet activities) and the capabilities to deny or degrade an adversary's ability to access, report, process, or disseminate information. These capabilities support MIS by providing access to digital media within the information environment to reach intended targets and denying TA information that does not support objectives.

MISO and Military Deception
MIS forces provide the JFC the ability to reduce the allocation of forces and resources required to deceive the adversary and facilitate mission accomplishment. MISO create and reinforce actions that are executed to deliberately mislead adversary military decision makers about US military capabilities, intentions, and operations. MILDEC operations that integrate MIS unit's targeting input provide the JFC with the ability to influence the adversary to take specific actions (or inactions), giving the joint force an advantage. MISO support to MILDEC operations must be carefully considered by the commander and staff, weighing the likely benefits of a deception operation against a potential short- and long-term loss of credibility with the media and local and regional audiences.

MISO and Operations Security
It is essential that MISO plans and messages are protected prior to execution through the proper use of information security, information assurance, physical security, and OPSEC. Additionally, it is essential during the effort to influence foreign audiences that MISO not reveal critical information or indicators of friendly operations to the adversary.

MISO and Electronic Warfare
EW platforms provide a means of disseminating MISO messages and shaping the information environment through the electronic dissemination of MISO products. The joint restricted frequency list deconflicts these two capabilities. When appropriate, EW platforms can also provide a means of denying enemy forces the ability to disseminate adversarial information. These platforms can also degrade the adversary's ability to see, report, and process information by jamming selected frequencies. EW validates the assessment of MISO effectiveness by providing information on threat responses to broadcasts.

MISO and Public Affairs
MISO are used to influence the attitudes, opinions, and behavior of foreign TAs in a manner favorable to US objectives.

Military PA forces plan, coordinate, and synchronize public information, command information, and community engagement activities and resources to support the commander's operational objectives. Through timely dissemination of factual information to international and domestic audiences, PA puts operational actions in context, facilitates the development of informed perceptions about military operations among information consumers, and undermines adversarial information efforts. PA operations and activities shall not focus on directing or manipulating public actions or opinion.

PA and MISO activities are separate and distinct, but they must support and reinforce each other, which requires coordination, synchronization, and occasionally deconfliction. These planning activities are generally accomplished in the IO working group, the IO cell, or other planning groups. In the event that formal planning groups are not established, informal coordination should be accomplished between these two capabilities as well as with other related capabilities. JFCs must ensure that appropriate coordination between MISO and PA activities are consistent with the DOD Principles of Information, policy or statutory limitation, and security.

PA is normally the source for official information for the media. Information disseminated by the joint force regardless of source or method of distribution will reach unintended audiences. Efforts of one capability must not undermine those of another. While PA will have no role in executing MISO, PA can use MISO products to educate the media about MISO missions, as appropriate.

IV. Example Joint MISO Activities

Ref: JP 3-61 (w/Chg 1), Public Affairs (Aug '16), fig. IV-1. p. IV-8.

Example MISO Activities (Across the ROMO)

MILITARY ENGAGEMENT, SECURITY COOPERATION, AND DETERRENCE	CRISIS RESPONSE AND LIMITED CONTINGENCY OPERATIONS	MAJOR OPERATIONS AND CAMPAIGNS
Modify the behavior of selected target audiences toward US and multinational capabilities	Mobilize popular support for US and multinational military operations	Explain US policies, aims, and objectives
Support the peacetime elements of US national policy objectives, national security strategy, and national military strategy	Gain and sustain popular belief in and support for US and multinational political systems (including ideology and infrastructure) and political, social, and economic programs	Arouse foreign public opinion or political pressures for, or against, a military operation
Support the geographic combatant commander's security strategy objectives	Attack the legitimacy and credibility of the adversary political systems	Influence the development of adversary strategy and tactics
Support the objectives of the country team	Publicize beneficial reforms and programs to be implemented after defeat of the adversary	Amplify economic and other nonviolent forms of sanctions against an adversary
Promote the ability of the host nation to defend itself against internal and external insurgencies and terrorism by fostering reliable military forces and encouraging empathy between host nation armed forces and the civilian populace	Shift the loyalty of adversary forces and their supporters to the friendly powers	Undermine confidence in the adversary leadership
	Deter adversary powers or groups from initiating actions detrimental to the interests of the US, its allies, or the conduct of friendly military operations	Lower the morale and combat efficiency of adversary soldiers
	Promote cessation of hostilities to reduce casualties on both sides, reduce collateral damage, and enhance transition to post-hostilities	Increase the psychological impact of US and multinational combat power
		Support military deception and operations security
		Counter hostile information activities

US MISO are developed and executed through a multiphase approach. The joint MISO process is a standard framework by which MISO assets and critical enablers plan, execute, and evaluate MISO with proficiency and consistency throughout major campaigns, operations, and peacetime engagements. The integration and execution of MISO hinge upon the proper implementation of this process.

The joint MISO process consists of seven phases: planning; target audience analysis (TAA); series development; product development and design; approval; production, distribution, dissemination; and evaluation. Each of these phases is designed to apply to any type or level of operation. Collectively, the phases address important considerations and include the necessary activities for the proper integration of MISO with the CCDR's military strategy and mission.

V. Operations Security (OPSEC)

Ref: JP 3-13.3, Operations Security (Jan '16) and ATP 3-13.3, Army Operations Security for Division and Below (Jul '19).

Joint forces often display personnel, organizations, assets, and actions to public view and to a variety of adversary intelligence collection activities, including sensors and systems. Joint forces can be under observation at their peacetime bases and locations, in training or exercises, while moving, or when deployed conducting actual operations. The actions or behavior of military family members and businesses associated with or supporting military operations are also subject to observation by adversaries, which could equally be associated with activities or operations of the joint force. Frequently, when a force performs a particular activity or operation a number of times, it establishes a pattern of behavior. Within this pattern, certain unique, particular, or special types of information might be associated with an activity or operation. Even though this information may be unclassified, it can expose US military operations to observation and/or attack. Commanders ensure OPSEC is practiced during all phases of operations. OPSEC is a capability that identifies and controls critical information, indicators of friendly force actions attendant to military operations, and incorporates countermeasures to reduce the risk of an adversary exploiting vulnerabilities. In addition, the adversary could compile and correlate enough information to predict and counter US operations.

I. Purpose of Operations Security

The purpose of OPSEC is to **reduce the vulnerability** of US and multinational forces to successful adversary exploitation of critical information. OPSEC applies to all activities that prepare, sustain, or employ forces.

The **OPSEC process** is a systematic method used to identify, control, and protect critical information and subsequently analyze friendly actions associated with military operations and other activities to:

- Identify those actions that may be observed by adversary intelligence systems.
- Determine what specific indications could be collected, analyzed, and interpreted to derive critical information in time to be useful to adversaries.
- Select countermeasures that eliminate or reduce vulnerability or indicators to observation and exploitation.
  - Avoid drastic changes as OPSEC countermeasures are implemented. Changes in procedures alone may indicate to the adversary that there is an operation or exercise starting.
  - Prevent the display or collection of critical information, especially during preparation for and execution of actual operations.
  - Avoid patterns of behavior, whenever feasible, to preclude the possibility of adversary intelligence constructing an accurate model.
- Preserve a commander's decision cycle and allow options for military actions.

Commanders cannot limit their protection efforts to a particular operational area or threat. With continuing rapid advancement and global use of communications systems and information technology, easily obtainable technical collection tools, and the growing use of the Internet and various social and mass media outlets, the ability to collect critical information virtually from anywhere in the world and threaten US military operations continues to expand. To prevent or reduce successful adversary collection and exploitation of US critical information, the commander should formulate a prudent, practical, timely, and effective OPSEC program.

II. Characteristics of Operations Security

OPSEC's most important characteristic is that it is a capability that employs a process. OPSEC is not a collection of specific rules and instructions. It is an analytical, planning, and executional process that can be applied to any operation or activity for the purpose of denying critical information to an adversary.

Unlike security programs that seek to protect classified information and controlled unclassified information (CUI), OPSEC identifies, controls, and protects unclassified critical information that is associated with specific military operations and activities. While some of the critical information in an OPSEC program may be CUI, most of the critical information is situation dependent. OPSEC and security programs must be closely coordinated to ensure appropriate aspects of military operations are protected. OPSEC and other security programs (i.e., information security, physical security, personnel security, industrial security, acquisition security, emissions security, cybersecurity, communications security [COMSEC], etc.) are complementary and should not be confused as being the same.

In OPSEC usage, an **indicator** is data derived from friendly detectable actions and open-source information that adversaries can interpret and piece together to reach conclusions or estimates of friendly intentions, capabilities, or activities. Selected indicators can be developed into an analytical model or profile of how a force prepares and how it operates. An **indication** is an observed specific occurrence or instance of an indicator. **OPSEC indicators** are friendly detectable actions and open-source information that can be interpreted or pieced together by an adversary to derive **critical information**.

Adversary intelligence personnel continuously analyze and interpret collected information to validate and/or refine the model. As adversary analysts apply more information to the analytical model, the likelihood increases that the analytical model will replicate the observed force. Thus, current and future capabilities and courses of action (COAs) can be revealed and compromised. **Critical information** consists of specific facts about friendly intentions, capabilities, and activities needed by adversaries to plan and act effectively so as to guarantee failure or unacceptable consequences for friendly mission accomplishment. Critical information can be either classified or unclassified.

III. OPSEC and Intelligence

Intelligence plays a key role in the OPSEC process. **Joint intelligence preparation of the operational environment (JIPOE)** is the analytical process used by joint intelligence organizations to produce intelligence assessments, estimates, and other intelligence products in support of the joint force commander's (JFC's) decision-making process. JIPOE's main focus is to provide predictive intelligence designed to help the JFC discern the adversary's probable intent and most likely future COA.

*Refer to Joint/Interagency SMARTbook 1: Joint Strategic & Operational Planning (Planning for Planners), 3rd Ed. (JIA1-3). JIA1-3 has an entire chapter (22 pages) covering **Joint Intelligence Preparation of the Operational Environment (JIPOE)**. JIA1-3 is the third edition of our Joint/Interagency SMARTbook 1: Joint Strategic & Operational Planning (Planning for Planners), completely reorganized and updated with the latest joint publications for 2023.*

IV. Implement Operations Security (OPSEC)

Ref: ADP 3-37, Protection (Jul '19), pp. 2-14 to 2-15.

Operations security is a capability that identifies and controls critical information, indicators of friendly force actions attendant to military operations, and incorporates countermeasures to reduce the risk of an adversary exploiting vulnerabilities (JP 3-13.3). Effective and disciplined operations security (OPSEC) is employed during decisive action. Units routinely employ OPSEC to protect essential elements of friendly information (EEFI). This helps to prevent enemy or adversary reconnaissance and other information collection capabilities from gaining an advantage because the threat has knowledge of identifiable or observable unit-specific information. OPSEC may also be used to—

- Identify actions that can be observed by enemy or adversary intelligence systems.
- Determine indicators of hostile intelligence that systems might obtain and which could be interpreted or pieced together to derive critical information in time to be useful to adversaries or enemies.
- Select countermeasures that eliminate or reduce vulnerability or indicators to observation and exploitation:
    - Avoid drastic changes as OPSEC countermeasures are implemented. Changes in procedures alone may alert the adversary that an operation or exercise is starting.
    - Prevent the display or collection of critical information, especially during the preparation for, and the execution of, actual operations.
    - Avoid patterns of behavior, when feasible, to preclude the possibility of adversary intelligence constructing an accurate model.
- Preserve a commander's decision cycle and allow options for military actions.

OPSEC applies to all operations. OPSEC is a force multiplier that can maximize operational effectiveness by saving lives and resources when integrated into operations, activities, plans, exercises, training, and capabilities. Good field craft and the disciplined enforcement of camouflage and concealment are essential to OPSEC. The unit OPSEC officer coordinates additional OPSEC measures with other staff and command elements and synchronizes with adjacent units. The OPSEC officer develops OPSEC measures during the military decisionmaking process (MDMP). The assistant chief of staff, intelligence, assists the OPSEC process by comparing friendly OPSEC indicators with enemy or adversary intelligence collection capabilities.

OPSEC, integrated and synchronized in combination with other protection measures, may be employed with deception to ensure that only desired events reach the enemy and supported operations are concealed. At times, unit commanders employ deception in support of OPSEC to create multiple false indicators that confuse enemy or adversary forces operating in the unit's area of operations, making unit intentions harder to interpret. Deception in support of OPSEC uses controlled information about friendly force capabilities, activities, and intentions to shape perceptions. It targets and counters intelligence, surveillance, and reconnaissance capabilities to distract intelligence collection away from, or provide cover for, unit operations. Deception in support of OPSEC is a relatively easy countermeasure to use and is appropriate for use at battalion-level and below.

To be successful, OPSEC and deception requirements must achieve balance (refer to ATP 3-13.3 for additional information).

V. The Operations Security Process
Ref: JP 3-13.3, Operations Security (Jan '16), chap. II.

The OPSEC process is applicable across the range of military operations. Use of the process ensures that the resulting OPSEC countermeasures address all significant aspects of the particular situation and are balanced against operational requirements. OPSEC is a continuous process. The OPSEC process (Figure II-1) consists of five distinct actions: identification of critical information, analysis of threats, analysis of vulnerabilities, assessment of risk, and application of appropriate OPSEC countermeasures. These OPSEC actions are applied continuously during OPSEC planning. In dynamic situations, however, individual actions may be reevaluated at any time. New information about the adversary's intelligence collection capabilities, for instance, would require a new analysis of threats.

An understanding of the following terms is required before the process can be explained.

Critical Information
These are specific facts about friendly intentions, capabilities, and activities needed by adversaries to plan and act effectively against friendly mission accomplishment.

OPSEC Indicators
Friendly detectable actions and open-source information that can be interpreted or pieced together by an adversary to derive critical information.

OPSEC Vulnerability
A condition in which friendly actions provide OPSEC indicators that may be obtained and accurately evaluated by an adversary in time to provide a basis for effective adversary decision making.

A. Identify Critical Information
The identification of critical information is a key part of the OPSEC process because it focuses the remainder of the OPSEC process on protecting vital information rather than attempting to protect all unclassified information. Critical information answers key questions likely to be asked by adversaries about specific friendly intentions, capabilities, and activities necessary for adversaries to plan and act effectively against friendly mission accomplishment. There are many areas within an organization where elements of critical information can be obtained. Personnel from outside the organization may also handle portions of its critical information. Therefore it is important to have personnel from each staff section and component involved in the process of identifying critical information. The critical information items should be consolidated into a list known as a CIL.

Critical information is listed in tab C (Operations Security) to appendix 3 (Information Operations) to annex C (Operations) of an OPLAN or OPORD. Generic CILs (Figure II-2) can be developed beforehand to assist in identifying the specific critical information.

B. Threat Analysis
This action involves the research and analysis of intelligence, CI, and open-source information to identify the likely adversaries to the planned operation.

The operations planners, working with the intelligence and CI staffs and assisted by the OPSEC program manager, seek answers to the following threat questions:

- Who is the adversary? (Who has the intent and capability to take action against the planned operation?)

- What are the adversary's goals? (What does the adversary want to accomplish?)

- What is the adversary's COA for opposing the planned operation? (What actions might the adversary take? Include the most likely COA and COA most dangerous to friendly forces and mission accomplishment.)
- What critical information does the adversary already know about the operation? (What information is too late to protect?)
- What are the adversary's intelligence collection capabilities?
- Who are the affiliates of the adversary, and will they share information?

C. Vulnerability Analysis

The purpose of this action is to identify an operation's or activity's vulnerabilities. It requires examining each aspect of the planned operation to identify any OPSEC indicators or vulnerabilities that could reveal critical information and then comparing those indicators or vulnerabilities with the adversary's intelligence collection capabilities identified in the previous action. A vulnerability exists when the adversary is capable of collecting critical information, correctly analyzing it, and then taking timely action. The adversary can then exploit that vulnerability to obtain an advantage.

Continuing to work with the intelligence personnel, the operations planners seek answers to the following vulnerability questions:

(1) What indicators (friendly actions and open-source information) of critical information not known to the adversary will be created by the friendly activities that will result from the planned operation?

(2) What indicators can the adversary actually collect?

(3) What indicators will the adversary be able to use to the disadvantage of friendly forces? (Can the adversary analyze the information, make a decision, and take appropriate action in time to interfere with the planned operation?)

(4) Will the application of OPSEC countermeasures introduce more indicators that the adversary will be able to collect?

Refer to JP 3-13.3, app. A, "Operations Security Indicators," for a detailed discussion of OPSEC indicators.

D. Risk Assessment

This action has three components. First, planners analyze the vulnerabilities identified in the previous action and identify possible OPSEC countermeasures for each vulnerability. Second, the commander and staff estimate the impact to operations such as cost in time, resources, personnel or interference with other operations associated with implementing each possible OPSEC countermeasure versus the potential harmful effects on mission accomplishment resulting from an adversary's exploitation of a particular vulnerability. Third, the commander and staff select specific OPSEC countermeasures for execution based upon a risk assessment done by the commander and staff.

OPSEC countermeasures reduce the probability of the adversary either observing indicators or exploiting vulnerabilities, being able to correctly analyze the information obtained, and being able to act on this information in a timely manner.

OPSEC countermeasures can be used to prevent the adversary from detecting an indicator or exploiting a vulnerability, provide an alternative analysis of a vulnerability or an indicator (prevent the adversary from correctly interpreting the indicator), and/or attack the adversary's collection system.

OPSEC countermeasures include, among other actions, cover, concealment, camouflage, deception, intentional deviations from normal patterns, and direct strikes against the adversary's intelligence system.

More than one possible measure may be identified for each vulnerability. Conversely, a single measure may be used for more than one vulnerability.

VI. Operations Security Indicators
Ref: ATP 3-13.3, Army Operations Security for Division and Below (Jul '19).

The indicator's signature is a characteristic that serves to set the indicator apart. A signature makes the indicator identifiable or causes it to stand out. Uniqueness and stability are properties of a signature. Uncommon or unique features reduce the ambiguity of an indicator.

Association. Association is the process of forming mental connections to an indicator. It is the key to interpretation. An enemy compares current data with previously gathered information to identify possible relationships. Continuity of actions, objects, or other indicators, which register as patterns, provides another association. For example, the presence of special operations aviation aircraft, such as the MH-6, MH-60, and MH-47, may be indicators of other special operations forces operating in the area. Certain items of equipment particular to specific units are indicators of the potential presence of related equipment. For instance, the sighting of an M-88A2 Hercules Recovery Vehicle likely indicates the presence of an armored unit equipped with M1A2-series tanks, as the M-88A2 is rated to recover and tow the M1A2-series tanks. Such continuity can result from repetitive practices or sequencing instead of from planned procedures. When detecting some components of symmetrically-arrayed organizations, the enemy can assume the existence of the rest. As another example, the adversary would suspect the presence of an entire infantry battalion when intelligence detects the headquarters company and one line company. When evaluated as a whole, the pattern can be a single indicator, which simplifies the enemy's analysis.

Profile. A profile is accumulated data that portray the significant features of an indicator. Profiles are linked to functional activity, which has a profile comprising unique indicators, patterns, and associations. This profile, in turn, has several sub-profiles for the functional activities needed to deploy the particular mission aircraft (for example, fuels, avionics, munitions, communications, air traffic control, supply, personnel, and transportation). If a functional profile does not appear to change from one operation to the next, it is difficult for an enemy to interpret. However, if it is distinct, the profile may be the key or only indicator needed to understand the operation. Unique profiles reduce the time needed to make accurate situational assessments. They are primary warning tools because they provide a background for contrasts.

Contrast. Contrast is the change in an indicator's established profile. The key to obtaining the contrast of an indicator lies in how it differs from what has been shown previously. Contrasts are the simplest and most reliable means of detection because they only need to be recognized, not understood. One question prompts several additional ones concerning contrasts in profile. The nature of the indicator's exposure is an important aspect when seeking profile contrasts. For example, if the adversary identifies items specific to special operations aviation at an airfield, this will contrast with what is "normal" at the airfield and will indicate the deployment of special operations aircraft to the airfield without having actually observed them.

Exposure. Exposure is the condition of being presented to view or made known—the condition of being unprotected. For an OPSEC indicator, exposure increases according to the duration, repetition, and timing of its appearance. The exposure of an indicator often reveals its relative importance and meaning. Limited duration and repetition reduces detailed observation and associations. An indicator that appears for a short time will likely fade into the background of insignificant anomalies. An indicator that appears over a long period of time, however, becomes part of a profile. Indicators exposed repeatedly present the biggest danger. Operations conducted the same way several times with little or no variation provide an adversary the information needed to determine where, when, how, and with what to attack. Repetitive operations cost many lives in wartime.

VI. Cyberspace and the Electromagnetic Spectrum

Ref: ATP 3-13.1, The Conduct of Information Operations (Oct '18), pp. 3-3 to 3-4 and FM 3-12, Cyberspace Operations & Electromagnetic Warfare (Aug '21), chap. 1.

Cyberspace operations and electromagnetic warfare (EW) play an essential role in the Army's conduct of unified land operations as part of a joint force and in coordination with unified action partners.

Cyberspace Operations *(See pp. 3-47 to 3-54.)*
Cyberspace operations are the employment of cyberspace capabilities where the primary purpose is to achieve objectives in or through cyberspace (JP 3-0).

Electromagnetic Warfare (EW) *(See pp. 3-55 to 3-60.)*
Electromagnetic warfare (EW) is a military action involving the use of electromagnetic and directed energy to control the **electromagnetic spectrum** or to attack the enemy (JP 3-85).

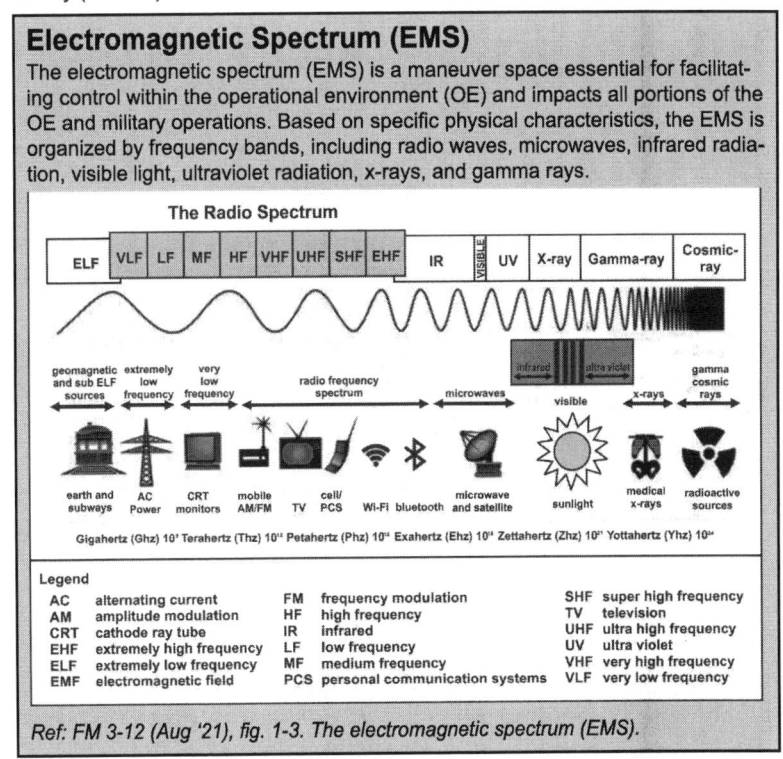

Electromagnetic Spectrum (EMS)
The electromagnetic spectrum (EMS) is a maneuver space essential for facilitating control within the operational environment (OE) and impacts all portions of the OE and military operations. Based on specific physical characteristics, the EMS is organized by frequency bands, including radio waves, microwaves, infrared radiation, visible light, ultraviolet radiation, x-rays, and gamma rays.

The Radio Spectrum

| ELF | VLF | LF | MF | HF | VHF | UHF | SHF | EHF | IR | VISIBLE | UV | X-ray | Gamma-ray | Cosmic-ray |

geomagnetic and sub ELF sources — extremely low frequency — very low frequency — radio frequency spectrum — microwaves — infrared — visible — ultra violet — x-rays — gamma cosmic rays

earth and subways | AC Power | CRT monitors | mobile AM/FM | TV | cell/PCS | Wi-Fi bluetooth | microwave and satellite | sunlight | medical x-rays | radioactive sources

Gigahertz (Ghz) 10^9 Terahertz (Thz) 10^{12} Petahertz (Phz) 10^{15} Exahertz (Ehz) 10^{18} Zettahertz (Zhz) 10^{21} Yottahertz (Yhz) 10^{24}

Legend

AC	alternating current	FM	frequency modulation	SHF	super high frequency
AM	amplitude modulation	HF	high frequency	TV	television
CRT	cathode ray tube	IR	infrared	UHF	ultra high frequency
EHF	extremely high frequency	LF	low frequency	UV	ultra violet
ELF	extremely low frequency	MF	medium frequency	VHF	very high frequency
EMF	electromagnetic field	PCS	personal communication systems	VLF	very low frequency

Ref: FM 3-12 (Aug '21), fig. 1-3. The electromagnetic spectrum (EMS).

Cyberspace is one of the five domains of warfare and uses a portion of the electromagnetic spectrum (EMS) for operations, for example, Bluetooth, Wi-Fi, and satellite transport. Therefore, cyberspace operations and EW require frequency assignment, management, and coordination performed by spectrum management operations.

Cyberspace Electromagnetic Activities (CEMA) & Information Advantage

Ref: FM 3-12, Cyberspace Operations and Electromagnetic Warfare (Aug '21), fig. 1-4.

Cyberspace and the EMS are critical for success in today's operational environment (OE). U.S. and adversary forces alike rely heavily on cyberspace and EMS-dependent technologies for command and control, information collection, situational understanding, and targeting. Achieving relative superiority in cyberspace and the EMS gives commanders an advantage over adversaries and enemies. By conducting cyberspace operations and EW, commanders can limit adversaries' available courses of action, diminish their ability to gain momentum, degrade their command and control, and degrade their ability to operate effectively in the other domains.

Commanders must leverage cyberspace and EW capabilities using a combined arms approach to seize, retain, and exploit the operational initiative. Effective use of cyberspace operations and EW require commanders and staffs to conduct **cyberspace electromagnetic activities (CEMA).** Cyberspace electromagnetic activities is the process of planning, integrating, and synchronizing cyberspace operations and electromagnetic warfare in support of unified land operations (ADP 3-0). By integrating and synchronizing cyberspace operations and EW, friendly forces gain an **information advantage** across multiple domains and lines of operations.

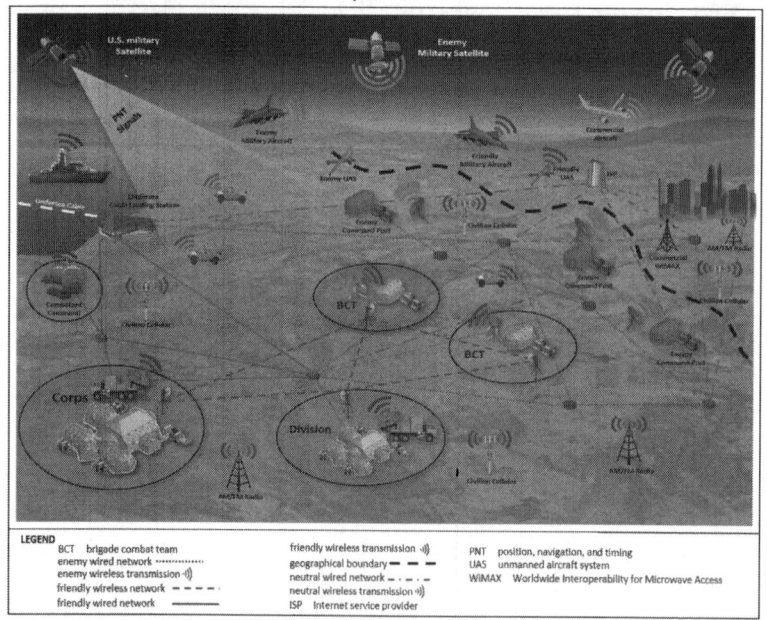

LEGEND
BCT brigade combat team
enemy wired network
enemy wireless transmission
friendly wireless network
friendly wired network
friendly wireless transmission
geographical boundary
neutral wired network
neutral wireless transmission
ISP Internet service provider
PNT position, navigation, and timing
UAS unmanned aircraft system
WiMAX Worldwide Interoperability for Microwave Access

CYBER1-1: The Cyberspace Operations & Electronic Warfare SMARTbook (w/SMARTupdate 1*) topics and chapters include cyber intro (global threat, contemporary operating environment, information as a joint function), joint cyberspace operations (CO), cyberspace operations (OCO/DCO/DODIN), electromagnetic warfare (EW) operations, cyber & EW (CEMA) planning, spectrum management operations (SMO/JEMSO), DoD information network (DODIN) operations, acronyms/abbreviations, and glossary of cyber terms.

VI(a). Cyberspace Operations (CO)

Ref: FM 3-12, Cyberspace Operations and Electromagnetic Warfare (Aug '21), chap. 2.

Cyberspace operations and electromagnetic warfare (EW) can benefit from synchronization with other Army capabilities using a combined arms approach to achieve objectives against enemy forces. Cyberspace operations and EW can provide commanders with positions of relative advantage in the multi-domain fight. Effects that bleed over from the cyberspace domain into the physical domain can be generated and leveraged against the adversary. A cyberspace capability is a device or computer program, including any combination of software, firmware, or hardware, designed to create an effect in or through cyberspace (JP 3-12).

Electromagnetic Spectrum Superiority

Electromagnetic spectrum superiority is the degree of control in the electromagnetic spectrum that permits the conduct of operations at a given time and place without prohibitive interference, while affecting the threat's ability to do the same (JP 3-85). Electromagnetic warfare (EW) creates effects in the EMS and enables commanders to gain EMS superiority while conducting Army operations. EW capabilities consist of the systems and weapons used to conduct EW missions to create lethal and non-lethal effects in and through the EMS.

See pp. 3-55 to 3-60, Electromagnetic Warfare (EW), for further discussion.

I. Cyberspace Operations (CYBER1-1, chap. 2.)

The joint force and the Army divide cyberspace operations into three categories based on the portion of cyberspace in which the operations take place and the type of cyberspace forces that conduct those operations. Each of type of cyberspace operation has varying associated authorities, approval levels, and coordination considerations. An Army taxonomy of cyberspace operations is depicted in figure 2-1, below. The three types of cyberspace operations are—

Cyberspace Operations

- **A** DODIN Operations
- **B** Defensive Cyberspace Operations (DCO)
- **C** Offensive Cyberspace Operations (OCO)

The Army conducts DODIN operations on internal Army and DOD networks and systems using primarily signal forces. The Army employs cyberspace forces to conduct DCO which includes two further sub-divisions—DCO-IDM and defensive cyberspace operations-response actions (DCO-RA). Cyberspace forces conduct DCO-IDM within the DODIN boundary, or on other friendly networks when authorized, in order to defend those networks from imminent or ongoing attacks. At times cyberspace forces may also take action against threat cyberspace actors in neutral or adversary

Cyberspace Domain

Ref: FM 3-12, Cyberspace Operations and Electromagnetic Warfare (Aug '21), pp. 1-5 to 1-7.

Cyberspace is a global domain within the information environment consisting of the interdependent networks of information technology infrastructures and resident data, including the Internet, telecommunications networks, computer systems, and embedded processors and controllers (JP 3-12). Cyberspace operations require the use of links and nodes located in other physical domains to perform logical functions that create effects in cyberspace that then permeate throughout the physical domains using both wired networks and the EMS.

The use of cyberspace is essential to operations. The Army conducts cyberspace operations and supporting activities as part of both Army and joint operations. Because cyberspace is a global communications and data-sharing medium, it is inherently joint, inter-organizational, multinational, and often a shared resource, with signal and intelligence maintaining significant equities. Friendly, enemy, adversary, and host-nation networks, communications systems, computers, cellular phone systems, social media websites, and technical infrastructures are all part of cyberspace.

To aid the planning and execution of cyberspace operations, cyberspace is sometimes visualized in three layers. These layers are interdependent, but each layer has unique attributes that affect operations. Cyberspace operations generally traverse all three layers of cyberspace but may target effects at one or more specific layers. Planners must consider the challenges and opportunities presented by each layer of cyberspace as well as the interactions amongst the layers. Figure 1-2 on page 1-6 depicts the relationship between the three cyberspace layers. The three cyberspace layers are—

- The physical network layer.
- The logical network layer.
- The cyber-persona layer.

Physical Network Layer

The physical network layer consists of the information technology devices and infrastructure in the physical domains that provide storage, transport, and processing of information within cyberspace, to include data repositories and the connections that transfer data between network components (JP 3-12). Physical network components include the hardware and infrastructure such as computing devices, storage devices, network devices, and wired and wireless links. Components of the physical network layer require physical security measures to protect them from damage or unauthorized access, which, if left vulnerable, could allow a threat to gain access to both systems and critical data.

Every physical component of cyberspace is owned by a public or private entity. The physical layer often crosses geo-political boundaries and is one of the reasons that cyberspace operations require multiple levels of joint and unified action partner coordination. Cyberspace planners use knowledge of the physical location of friendly, neutral, and adversary information technology systems and infrastructures to understand appropriate legal frameworks for cyberspace operations and to estimate impacts of those operations. Joint doctrine refers to portions of cyberspace, based on who owns or controls that space, as either blue, gray, or red cyberspace (refer to JP 3-12). This publication refers to these areas as friendly, neutral, or enemy cyberspace respectively.

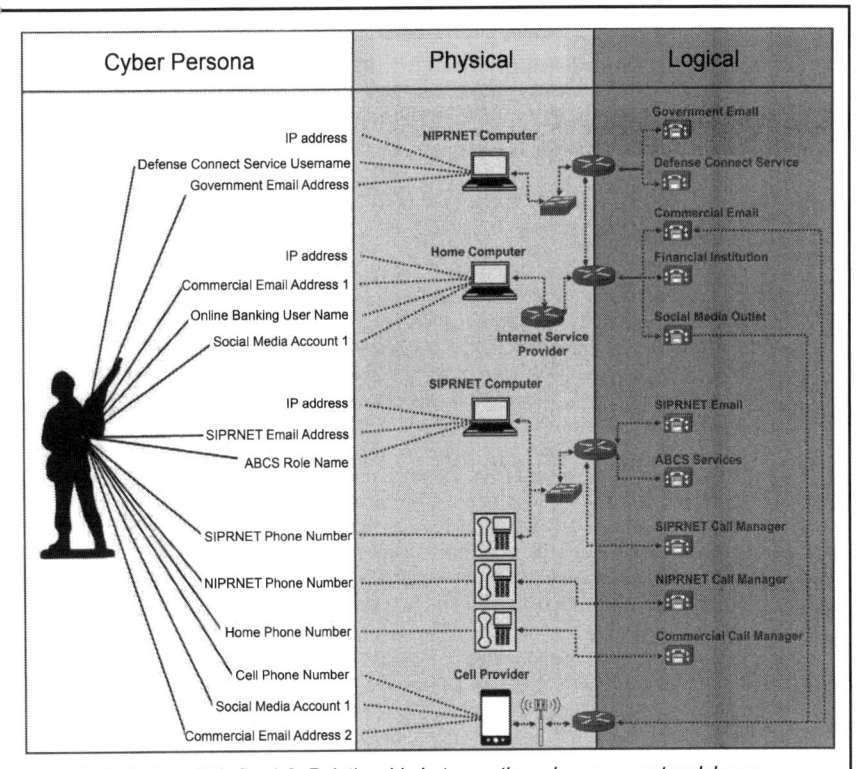

Ref: FM 3-12 (Aug '21), fig. 1-2. Relationship between the cyberspace network layers.

Logical Network Layer

The logical network layer consists of those elements of the network related to one another in a way that is abstracted from the physical network, based on the logic programming (code) that drives network components (i.e., the relationships are not necessarily tied to a specific physical link or node, but to their ability to be addressed logically and exchange or process data) (JP 3-12). Nodes in the physical layer may logically relate to one another to form entities in cyberspace not tied to a specific node, path, or individual. Web sites hosted on servers in multiple physical locations where content can be accessed through a single uniform resource locator or web address provide an example. This may also include the logical programming to look for the best communications route, instead of the shortest physical route, to provide the information requested.

Cyber-Persona Layer

The cyber-persona layer is a view of cyberspace created by abstracting data from the logical network layer using the rules that apply in the logical network layer to develop descriptions of digital representations of an actor or entity identity in cyberspace, known as a cyber-persona (JP 3-12). Cyber-personas are not confined to a single physical or logical location and may link to multiple physical and logical network layers. When planning and executing cyberspace operations, staffs should understand that one actor or entity (user) may have multiple cyber-personas, using multiple identifiers in cyberspace. These various identifiers can include different work and personal emails and different identities on different Web forums, chatrooms, and social network sites. For example, an individual's account on a social media website, consisting of the username and digital information associated with that username, may be just one of that individual's cyber-personas.

networks in defense of the DODIN or friendly networks. These types of actions, called DCO-RA, require additional authorities and coordination measures. Lastly, cyberspace forces deliberately target threat capabilities in neutral, adversary, and enemy-held portions of cyberspace by conducting OCO. Cyberspace forces may include joint forces from the DOD cyber mission forces or Army-retained cyberspace forces.

See pp. 2-27 to 2-36 for discussion of cyberspace forces.

A. Department of Defense Information Network Operations (DODIN)

The Department of Defense information network is the set of information capabilities and associated processes for collecting, processing, storing, disseminating, and managing information on demand to warfighters, policy makers, and support personnel, whether interconnected or stand-alone. Also called DODIN (JP 6-0). This includes owned and leased communications and computing systems and services, software (including applications), data, security services, other associated services, and national security systems. Department of Defense information network operations are operations to secure, configure, operate, extend, maintain, and sustain Department of Defense cyberspace to create and preserve the confidentiality, availability, and integrity of the Department of Defense information network. Also called DODIN operations (JP 3-12). DODIN operations provide authorized users at all echelons with secure, reliable end-to-end network and information system availability. DODIN operations allow commanders to effectively communicate, collaborate, share, manage, and disseminate information using information technology systems.

Signal forces install tactical networks, conduct maintenance and sustainment activities, and security evaluation and testing. Signal forces performing DODIN operations may also conduct limited DCO-IDM. Since both cyberspace security and defense tasks are ongoing, standing orders for DODIN operations and DCO-IDM cover most cyberspace security and initial cyberspace defense tasks.

The Army secures the DODIN-A using a layered defense approach. Layered defense uses multiple physical, policy, and technical controls in to guard against threats on the network. Layering integrates people, technology, and operational capabilities to establish security barriers across multiple layers of the DODIN-

A. Various types of security barriers include—

- Antivirus software.
- Firewalls.
- Anti-spam software.
- Communications security.
- Data encryption.
- Password protection.
- Physical and technical barriers.
- Continuous security training.
- Continuous network monitoring.

Security barriers are protective measures against acts that may impair the effectiveness of the network, and therefore the mission command system. Additionally, layering includes perimeter security, enclave security, host security, physical security, personnel security, and cybersecurity policies and standards. Layering protects the cyberspace domain at the physical, logical, and administrative control levels.

B. Defensive Cyberspace Operations (DCO)

Defensive cyberspace operations are missions to preserve the ability to utilize blue cyberspace capabilities and protect data, networks, cyberspace-enabled devices,

Cyberspace Operations (Missions & Actions)

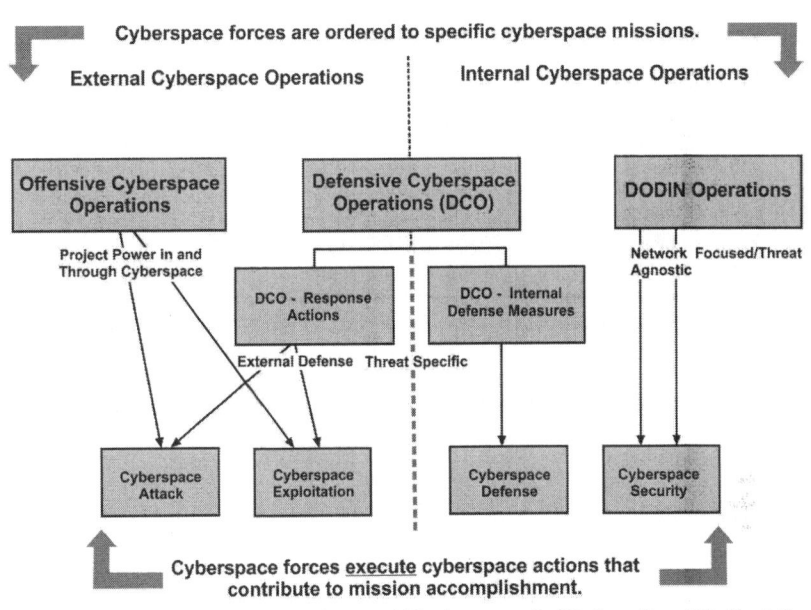

Cyberspace forces are ordered to specific cyberspace missions.

External Cyberspace Operations | **Internal Cyberspace Operations**

- **Offensive Cyberspace Operations** — Project Power in and Through Cyberspace
- **Defensive Cyberspace Operations (DCO)**
  - DCO - Response Actions (External Defense)
  - DCO - Internal Defense Measures (Threat Specific)
- **DODIN Operations** — Network Focused/Threat Agnostic

Actions:
- Cyberspace Attack
- Cyberspace Exploitation
- Cyberspace Defense
- Cyberspace Security

Cyberspace forces **execute** cyberspace actions that contribute to mission accomplishment.

Ref: FM 3-12, Cyberspace Operations and Electromagnetic Warfare (Aug '21), fig. 2-2. Cyberspace operations missions and actions.

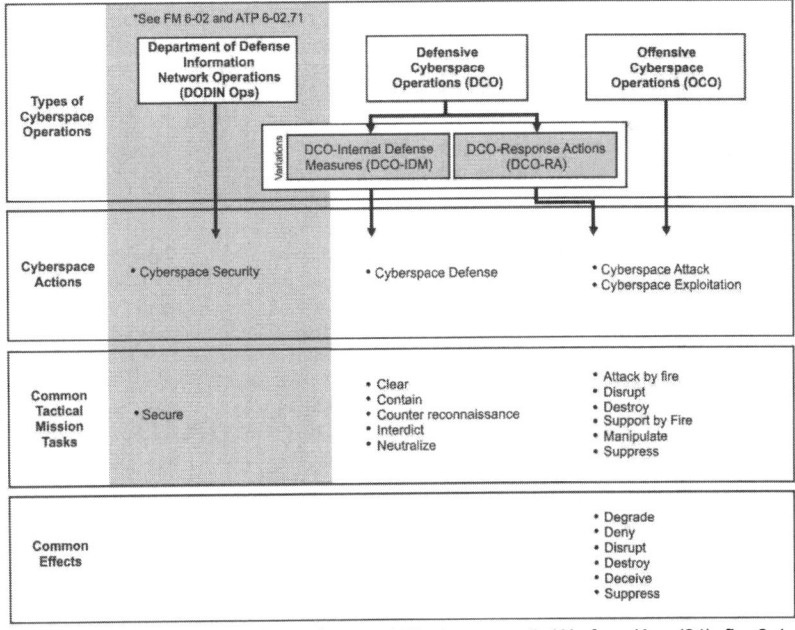

	Department of Defense Information Network Operations (DODIN Ops)*	Defensive Cyberspace Operations (DCO)		Offensive Cyberspace Operations (OCO)
Types of Cyberspace Operations		DCO-Internal Defense Measures (DCO-IDM)	DCO-Response Actions (DCO-RA)	
Cyberspace Actions	• Cyberspace Security	• Cyberspace Defense		• Cyberspace Attack • Cyberspace Exploitation
Common Tactical Mission Tasks	• Secure	• Clear • Contain • Counter reconnaissance • Interdict • Neutralize		• Attack by fire • Disrupt • Destroy • Support by Fire • Manipulate • Suppress
Common Effects				• Degrade • Deny • Disrupt • Destroy • Deceive • Suppress

*See FM 6-02 and ATP 6-02.71

Ref: FM 3-12, Cyberspace Operations and Electromagnetic Warfare (Aug '21), fig. 2-1. Cyberspace operations taxonomy.

and other designated systems by defeating on-going or imminent malicious cyberspace activity (JP 3-12). The term blue cyberspace denotes areas in cyberspace protected by the United States, its mission partners, and other areas the Department of Defense may be ordered to protect. DCO are further categorized based on the location of the actions in cyberspace as—

Defensive Cyberspace Operations-Internal Defensive Measures (DCO-IDM)

Defensive cyberspace operations-internal defensive measures are operations in which authorized defense actions occur within the defended portion of cyberspace (JP 3-12). DCO-IDM is conducted within friendly cyberspace. DCO-IDM involves actions to locate and eliminate cyber threats within friendly networks. Cyberspace forces employ defensive measures to neutralize and eliminate threats, allowing reestablishment of degraded, compromised, or threatened portions of the DODIN. Cyberspace forces conducting DCO-IDM primarily conduct cyberspace defense tasks, but may also perform some tasks similar to cyberspace security.

Cyberspace defense includes actions taken within protected cyberspace to defeat specific threats that have breached or are threatening to breach cyberspace security measures and include actions to detect, characterize, counter, and mitigate threats, including malware or the unauthorized activities of users, and to restore the system to a secure configuration. (JP 3-12). Cyberspace forces act on cues from cybersecurity or intelligence alerts of adversary activity within friendly networks. Cyberspace defense tasks during DCO-IDM include hunting for threats on friendly networks, deploying advanced countermeasures, and responding to eliminate these threats and mitigate their effects.

Defensive Cyberspace Operations-Response Actions (DCO-RA)

Defensive cyberspace operation-response actions are operations that are part of a defensive cyberspace operations mission that are taken external to the defended network or portion of cyberspace without permission of the owner of the affected system (JP 3-12). DCO-RA take place outside the boundary of the DODIN. Some DCO-RA may include actions that rise to the level of use of force and may include physical damage or destruction of enemy systems. DCO-RA consist of conducting cyberspace attacks and cyberspace exploitation similar to OCO. However, DCO-RA use these actions for defensive purposes only, unlike OCO that is used to project power in and through cyberspace.

Decisions to conduct DCO-RA depend heavily on the broader strategic and operational contexts such as the existence or imminence of open hostilities, the degree of certainty in attribution of the threat; the damage the threat has or is expected to cause, and national policy considerations. DCO-RA are conducted by national mission team(s) and require a properly coordinated military order, coordination with interagency and unified action partners, and careful consideration of scope, rules of engagement, and operational objectives.

C. Offensive Cyberspace Operations (OCO)

Offensive cyberspace operations are missions intended to project power in and through cyberspace (JP 3-12). Cyberspace forces conduct OCO outside of DOD networks to achieve positions of relative advantage through cyberspace exploitation and cyberspace attack actions in support of commanders' objectives. Commanders must integrate OCO within the combined arms scheme of maneuver throughout the operations process to achieve optimal effects.

The Army provides cyberspace forces trained to perform OCO across the range of military operations to the joint force. Army forces conducting OCO do so under the authority of a joint force commander. Refer to Appendix C for information on integrating with unified action partners. Joint forces may provide OCO support to corps and below Army commanders in response to requests through the joint targeting process. Refer to Appendix D for more information on joint cyberspace forces. Targets for cyberspace effects may require extended planning time, extended approval time, as well as synchronization and deconfliction with partners external to the DOD. Chapter 4 covers targeting considerations in detail.

II. Cyberspace Actions

Execution of these cyberspace operations entails one or more specific tasks, which joint cyberspace doctrine refers to as cyberspace actions (refer to JP 3-12), and the employment of one or more cyberspace capabilities. Figure 2-2 on page 2-6 depicts the relationships between the types of cyberspace operations and their associated actions, the location of those operations in cyberspace, and the forces that conduct those operations. The four cyberspace actions are—

A. Cyberspace Security

Cyberspace security is actions taken within protected cyberspace to prevent unauthorized access to, exploitation of, or damage to computers, electronic communications systems, and other information technology, including platform information technology, as well as the information contained therein, to ensure its availability, integrity, authentication, confidentiality, and nonrepudiation (JP 3-12). These preventive measures include protecting the information on the DODIN, ensuring the information's availability, integrity, authenticity, confidentiality, and nonrepudiation. Cyberspace security is generally preventative in nature, but also continues throughout DCO-IDM and incident responses in instances where a cyberspace threat compromises the DODIN. Some common types of cyberspace security actions include—

- Password management.
- Software patching.
- Encryption of storage devices.
- Mandatory cybersecurity training for all users.
- Restricting access to suspicious websites.
- Implementing procedures to define the roles, responsibilities, policies, and administrative functions for managing DODIN operations.

B. Cyberspace Defense

Cyberspace defense are actions taken within protected cyberspace to defeat specific threats that have breached or are threatening to breach cyberspace security measures and include actions to detect, characterize, counter, and mitigate threats, including malware or the unauthorized activities of users, and to restore the system to a secure configuration. (JP 3-12)

C. Cyberspace Exploitation

Cyberspace exploitation consists of actions taken in cyberspace to gain intelligence, maneuver, collect information, or perform other enabling actions required to prepare for future military operations (JP 3-12). These operations must be authorized through mission orders and are part of OCO or DCO-RA actions in gray or red cyberspace that do not create cyberspace attack effects, and are often intended to remain clandestine. Cyberspace exploitation includes activities to support operational preparation of the environment for current and future operations by gaining and maintaining access to networks, systems, and nodes of military value; maneuvering to positions of advantage within cyberspace; and positioning cyberspace capabilities to facilitate follow-on actions. Cyberspace exploitation actions are deconflicted with other United States Government departments and agencies in accordance with national policy.

D. Cyberspace Attack (See p. 1-50.)

Ref: FM 3-12, Cyberspace Operations and Electromagnetic Warfare (Aug '21), p. 2-7.

Cyberspace attack actions taken in cyberspace that create noticeable denial effects (i.e., degradation, disruption, or destruction) in cyberspace or manipulation that leads to denial effects in the physical domains (JP 3-12). A cyberspace attack creates effects in and through cyberspace and may result in physical destruction. Modification or destruction of cyberspace capabilities that control physical processes can lead to effects in the physical domains. Some illustrative examples of common effects created by a cyberspace attack include—

Deny
To prevent access to, operation of, or availability of a target function by a specified level for a specified time (JP 3-12). Cyberspace attacks deny the enemy's ability to access cyberspace by hindering hardware and software functionalities for a specific duration of time.

Degrade
To deny access to, or operation of, a target to a level represented as a percentage of capacity. Level of degradation is specified. If a specific time is required, it can be specified (JP 3-12).

Disrupt
To completely but temporarily deny access to, or operation of, a target for a period of time. A desired start and stop time are normally specified. Disruption can be considered a special case of degradation where the degradation level is 100 percent (JP 3-12). Commanders can use cyberspace attacks that temporarily but completely deny an enemy's ability to access cyberspace or communication links to disrupt decision making, ability to organize formations, and conduct command and control. Disruption effects in cyberspace are usually limited in duration.

Destroy
To completely and irreparably deny access to, or operation of, a target. Destruction maximizes the time and amount of denial. However, destruction is scoped according to the span of a conflict, since many targets, given enough time and resources, can be reconstituted (JP 3-12). Commanders can use cyberspace attacks to destroy hardware and software beyond repair where replacement is required to restore system function. Destruction of enemy cyberspace capabilities could include irreversible corruption to system software causing loss of data and information, or irreparable damage to hardware such as the computer processor, hard drive, or power supply on a system or systems on the enemy's network.

Manipulate
Manipulation, as a form of cyberspace attack, controls or changes information, information systems, and/or networks in gray or red cyberspace to create physical denial effects, using deception, decoying, conditioning, spoofing, falsification, and other similar techniques. It uses an adversary's information resources for friendly purposes, to create denial effects not immediately apparent in cyberspace (JP 3-12). Commanders can use cyberspace attacks to manipulate enemy information or information systems in support of tactical deception objectives or as part of joint military deception. Refer to FM 3-13.4 for information on Army support to military deception.

Note. Cyberspace attacks are types of fires conducted during DCO-RA and OCO actions and are limited to cyber mission force(s) engagement. They require coordination with other United States Government departments and agencies and careful synchronization with other lethal and non-lethal effects through established targeting processes.

I. Electromagnetic Warfare (EW)

Ref: FM 3-12, Cyberspace Operations and Electromagnetic Warfare (Aug '21), pp. 2-8 to 2-15.

I. Electromagnetic Warfare (EW) *(CYBER1-1, chap. 3.)*

Electromagnetic Warfare (EW) is military action involving the use of electromagnetic and directed energy to control the electromagnetic spectrum or to attack the enemy. EW consists of three functions: electromagnetic attack, electromagnetic protection, and electromagnetic support.

Modern militaries rely on communications equipment using broad portions of the electromagnetic spectrum (EMS) to conduct military operations allowing forces to talk, transmit data, and provide navigation and timing information, and command and control troops worldwide. They also rely on the EMS for sensing and awareness of the OE. The Army conducts electromagnetic warfare (EW) to gain and maintain positions of relative advantage within the EMS. The Army's contribution to electromagnetic spectrum operations is accomplished by integrating and synchronizing EW and spectrum management operations.

Electromagnetic Warfare (EW)

- **A** Electromagnetic Attack (EA)
- **B** Electromagnetic Protection (EP)
- **C** Electromagnetic Support (ES)

- ***** Electromagnetic Warfare Reprogramming

The three divisions often mutually support each other in operations. For example, radar-jamming EA can serve a protection function for friendly forces to penetrate defended airspace; it can also prevent an adversary from having a complete operating picture.

CYBER1-1: The Cyberspace Operations & Electronic Warfare SMARTbook (w/SMARTupdate 1*) topics and chapters include cyber intro (global threat, contemporary operating environment, information as a joint function), joint cyberspace operations (CO), cyberspace operations (OCO/DCO/DODIN), electromagnetic warfare (EW) operations, cyber & EW (CEMA) planning, spectrum management operations (SMO/JEMSO), DoD information network (DODIN) operations, acronyms/abbreviations, and glossary of cyber terms.

A. Electromagnetic Attack (EA) *(See p. 1-50.)*

Army forces conduct both offensive and defensive EA to fulfill the commander's objectives in support of the mission. EA projects power in and through the EMS by implementing active and passive actions to deny enemy capabilities and equipment, or by employing passive systems to protect friendly capabilities. Electromagnetic attack is a division of electromagnetic warfare involving the use of electromagnetic energy, directed energy, or antiradiation weapons to attack personnel, facilities, or equipment with the intent of degrading, neutralizing, or destroying enemy combat capability and considered a form of fires (JP 3-85). EA requires systems or weapons that radiate electromagnetic energy as active measures and systems that do not radiate or re-radiate electromagnetic energy as passive measures.

Offensive EA

Offensive EA prevents or reduces an enemy's effective use of the EMS by employing jamming and directed energy weapon systems against enemy spectrum-dependent systems and devices. Offensive EA systems and capabilities include—

- Jammers.
- Directed energy weaponry.
- Self-propelled decoys.
- Electromagnetic deception.
- Antiradiation missiles.

Defensive EA

Defensive EA protects against lethal attacks by denying enemy use of the EMS to target, guide, and trigger weapons that negatively impact friendly systems. Defensive EA supports force protection, self-protection and OPSEC efforts by degrading, neutralizing, or destroying an enemy's surveillance capabilities against protected units. Defensive EA systems and capabilities include—

- Expendables (flares and active decoys).
- Jammers.
- Towed decoys.
- Directed energy infrared countermeasure systems.
- Radio controlled improvised explosive device (RCIED) systems.
- Counter Unmanned Aerial Systems (C-UAS).

Electromagnetic Attack (EA) Effects

EA effects available to the commander include—

- **Destroy.** Destruction makes the condition of a target so damaged that it can neither function nor be restored to a usable condition in a timeframe relevant to the current operation. When used in the EW context, destruction is the use of EA to eliminate targeted enemy personnel, facilities, or equipment (JP 3-85).
- **Degrade.** Degradation reduces the effectiveness or efficiency of an enemy EMS-dependent system. The impact of degradation may last a few seconds or remain throughout the entire operation (JP 3-85).
- **Disrupt.** Disruption temporarily interrupts the operation of an enemy EMS dependent system (JP 3-85).
- **Deceive.** Deception measures are designed to mislead the enemy by manipulation, distortion, or falsification of evidence to induce them to react in a manner prejudicial to their interests. Deception in an EW context presents enemy operators and higher-level processing functions with erroneous inputs, either directly through the sensors themselves or through EMS-based networks such as voice communications or data links (JP 3-85).

II. Spectrum Management

Ref: FM 3-12, Cyberspace Operations and Electromagnetic Warfare (Aug '21), p. 1-34 to 1-35.

Spectrum management is the operational, engineering, and administrative procedures to plan, coordinate, and manage use of the electromagnetic spectrum and enables cyberspace, signal and EW operations. Spectrum management includes frequency management, host nation coordination, and joint spectrum interference resolution. Spectrum management enables spectrum-dependent capabilities and systems to function as designed without causing or suffering unacceptable electromagnetic interference. Spectrum management provides the framework to utilize the electromagnetic spectrum in the most effective and efficient manner through policy and procedure.

Electromagnetic Interference (EMI)
Electromagnetic interference is any electromagnetic disturbance, induced intentionally or unintentionally, that interrupts, obstructs, or otherwise degrades or limits the effective performance of electronics and electrical equipment (JP 3-13.1). It can be induced intentionally, as in some forms of EW, or unintentionally, because of spurious emissions and responses, intermodulation products, and other similar products.

Frequency Interference Resolution
Interference is the radiation, emission, or indication of electromagnetic energy (either intentionally or unintentionally) causing degradation, disruption, or complete obstruction of the designated function of the electronic equipment affected. The reporting end user is responsible for assisting the spectrum manager in tracking, evaluating, and resolving interference. Interference resolution is performed by the spectrum manager at the echelon receiving the interference. The spectrum manager is the final authority for interference resolution. For interference affecting satellite communications, the Commander, Joint Functional Component Command for Space is the supported commander and final authority of satellite communications interference.

Spectrum Management Operations (SMO)
SMO are the interrelated functions of spectrum management, frequency assignment, host nation coordination, and policy that together enable the planning, management, and execution of operations within the electromagnetic operational environment during all phases of military operations. The SMO functional area is ultimately responsible for coordinating EMS access among civil, joint, and multinational partners throughout the operational environment. The conduct of SMO enables the commander's effective use of the EMS. The spectrum manager at the tactical level of command is the commander's principal advisor on all spectrum related matters.

Electromagnetic Warfare Coordination
The spectrum manager should be an integral part of all EW planning. The SMO assists in the planning of EW operations by providing expertise on waveform propagation, signal, and radio frequency theory for the best employment of friendly communication systems to support the commander's objectives. The advent of common user "jammers" has made this awareness and planning critical for the spectrum manager. In addition to jammers, commanders and staffs must consider non-lethal weapons that use electromagnetic radiation. Coordination for EW will normally occur in the CEMA section. It may occur in the EW cell if it is operating under a joint construct or operating at a special echelon.

B. Electromagnetic Protection (EP)

Ref: FM 3-12, Cyberspace Operations and Electromagnetic Warfare (Aug '21), pp. 2-11 to 2-13.

Electromagnetic protection is the division of electromagnetic warfare involving actions taken to protect personnel, facilities, and equipment from any effects of friendly, neutral, or enemy use of the electromagnetic spectrum that degrade, neutralize, or destroy friendly combat capability (JP 3-85). EP measures eliminate or mitigate the negative impact resulting from friendly, neutral, enemy, or naturally occurring EMI.

Electromagnetic Protection (EP) Tasks

Adversaries are heavily invested in diminishing our effective use of the electromagnetic spectrum. It is crucial we understand the enemy threat and our vulnerabilities to our systems, equipment and personnel. Effective EP measures will minimize natural phenomena and mitigate the enemy's ability to conduct ES and EA actions against friendly forces successfully.

EP tasks include—
- Electromagnetic environmental effects deconfliction.
- Electromagnetic compatibility.
- Electromagnetic hardening.
- Emission control.
- Electromagnetic masking.
- Preemptive countermeasures.
- Electromagnetic security.
- Wartime reserve modes.

Electromagnetic Environmental Effects Deconfliction.

Electromagnetic vulnerability is the characteristics of a system that cause it to suffer a definite degradation (incapability to perform the designated mission) as a result of having been subjected to a certain level of electromagnetic environmental effects (JP 3-85). Any system operating in the EMS is susceptible to electromagnetic environmental effects. Any spectrum-dependent device exposed to or having electromagnetic compatibility issues within an EMOE may result in the increased potential for such electromagnetic vulnerability as safety, interoperability, and reliability issues. Electromagnetic vulnerability manifests when spectrum-dependent devices suffer levels of degradation that render them incapable of performing operations when subjected to electromagnetic environmental effects.

Electromagnetic compatibility, EMS deconfliction, electromagnetic pulse, and EMI mitigation reduce the impact of electromagnetic environmental effects. Recognizing the different types of electromagnetic radiation hazards allows planners to use appropriate measures to counter or mitigate electromagnetic environmental effects. Electromagnetic radiation hazards include— hazards of electromagnetic radiation to personnel, hazards of electromagnetic radiation to ordnance, and hazards of electromagnetic radiation to fuels. Electromagnetic environmental effects can also occur from natural phenomena such as lightning and precipitation static.

Electromagnetic Compatibility.
Electromagnetic compatibility is the ability of systems, equipment, and devices that use the electromagnetic spectrum to operate in their intended environments without causing or suffering unacceptable or unintentional degradation because of electromagnetic radiation or response (JP 3-85). The CEMA spectrum manager assists the G-6 or S-6 spectrum manager with implementing electromagnetic compatibility to mitigate electromagnetic vulnerabilities by applying sound spectrum planning, coordination, and management of the EMS. Operational forces have minimal ability to mitigate

electromagnetic compatibility issues. Instead, they must document identified electromagnetic compatibility issues so that the Service component program management offices may coordinate the required changes necessary to reduce compatibility issues.

Electromagnetic Hardening. Electromagnetic hardening consists of actions taken to protect personnel, facilities, and/or equipment by blanking, filtering, attenuating, grounding, bonding, and/or shielding against undesirable effects of electromagnetic energy (JP 3-85). Electromagnetic hardening can protect friendly spectrum-dependent devices from the impact of EMI or threat EA such as lasers, high-powered microwave, or electromagnetic pulse. An example of electromagnetic hardening includes installing electromagnetic conduit consisting of conductive or magnetic materials to shield against undesirable effects of electromagnetic energy.

Emission Control. Emission control is the selective and controlled use of electromagnetic, acoustic, or other emitters to optimize command and control capabilities while minimizing, for operations security: a. detection by enemy sensors, b. mutual interference among friendly systems, and/or c. enemy interference with the ability to execute a military deception plan (JP 3-85). emission control enables OPSEC by—
- Decreasing detection probability and countering detection range by enemy sensors.
- Identifying and mitigating EMI among friendly spectrum-dependent devices
- Identifying enemy EMI that allows execution of military deception planning.

Emission control enables electromagnetic masking by integrating intelligence, and EW to adjust spectrum management and communications plans. A practical and disciplined emission control plan, in conjunction with other EP measures, is a critical aspect of good OPSEC. *Refer to ATP 3-13.3 for OPSEC techniques at division and below.*

Electromagnetic Masking. Electromagnetic masking is the controlled radiation of electromagnetic energy on friendly frequencies in a manner to protect the emissions of friendly communications and electronic systems against enemy electromagnetic support measures/signals intelligence without significantly degrading the operation of friendly systems (JP 3-85). Electromagnetic masking disguises, distorts, or manipulates friendly electromagnetic radiation to conceal military operations information or present false perceptions to adversary commanders. Electromagnetic masking is an essential component of military deception, OPSEC, and signals security.

Preemptive Countermeasures. Countermeasures consist of that form of military science that, by the employment of devices and/or techniques, has as its objective the impairment of the operational effectiveness of enemy activity (JP 3-85). Countermeasures can be passive (non-radiating or reradiating electromagnetic energy) or active (radiating electromagnetic energy) and deployed preemptively or reactively. Preemptive deployment of passive countermeasures are precautionary procedures to disrupt an enemy attack in the EMS through the use of passive devices such as chaff which reradiates, or the use of radio frequency absorptive material which impedes the return of the radio frequency signal.

Electromagnetic Security. Electromagnetic security is the protection resulting from all measures designed to deny unauthorized persons information of value that might be derived from their interception and study of noncommunications electromagnetic radiation (e.g., radar) (JP 3-85). Changing the modulation and characteristics of electromagnetic frequencies used for radars make it difficult for a threat to intercept and study radar signals.

Wartime Reserve Modes. Wartime reserve modes are characteristics and operating procedures of sensor, communications, navigation aids, threat recognition, weapons, and countermeasure systems that will contribute to military effectiveness if unknown to or misunderstood by opposing commanders before they are used, but could be exploited or neutralized if known in advance (JP 3-85). Wartime reserve modes are held deliberately in reserve for wartime or emergency use.

C. Electromagnetic Support (ES)

Electromagnetic support refers to the division of electromagnetic warfare involving actions tasked by, or under the direct control of, an operational commander to search for, intercept, identify, and locate or localize sources of intentional and unintentional radiated electromagnetic energy for immediate threat recognition, targeting, planning, and conduct of future operations (JP 3-85). In multi-domain operations, commanders work to dominate the EMS and shape the operational environment by detecting, intercepting, analyzing, identifying, locating, and affecting (deny, degrade, disrupt, deceive, destroy, and manipulate) adversary electromagnetic systems that support military operations. Simultaneously, they also work to protect and enable U.S. and Allied forces' freedom of action in and through the EMS.

The purpose of ES is to acquire adversary combat information in support of a commander's maneuver plan. Combat information is unevaluated data, gathered by or provided directly to the tactical commander which, due to its highly perishable nature or the criticality of the situation, cannot be processed into tactical intelligence in time to satisfy the user's tactical intelligence requirements (JP 2-01). Combat information used for planning or conducting combat operations, to include EA missions, is acquired under Command authority; however, partner nation privacy concerns must be taken into account. Decryption of communications is an exclusively SIGINT function and may only be performed by SIGINT personnel operating under Director, National Security Agency and Chief, National Security Service SIGINT operational control (DODI O-3115.07).

ES supports operations by obtaining EMS-derived combat information to enable effects and planning. Combat information is collected for immediate use in support of threat recognition, current operations, targeting for EA or lethal attacks, and support the commander's planning of future operations. Data collected through ES can also support SIGINT processing, exploitation, and dissemination to support the commander's intelligence and targeting requirements and provide situational understanding. Data and information obtained through ES depend on the timely collection, processing, and reporting to alert the commander and staff of potential critical combat information.

Electromagnetic Support (ES) Tasks

When conducting electromagnetic support, commanders employ EW platoons located in the brigade, combat team (BCT) military intelligence company (MICO) to support with information collection efforts, survey of the EMS, integration and multisource analysis by providing indications and warning, radio frequency direction finding and geolocation of threat emissions.

ES tasks include—

- Electromagnetic Reconnaissance.
- Threat Warning.
- Direction finding.

* Electromagnetic Warfare Reprogramming

Electromagnetic warfare reprogramming is the deliberate alteration or modification of electromagnetic warfare or target sensing systems, or the tactics and procedures that employ them, in response to validated changes in equipment, tactics, or the electromagnetic environment (JP 3-85). The purpose of EW reprogramming is to maintain or enhance the effectiveness of EW and targeting sensing systems. EW reprogramming includes changes to EW and targeting sensing software (TSS) equipment such as self-defense systems, offensive weapons systems, and intelligence collection systems. EW consists of three distinct divisions: EA, EP, and ES, which are supported by EW reprogramming activities.

For more information on EW reprogramming, refer to FM 3-12, app. F.

VII. Space Operations

Ref: JP 3-14 (w/Chg 1), Space Operations (Oct '20).

Access to space is vital to the collective security of the United States and its allies and partners. The Department of Defense (DOD) space policy is focused on deterring adversaries, defending against threats, and pursuing resilient space architectures that contribute to achieving space mission assurance and objectives. Further, the United States must sustain the ability to attribute malicious or irresponsible actions that jeopardize the viability of space for all. Sustained space access is vital to the collective security of the United States and its allies and partners.

Space Domain

The space domain is the area above the altitude where atmospheric effects on airborne objects become negligible. United States Space Command (USSPACECOM) area of responsibility (AOR) is the area surrounding the Earth at altitudes equal to, or greater than, 100 kilometers (54 nautical miles) above mean sea level. Like the air, land, and maritime domains, space is a physical domain within which military, civil, and commercial activities are conducted. The relationship between space and cyberspace is unique in that many space operations depend on cyberspace, and a critical portion of cyberspace can only be provided via space operations.

Proper planning and execution of military operations in space enables activities such as intelligence collection; early warning; environmental monitoring; satellite communications (SATCOM); and positioning, navigation, and timing (PNT). Activities conducted in space support freedom of action throughout the operational environment (OE), and operations in other domains may create effects in space.

I. Space Operations

Space operations are those operations impacting or directly utilizing space-and ground-based capabilities to enhance the potential of the United States and multinational partners. Joint space forces are the space and terrestrial systems, equipment, facilities, organizations, and personnel, or combination thereof, necessary to conduct space operations. Space systems consist of three related segments: space, link, and ground.

- The **ground segment** consists of ground-based facilities and equipment supporting command and control (C2) of space segment resources, as well as ground-based processing equipment, Earth terminals or user equipment, space situational awareness (SSA) sensors, and the interconnectivity between the facilities in which this equipment is housed.

- The **link segment** consists of signals connecting ground and space segments through the electromagnetic spectrum (EMS). This link normally includes telemetry, tracking, and commanding (TT&C) signals necessary for controlling the spacecraft and payload. Separate from the TT&C signals, the satellite payload may contribute to the link segment through the use of SATCOM signals between two terminals on the ground or a PNT signal enhancing air, ground, and naval maneuver.

- The **space segment** involves the operational spacecraft within the space domain.

II. Space Capabilities
Ref: JP 3-14 (w/Chg 1), Space Operations (Oct '20), chap. 2.

Due to the complexities of the operational environment (OE) and the required integration and coordination between elements of the joint force, a shared understanding of selected aspects of specific space capabilities is essential to foster and enhance unified action.

Space Situational Awareness (SSA)
Space situational awareness (SSA) is the requisite foundational, current, and predictive knowledge and characterization of space objects and the OE upon which space operations depend—including physical, virtual, information, and human dimensions—as well as all factors, activities, and events of all entities conducting, or preparing to conduct, space operations. Space surveillance capabilities include a mix of space-based and ground-based sensors. SSA is dependent on integrating space surveillance, collection, and processing; environmental monitoring; status of US and cooperative satellite systems; understanding of US and multinational space readiness; and analysis of the space domain.

Space Control
Space control includes offensive space control and defensive space control operations to ensure freedom of action in space and, when directed, defeat efforts to interfere with or attack US or allied space systems. Space control uses a broad range of response options to provide continued, sustainable use of space. Space control contributes to space deterrence by employing a variety of measures to assure the use of space; attributing enemy attacks; and being consistent with the right to self-defense, target-threat space capabilities.

See following page (p. 3-64) for further discussion of space control and superiority.

Positioning, Navigation, and Timing (PNT)
Military users depend on assured positioning, navigation, and timing (PNT) systems for precise and accurate geo-location, navigation, and time reference services. PNT information, whether from space-based global navigation satellite systems (GNSSs), such as Global Positioning System, or non-GNSS sources, is considered mission-essential for virtually every modern weapons system.

Intelligence, Surveillance, Reconnaissance
Space-based intelligence collection synchronizes and integrates sensors, assets, and systems for gathering data and information on an object or in an area of interest on a persistent, event-driven, or scheduled basis. Space-based intelligence, surveillance, and reconnaissance, which includes overhead persistent infrared (OPIR), is conducted by an organization's intelligence collection manager to ensure integrated, synchronized, and deconflicted operations of high-demand assets.

Satellite Communications (SATCOM)
Satellite communications (SATCOM) systems inherently facilitate beyond line-of-sight connectivity. Depending on its configuration, a robust SATCOM architecture provides either equatorial coverage (nonpolar) or high-latitude coverage (includes poles). This provides national and strategic leadership with a means to maintain situational awareness and convey their intent to the operational commanders responsible for conducting joint operations.

Environmental Monitoring
Terrestrial environmental monitoring provides information on meteorological and oceanographic factors that affect military operations. Space environmental monitoring provides data that supports forecasts, alerts, and warnings for the space environment that may

affect space capabilities, space operations, and their terrestrial users. Environmental monitoring support to joint operations gives the JFC awareness of the OE.

Missile Warning
The missile warning mission uses a mix of OPIR and ground-based radars. Missile warning supports the warning mission executed by North American Aerospace Defense Command to notify national leaders of a missile attack against North America, as well as attacks against multinational partners (via shared early warning) in other geographic regions. It also includes notification to combatant commands (CCMDs), multinational partners, and forward-deployed personnel of missile attack and the assessment of a missile attack if the applicable CCMD or multinational partner is unable to do so.

Nuclear Detonation Detection
Nuclear detonation detection capabilities provide persistent, global, and integrated sensors to provide surveillance coverage of critical regions of the globe and provide warning and assessment recommendations to the President, Secretary of Defense (SecDef), and CCDRs, indicating place, height of burst, and yield of nuclear detonations.

Spacelift
Spacelift is the ability to deliver payloads (satellites or other materials) into space.

Satellite Operations
Satellite operations maneuver, configure, operate, and sustain on-orbit spacecraft. In a conflict, satellite operations are critical to the command and control (C2), movement and maneuver, protection, and sustainment of space capabilities.

Space Operations & the Joint Functions

Space-based intelligence collection supports **C2** by providing information used to develop a shared understanding of the threat. A large percentage of the intelligence required to make decisions for employment of forces is obtained from spacecraft.

Spacecraft complement non-space-based **intelligence** sources by providing decision makers with timely, accurate data for information that can create a decisive advantage across the competition continuum.

Space operations support air, land, maritime, and cyberspace **fires** through intelligence, PNT, and communications capabilities. Space operations movement and maneuver include the deployment, repositioning, or re-orientation of on-orbit assets and joint space forces. These movements may support service optimization, protection from environmental hazards, passive defense from threats, or the positioning of assets to enable active defensive or offensive measures.

Protection in space operations includes all measures taken to ensure friendly space systems perform as designed by overcoming attempts to deny or manipulate them. Protection includes all measures to passively neutralize or mitigate threats and man-made and/or environmental hazards, to include enemy attack, terrestrial weather, space weather, on-orbit conjunctions, and non-hostile EMI.

Space operations **sustainment** is conducted through spacelift, satellite operations, force reconstitution, maintenance of a force of space operations personnel, and support to human space flight. Spacelift includes launch systems, launch facilities, and ground personnel capable of placing satellites on orbit.

Space supports the flow of information and decision making. It may also serve as an activity essential to the delivery of specific information in the information environment.

See p. 2-6 for further discussion of the joint functions from JP 3-0.

III. Unity of Effort

Ref: JP 3-14 (w/Chg 1), Space Operations (Oct '20), pp. I-3 to I-4.

Synergy throughout the OE is necessary for effective joint operations. Every joint operation requires synchronization of capabilities throughout the OE to support the commander's intent and CONOPS. Space capabilities must be thoroughly integrated into every aspect of joint planning long before operations begin. Lack of integration increases the fog and friction of war.

Joint forces rely on space capabilities such as weather data; missile warning; PNT; SSA; intelligence, surveillance, and reconnaissance (ISR); and SATCOM. As with the littorals, the air and space domains also have a transitional region as the Earth's atmosphere and effects of gravity taper at increasing altitudes. Similar to air, land, and maritime operations and forces, space operations and forces are interconnected with cyberspace through the EMS.

While space operations and cyberspace operations (CO) are distinct, they are interdependent. Operations in space enable many CO, and control of space systems' segments require use of cyberspace. Cyberspace provides a means for satellite control and spacecraft data transport. The Department of Defense information network (DODIN) end-to-end connectivity and security under CO treats SATCOM payloads like a communications transport medium, much like copper wire, microwave links, or fiber optics do in other physical domains. The transport layer is critical, and the linkages must be addressed during planning and operations to ensure cyberspace concerns are met.

Space Control

Joint space forces conduct space control operations to ensure freedom of action in space for the United States and its allies and partners and, when directed, to deny an adversary freedom of action in space. The purpose of these operations is to achieve space superiority.

Space Superiority

Space superiority is the degree of control in space of one force over any others that permits the conduct of its operations at a given time and place without prohibitive interference from terrestrial and space-based threats. The purpose and value of space superiority is to provide the freedom of action in space in the pursuit and defense of national security interests. The US ability to capitalize on and protect space systems, and to counter enemy capabilities, contributes to US space superiority. To establish and maintain space superiority, commanders require resilient space forces that have the skill and the experience to protect and defend space systems across the competition continuum and to deny the same to the enemy. Commanders should understand how to request space forces and determine whether the capabilities are continuously available or know when special authorization and coordination are required. The use of offensive and defensive operations in multiple portions of the OE may be necessary to maintain space superiority. To maintain space superiority, joint forces must have SSA and the knowledge, training, resources, and authorities needed to defend our ability to use space. This includes being prepared to prevent the threat from exploiting space against the United States and its allies and partners. Commander, United States Space Command (CDRUSSPACECOM), is responsible for establishing space superiority and may request specific contributions from other CCDRs.

Benefits from Access to Space

Space operations support and enable the joint force to conduct operations. These capabilities come from the unique characteristics of space, including a global perspective and lack of overflight restrictions, as well as the rapid revisit times provided by low Earth orbit (LEO) spacecraft and the persistence afforded by geosynchronous satellites. Space capabilities provide CCDRs with near-worldwide coverage and access to otherwise denied areas. Commanders should account for specific space characteristics to plan and operate effectively.

There are several distinct advantages to using space for operational purposes. Despite the challenges stemming from competition in space, these advantages are commonly applicable to space operations, whether military, civil, or commercial.

Freedom of Action
Since early in the Space Age, the United States has led the world in space capability and capacity, thereby achieving overmatch in space and significant freedom of action. However, competition from adversaries is beginning to limit US freedom of action.

Overflight
International law does not extend a nation's territorial boundaries into space. Unlike the rules for aircraft overflight, there are no overflight restrictions for spacecraft in outer space. Therefore, space-faring nations benefit from unrestricted space overflight. This characteristic makes space-based ISR, remote sensing, SATCOM, and PNT more responsive than terrestrial alternatives.

Global Perspective
Space has been characterized as "the ultimate high ground." Orbiting the Earth in 90 minutes, LEO satellites have fields of view spanning hundreds of miles. Geosynchronous Earth orbit (GEO) satellites can view 42% of the Earth's surface area. Space affords a global vantage point from which to assess large swaths of the land, oceans, and air for strategic-, operational-, and tactical-level applications.

Responsiveness
Space operations provide the ability to surge some types of capabilities, such as communications or ISR, on much faster timescales than ground-based or airborne capabilities. As priorities change, some space resources can be rapidly reallocated to the areas where they are needed most. As an example, in the event of increased operating tempo or loss of a satellite, available SATCOM bandwidth can be quickly reallocated to meet the highest-priority requirements. Such retaskings may result in a degradation of system performance and/or life span due to expenditure of propellant.

Multi-User Capacity
Space operations typically support multiple users and, in some cases, such as PNT, can provide service to an unlimited number of users. The joint force can have access to shared effects generated in space nearly anywhere on the globe, in near real time.

Speed, Reach, and Persistence
A spacecraft's orbital parameters (e.g., velocity, distance, and inclination) enable satellites not only to overfly vast areas in very short periods but also to enable continuous operation, creating effects at great distances, with persistent coverage.

IV. Army Space Capabilities

Ref: FM 3-14, Space Operations (Oct '19), p. 1-4 to 1-10.

All Services contribute to and use the body of knowledge residing within the ten codified space capabilities. Some capabilities are provided by other Services and Agencies, but they are required to support Army operations and contribute to the success of Army missions. Army space operations may receive space-related intelligence and environmental monitoring products through Army intelligence channels, directly from other Services, centers, or agencies.

The distinct space capabilities, effects, and products used by the Army, joint, allied forces, and partner nations are planned, developed, prepared, and made available to the force by Soldiers conducting Army space operations and space-enabled operations. The Soldiers conducting space operations and space-enabled operations may be assigned to space operations, signal, cyber, electronic warfare, intelligence operations, and other military operations specialties. Not all Army Soldiers who configure and use equipment reliant on space capabilities are designated as space operators. However, these Soldiers are instrumental to the Army's critical use of space capabilities in unified land operations.

Army space-related activities include:

Army Space Operations

Army space operations, duties, and responsibilities are centered on these eight codified joint space capabilities:

- Space situational awareness (SSA)
- Positioning, navigation, and timing (PNT)
- Space control - defensive space control (DSC), Offensive Space Control (OSC), and Navigation Warfare (NAVWAR).
- Satellite communications (SATCOM)
- Satellite operations
- Missile warning
- Environmental monitoring
- Space-based intelligence, surveillance, and reconnaissance

There are two other codified joint space capabilities—nuclear detonation detection and spacelift—but the Army has no involvement with those.

Army Space-Enabled Operations

Army space-enabled operations are not specifically codified in joint doctrine as space capabilities, but are combined, derived, or second order tasks and actions enabled by space capabilities. These include, but are not limited to,

- Joint friendly force tracking (FFT)
- Network transport of Department of Defense information network
- Commercial imagery
- National Reconnaissance Office overhead systems
- Army tactical exploitation of national capabilities (TENCAP) program
- National-to-Theater program interfaces
- Geospatial intelligence
- Integrated broadcast service
- Common interactive broadcast

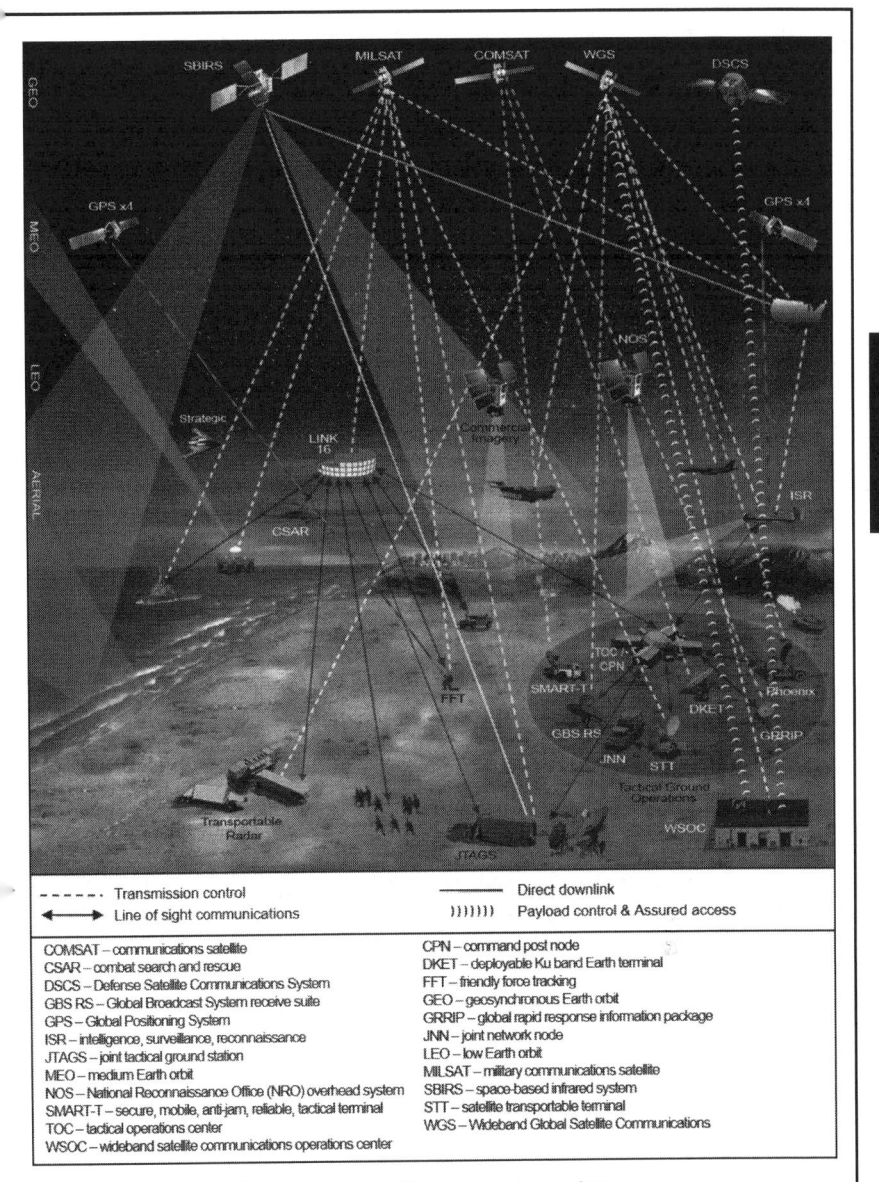

Ref: FM 3-14, fig. 1-1. Army space operations concept overview.

Space operations bring essential capabilities with unique tools to influence, enable, and enhance all mission areas in unified land operations. Unit commanders must have a clear understanding of the space capabilities available that contribute to mission operations and how best to utilize those assets.

The Army leverages space capabilities to support unified land operations from large scale combat operations to individual Soldiers at the tactical level of warfare. Space capabilities enhance the Army's ability to communicate, navigate, accurately target the enemy, protect and sustain our forces, and enable intelligence preparation of the battlefield (IPB).

V. Combined Space Tasking Order (CSTO)

Ref: JP 3-14 (w/Chg 1), Space Operations (Oct '20), p. IV-7.

Specific to space operations, the CFSCC produces plans/orders for the management of assigned space forces through the CSTO. The CSTO and special instructions (SPINS) direct space forces, assign tasks to meet joint force operational objectives, and synchronize space operations with other CCMD operations (see Figure IV-1).

The operational planning cycle includes inputs into the joint targeting cycle, as depicted in Figure IV-1. The space operations directive captures the CFSCC's guidance and intent. The space operations directive conveys prioritization and apportionment guidance focused on the applicable execution period. This is then used to form the master space plan. The master space plan is used to allocate resources to each desired effect and serves as the source to generate unit tasking and coordination within the CSTO and SPINS. The CSTO tasks execution and the SPINS provide amplifying guidance.

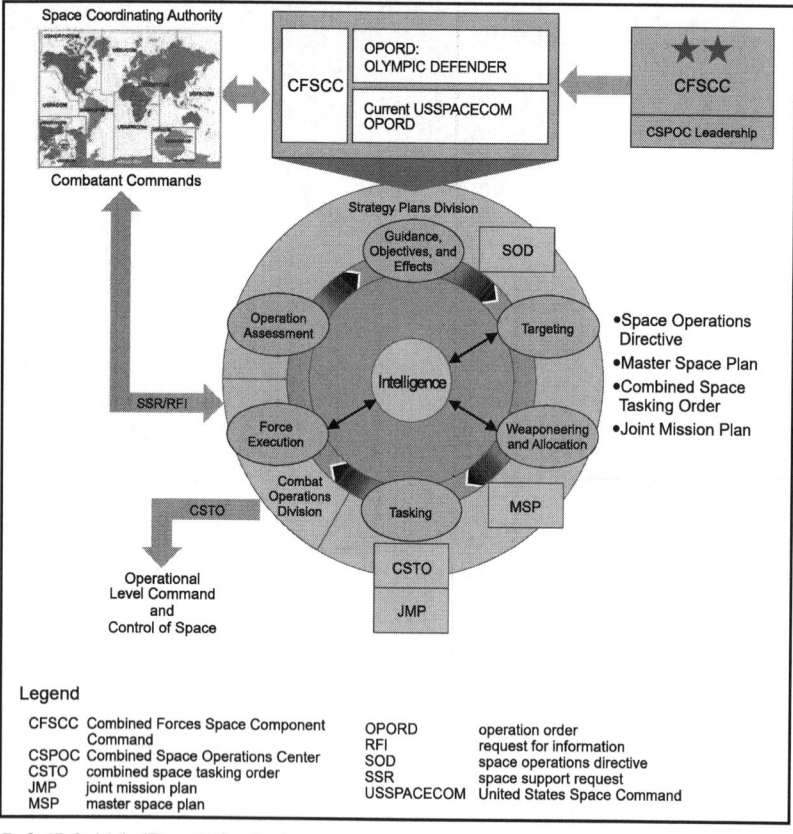

Ref: JP 3-14 (w/Chg 1), fig. IV-1. Combined Space Tasking Order Process.

The planning process may significantly compress during a crisis or to support major combat operations. In periods of conflict, the CSTO cycle may compress from a 30-day production cycle to synchronize with the supported CCDR's air tasking order cycle. The CSTO transmits the CFSCC's guidance and priorities for a short-duration timeframe, assigns tasks to meet operational objectives, and, when required, synchronizes and integrates CFSCC activities with other CCMD operations.

VIII. Additional Capabilities

Ref: * FM 3-13, Information Operations (Dec '16). (*See note p. 1-2.)

In addition to the specific IRCs covered on the previous pages, FM 3-13 discusses additional capabilities as outlined below.

Additional Capabilities

- **A** Integrated Joint Special Technical Operations (IJSTO)
- **B** Special Access Programs (SAP)
- **C** Personnel Recovery (PR)
- **D** Physical Attack
- **E** Physical Security
- **F** Presence, Profile, and Posture (PPP)
- **G** Soldier and Leader Engagement (SLE)
- **H** Police Engagement
- **I** Social Media

All unit operations, activities, and actions affect the information environment. Even if they primarily affect the physical dimension, they nonetheless also affect the informational and cognitive dimensions. For this reason, whether or not they are routinely considered an IRC, a wide variety of unit functions and activities can be adapted for the purposes of conducting information operations or serve as enablers to its planning, execution, and assessment.

See p. 3-1 for additional discussion.

The following additional capabilities are listed alphabetically, in part to reinforce the idea that they are co-equal in their potential contribution to the scheme of IO. Every IRC has one characteristic in common: a representative of each capability is a member of the IO working group. Some are habitual or core members while others attend on an as-needed basis. In either case, their participation is governed by the mission, current situation, and commander's discretion.

A. Integrated Joint Special Technical Operations (IJSTO)

Integrated joint special technical operations (IJSTO) are classified operations that harness specialized technical capabilities to gain a decisive advantage over an enemy or adversary. These technical capabilities can be information-related or, in some way, complement IO efforts. Therefore, IJSTO and IO must be deconflicted and synchronized through close coordination, primarily through the IO working group. According to JP 3-13, detailed information about IJSTO and its contribution to IO can be obtained from IJSTO planners at combatant command or Service component headquarters.

B. Special Access Programs (SAP)

Special access programs (known as SAPs) are sensitive acquisition, intelligence, or operations and support programs that impose need-to-know and access controls beyond those normally provided for access to confidential, secret, or top secret information (see DODD 5205.07 for more information on special access programs). As with IJSTO, detailed information related to special access programs can be obtained, when authorized, from the designated representative at combatant command or Service component headquarters.

C. Personnel Recovery (PR)

The core principle of Army personnel recovery (PR) is to recover isolated personnel before detention or capture through a systems-based approach that features proactive, integrated, rehearsed, and resourced measures and capabilities. To fulfill this principle, the Army has an obligation to train, equip, and protect its personnel (Soldier, DA Civilian, and contractor), prevent their capture and exploitation by adversaries, and reduce the potential for using isolated personnel as leverage against U.S. security objectives and national interests.

Note. PR did not appear in FM 3-13, 1 Dec 2017, but is added here, reinforcing the fact that any capability can serve as an IRC if it is affected by or affects the information environment.

D. Physical Attack *(See p. 1-50.)*

When synchronized as a part of information operations, physical attack—which includes physical maneuver, destruction, and lethal action—is the application of combat power to create desired effects in the information environment. Carefully applied force can play a major role in intimidation and deterrence and in obstructing a threat's ability to exercise command and control. It may include direct and indirect fires from ground, sea, and air platforms and direct actions by special operations forces. IO applications of physical attack to consider include—

- Preventing or degrading adversary reconnaissance and surveillance.
- Conducting physical attacks as deception events.
- Degrading the enemy's ability to process information.
- Degrading the enemy's ability to jam communications.
- Destroying command and control and communications systems.
- Reducing the enemy's ability to penetrate mission command systems.

As the list reveals, physical attack typically supports or complements other capabilities such as military deception, electronic warfare, or cyberspace operations.

When applying physical attack as a component of IO, consideration of second-and third-order effects, as well as consequence management, is a must. Total, or even partial, destruction of threat systems or capabilities or of indigenous capabilities co-opted by an enemy or adversary, may not be attainable or even desirable. For example, friendly forces may need to use threat command and control systems during the postconflict phase of military operations. Additionally, destroying indigenous capabilities may create animosity among the local populace, the effects of which are greater than any advantage gained over the threat.

E. Physical Security *(See p. 1-31.)*

Physical security is that part of security concerned with physical measures designed to safeguard personnel; to prevent unauthorized access to equipment, installations, material, and documents; and to safeguard them against espionage, sabotage, damage, and theft (JP 3-0). Physical security contributes directly to IO and each of its weighted efforts, most especially efforts to defend personnel, information, and systems that contribute to friendly decision making. Information, information-based processes, and information systems—such as mission command systems, weapon systems, and information infrastructures—are protected relative to the value of the information they contain and the risks associated with the compromise or loss of information. Physical security is a unit program directed by the commander and overseen by the operations staff officer.

Refer to AR 190-13 and AR 190-16 for details on physical security.

F. Presence, Profile, and Posture (PPP)

The mere presence of a force can significantly affect all audiences in the AO. Deploying, moving, or assigning forces to the right place at the right time can add substantial credibility to messages being delivered through other channels and provide a major contribution to deterrence. Whenever Soldiers or forces leave base or cross the line of departure, they do two things: collect information and send a message. If either collection or PPP are not a deliberate, coordinated effort, both the information coming back and the message sent appear haphazard and inconsistent. PPP is always in play, and the IO officer should always provide PPP guidance on behalf of the commander.

> **Presence** is the act of being physically present, although technology is increasingly enabling virtual presence. Presence can be menacing or reassuring, depending on the situation. Absence, or the lack of presence, can create perceptions that work for or against the unit's aims. Being very conscious and deliberate about being present or absent can be a powerful form of influence and should not be left to chance. Once units determine presence is required, or no choice exists but to be present, how they convey that presence is important. Both profile and posture address the way units, patrols, and Soldiers are present.
>
> **Profile** is about the degree of presence, both in terms of quantity and quality. Quantity is reflected in how much a unit is present, as in its footprint or task organization. Quality speaks to the nature of that presence, as in its current capability, as well as its reputation.
>
> The **posture** of a unit is an expression of its attitude. Whether active or passive, threatening or non-threatening, defensive or welcoming, posture dictates how units or Soldiers appear to others and how Soldiers act towards others. For example, the decision to wear soft caps instead of Kevlar helmets and body armor can considerably affect the perceptions and actions of adversaries and the local populace.
>
> The operations officer and IO officer or representative are the focal points for PPP. All leaders and Soldiers contribute to it.

G. Soldier and Leader Engagement (SLE) *(See p. 1-43.)*

Soldier and leader engagement is defined as interpersonal Service-member interactions with audiences in an area of operations (FM 3-13). These interactions can be dynamic, such as an impromptu meeting on the street or deliberate, such as a scheduled meeting. SLE can be in-person and face-to-face or conducted at a distance, facilitated by technology.

A primary purpose of SLE is to convey approved, pre-developed messages (to support approved public affairs or MISO themes) to enhance the credibility of unit personnel and legitimacy of unit operations. Key leader engagement is a subset of SLE.

The commander is the unit's chief engager and designates a staff focal point for planning, synchronizing, and assessing SLE, whether conducted by unit personnel or other IRCs, such as civil affairs, engineering, or military police forces; chaplains or religious affairs personnel; or medical personnel.

Chaplains and religious affairs personnel conduct SLEs at the commander's direction as the commander's principle advisor on religion, ethics, morals, and morale while maintaining their noncombatant status (see ATP 1-05.03 for details on religious support). By virtue of their roles as religious leaders, chaplains' very presence in an operational area opens avenues of approach for partnership.

H. Police Engagement

Police engagement occurs in all operational environments in which military police interact with elements external to their own organization. Police engagement is an IRC that occurs among police personnel, organizations, and populations for the purpose of maintaining social order. Military police and U.S. Army Criminal Investigative Command personnel engage local, host-nation, and coalition police partners; police agencies; civil leaders; and local populations for critical police information that can influence military operations or destabilize an AO. Ultimately police engagement aims to develop a routine and reliable interpersonal network through which police information can flow to military police. Based on the tactical situation, police engagement can be formal or informal. Police engagement may be a proactive activity as part of deliberate information gathering, targeting, or collection, or it can be conducted as a reactive response to an episodic event.

Refer to FM 3-39 for military police operations.

I. Social Media

The information environment spotlights the growing impact of social media. Although not listed in Table 3-1 because it is still an emergent IRC, social media has the potential to become a powerful capability for IO. Some possible applications include—

- Social media as a media channel, such as radio, newspapers, and television.
- Social media as an interactive medium for exerting influence.
- Social media as a means to communicate with an established network or networks.
- Social media as a near real-time sensor-to-sensor network.

Social media is rapidly expanding beyond the realm of public affairs, IO, or intelligence functions and becoming an integral component of operations, particularly those occurring in and through the information environment. Even as the institutional Army explores force modernization aspects of social media—such as doctrine, organizations, personnel, and training— commanders and staffs need to understand social media's impact and incorporate this understanding into planning and operations.

Chap 4
(Information) PLANNING

Ref: * FM 3-13, Information Operations (Dec '16), pp. 4-1 to 4-2. (*See note p. 1-2.)
See p. 1-60 for discussion of planning as related to information advantage (ADP 3-13) and p. 2-45 to 2-51 as related to operations in the information environment (JP 3-04).

Planning is the art and science of understanding a situation, envisioning a desired future, and laying out effective ways of bringing that future about (ADP 5-0). Planning helps commanders create and communicate a common vision between commanders, their staffs, subordinate commanders, and unified action partners. Planning results in a plan and orders that synchronize the action of forces in time, space, and purpose to achieve objectives and accomplish missions.

Commanders, supported by their staffs, ensure IO is fully integrated into the plan, starting with Army design methodology (ADM) and progressing through the military decisionmaking process (MDMP). The focal point for IO planning is the IO officer (or designated representative for IO). However, the entire staff contributes to planning products that describe and depict how IO supports the commander's intent and concept of operations. The staff also contributes to IO planning during IO working group meetings to include assessing the effectiveness of IO and refining the plan.

Commanders, supported by their staffs, ensure IO is fully integrated into the plan, starting with Army design methodology and progressing through the military decisionmaking process.

Army Design Methodology (ADM)
ADM helps commanders and staffs with the conceptual aspects of planning. These aspects include understanding, visualizing, and describing operations to include framing the problem and identifying an operational approach to solve the problem.

Military Decisionmaking Process (MDMP)
The MDMP helps commanders and staffs translate the commander's vision into an operations plan or operations order that synchronizes the actions of the force in time, space, and purpose to accomplish missions. Both the problem the commander needs to solve and the specific operation to advance towards its solution have significant information-related aspects.

See pp. 4-3 to 4-16 for discussion of commander, staff, and IO working group responsibilities for synchronizing information-related capabilities.

Planning activities occupy a continuum ranging from conceptual to detailed. **Conceptual planning** involves understanding operational environments and problems, determining the operation's end state, and visualizing an operational approach to attain that end state. **Detailed planning** translates the commander's operational approach into a complete and practical plan.

Refer to BSS7: The Battle Staff SMARTbook, 7th Ed., updated for 2023 to include FM 5-0 w/C1 (2022), FM 6-0 (2022), FMs 1-02.1/.2 (2022), and more. Focusing on planning & conducting multidomain operations (FM 3-0), BSS7 covers the operations process; commander/ staff activities; the five Army planning methodologies; integrating processes (IPB, information collection, targeting, risk management, and knowledge management); plans and orders; mission command, command posts, liaison; rehearsals & after action reviews; operational terms & military symbols.

A. IO & Army Design Methodology (ADM)

ADM is a methodology for applying critical and creative thinking to understand, visualize, and describe unfamiliar problems and approaches to solving them (ADP 5-0). By first framing an operational environment and associated problems, ADM enables commanders and staffs to think about the situation in depth. From this understanding, commanders and staffs develop a more informed approach to solve or manage identified problems. During operations, ADM supports organizational learning through reframing—a maturing of understanding that leads to a new perspective on problems or their resolution.

Problems typically facing Army forces and unified action partners, within a given area of operations, are human-centered. Human problems are driven by human decision making, which can be affected directly or indirectly through the use of IRCs, including effects produced by movement and maneuver. Therefore, the most essential part of ADM from an IO perspective is framing the current state of the information environment to determine key decision makers and the ways by which their decision process can be altered. This analysis identifies and creates understanding of decision makers' beliefs, motivations, grievances, biases, and preferred ways of communicating and obtaining information.

Framing the current state and desired future state of the information environment are key aspects of framing an operational environment and developing an operational approach. The operational approach provides a guide for more detailed IO planning, to include determining the effects necessary to bring about the desired end state in the information environment and the required combinations of IRCs needed to produce these effects.

Commanders typically employ a combination of direct and indirect approaches to defeating the enemy. A direct approach attacks the threat's center of gravity or principal strength by applying combat power against it. An indirect approach attacks the enemy's center of gravity by applying combat power against a series of decisive points that iteratively lead to the defeat of the center of gravity while avoiding the enemy's strengths. IO contributes to both approaches, especially when the threat's center of gravity or principal strength is information-related.

Refer to BSS7: The Battle Staff SMARTbook, 7th Ed. for a comprehensive discussion of various techniques used in framing the operational environment, framing the problem, developing an operational approach, and reframing (ATP 5-0.1, Army Design Methodology).

B. IO & the Military Decisionmaking Process (MDMP)

Commanders use the MDMP to understand the situation and mission confronting them and make informed decisions resulting in an operations plan or order for execution. Their personal interest and involvement is essential to ensuring that IO planning is integrated into MDMP from the beginning and effectively supports mission accomplishment.

IO planning is integral to several other processes, to include intelligence preparation of the battlefield (IPB) and targeting. The G-2 (S-2) and fire support representatives participate in the IO working group and coordinate with the IO officer to integrate IO with their activities and the overall operation.

See pp. 4-17 to 4-34 for discussion of information environment analysis (IO & IPB) and chapter 7 for discussion of fires and targeting.

Commanders use their mission statement for the overall operation, the IO mission statement, scheme of IO, IO objectives, and IRC tasks to describe and direct IO.

See pp. 4-35 to 4-60 for discussion of IO & the Military Decisionmaking Process (MDMP).

I. Synchronization of Info-Related Capabilities

Ref: * ATP 3-13.1, Conduct of Information Operations (Oct '18), chap. 4. (*See note p. 1-2.)

Creating effects in the information environment is not random. Units synchronize and sequence IRCs so that they actively contribute to fulfilling the unit's mission in accordance with the commander's intent and concept of operations. Mission command places responsibility for IRC synchronization on the staff; however, without the commander's direct involvement, stated intent, guidance, concept of operations, and narrative, the staff will fail to achieve desired and required operational outcomes.

I. Commanders' Responsibilities

Commanders drive the conduct of IO and are their unit's key informers and influencers. Their influence is a function of their position, authority, decisions, personal actions, and the combat power their unit generates. Every action they take, operation they lead, capability they employ, and word or image they convey sends a message. Ultimately, they have the responsibility to align and combine each message into a comprehensive and compelling narrative while ensuring their unit fulfills this narrative. Their narrative explains the why of military operations.

Commanders (and subordinate leaders) are responsible for driving the conduct of IO through their narrative, stated intent, guidance, concept of operations, and risk assessment to achieve desired and required operational outcomes.

See following pages (pp. 4-4 to 4-5) for an overview and further discussion.

II. Staff Responsibilities

The staff has responsibility for conducting IO through synchronizing IRCs. As the staff lead for IO, the IO officer or designated representative develops a range of products and chairs the IO working group. The **IO working group is the primary mechanism for synchronization** and produces several outputs that drive the unit's efforts in the information environment.

Key IO Planning Tools and Outputs

Key staff outputs include the:
- IO running estimate (*See pp. 4-6 to 4-7.*)
- Logic of the effort (*See p. 4-8.*)
- Commander's critical information requirements (CCIRs) and CCIRs and essential elements of friendly information (EEFIs) (*See p. 4-9.*)
- Combined Information Overlay (CIO) (*See pp. 4-32 to 4-33.*)

Information operations input to base orders and plans include:
- Mission Statement (*See p. 4-11.*)
- Scheme of information operations (*See pp. 4-12 to 4-13.*)
- IO Objectives & IRC tasks (*See pp. 4-14 to 4-15.*)
- IO Synchronization Matrix (*See p. 4-16.*)
- Battle drills (*See pp. 4-65 to 4-68.*)
- Other products as needed

See p. 6-3 for related discussion of the IO working group inputs and outputs (fig. 4-1) and chap. 7 for fires and targeting products.

Commander's Responsibilities (Overview)
Ref: *ATP 3-13.1, The Conduct of Information Operations (Oct '18), pp. 4-1 to 4-3.

Commanders drive the conduct of IO and are their unit's key informers and influencers. Their influence is a function of their position, authority, decisions, personal actions, and the combat power their unit generates. Every action they take, operation they lead, capability they employ, and word or image they convey sends a message. Ultimately, they have the responsibility to align and combine each message into a comprehensive and compelling narrative while ensuring their unit fulfills this narrative. Their narrative explains the why of military operations.

The why of operations comes down to establishing credibility and legitimacy. No matter the unit's mission, credibility and legitimacy are essential to success. Both credibility and legitimacy build on the Army bedrock of trust. Credible units match or align their actions with their messages (words and images). Trusted Army leaders and units fulfill commitments, are consistent in what they do, and ensure follow through. Legitimacy maintains legal and moral authority in the conduct of operations. Legitimacy, which can be a decisive factor in operations, is based on the actual and perceived legality, morality, and rightness of the actions from interested audiences' point of view. These audiences include American national leadership and domestic audiences; foreign governments, leaders, and civilian populations in the operational area; threats and adversaries; and other nations and organizations around the world.

Commanders (and subordinate leaders) are responsible for driving the conduct of IO through their:

- Narrative
- Stated intent
- Guidance
- Concept of operations
- Risk assessment

Commander's Narrative

Aligned and synchronized actions and messages help create and convey a credible narrative comprising legitimate actions. To build trust, enable unity of effort, and strengthen legitimacy, commanders, leaders, and IO professionals demonstrate their character, competence, and commitment through their decisions and actions.

A narrative is an overarching expression of context and desired results (JDN 2-13). It focuses primarily on shaping perceptions of relevant audiences in the AO. Not only does it provide rationale to audiences affected by military operations but the narrative serves as a guide to units so that their actions (deeds), words, and images appropriately align. The final result: a unit whose actions support and reinforce the narrative and ensure its consistency, viability, and effectiveness.

The IO officer plays a significant role in assisting the commander to craft the narrative. As the unit's effects coordinator for IRCs, the IO officer advises the commander on ways IRCs can affect operations and ways operations can affect the information and operational environments. An effective narrative helps shape both environments by creating or facilitating conditions favorable to the commander's intent, especially in bolstering confidence in the U.S.'s or coalition's mission and creating an alternative to the enemy's or adversary's narrative.

Commanders typically develop formal, explicit narratives at the strategic and possibly operational levels and convey them downward, within which subordinate units nest their messages, actions, and activities. Yet even the lowest-level commanders or leaders consciously envision how their units' actions, words, and images either support or confound the approved narrative. These leaders then tailor and adapt unit actions and messages to their AOs. If necessary, subordinate commanders get clarification from higher headquarters.

Commander's Intent
Commander's intent is a clear and concise expression of the purpose of the operation and the desired military end state that supports mission command, provides focus to the staff, and helps subordinate and supporting commanders act to achieve the commander's desired results without further orders, even when the operation does not unfold as planned (JP 3-0). Mission command requires commanders to convey a clear commander's intent for operations in which multiple operational and mission variables interact with the lethal application of ground combat power. Such dynamic interactions—many of which occur in the information environment—often compel subordinate commanders to make difficult decisions in unforeseen circumstances. Commander's intent is also essential for exercising disciplined initiative, which is particularly critical to executing a range of IO actions and activities. Such actions and activities can include military deception, SLE, and PPP.

Commander's Guidance (Initial and Subsequent)
Commander's planning guidance conveys the essence of the commander's visualization and may be broad or detailed. It outlines an operational approach—a broad description of the mission, operational concepts, tasks, and actions required to accomplish the mission (JP 5-0) and discusses COAs the commander initially favors from those the staff should not consider. It broadly describes when, where, and how the commander intends to employ combat power to accomplish the mission within the higher commander's intent. In terms of IO, if commanders determine that an information-related line of effort is decisive or that attaining an IO objective requires a significant lead time, they will issue relevant instructions as part of their guidance. In this guidance, they may frame their narrative and subordinating themes and messages, request information about the information environment, identify key leaders with whom they must engage, and discuss how IRCs will support COAs.

Concept of Operations
The concept of operations describes how the commander or leader envisions an operation unfolding from its start to its conclusion or end state. It determines how accomplishing each task leads to executing the next. It identifies the best ways to use available terrain (both physical and virtual) and employs unit strengths against enemy weaknesses. As a line of effort that supports the overall operation—as well as specific lines of operation or effort—IO is an essential element of any concept of operations. IO's contribution to the concept of operation is expressed in its scheme of IO

See pp. 4-12 to 4-13 for more information on scheme of IO.

Risk Assessment
Commanders and their staffs, as trusted Army professionals, incorporate ethical risk assessments in their planning and conduct of operations. These assessments seek to—
- Mitigate unnecessary risk to personnel and mission accomplishment, friendly and allied forces, and noncombatants.
- Avoid improper use of resources and assets.
- Avert decisions and actions that may produce short-term tactical benefits to operations but long-term negative strategic consequences.

A. IO Running Estimate (See also pp. 4-37.)

Ref: *ATP 3-13.1, The Conduct of Information Operations (Oct '18), pp. 4-3 to 4-6.

A running estimate is the continuous assessment of the current situation used to determine if the current operation is proceeding according to the commander's intent and if planned future operations are supportable (ADP 5-0). Running estimates help the IO officer record and track pertinent information about the information environment leading to a basis for recommendations to the commander. The IO officer uses the running estimate to assist with completion of each step of the MDMP. An effective running estimate is as comprehensive as possible within the time available but also organized so that the information is easily communicated and processed. Normally, the running estimate provides enough information to draft the applicable IO sections of warning orders as required during planning and, ultimately, to draft applicable IO sections of the operation order or operation plan. Running estimates enable planning officers to track and record pertinent information and provide recommendations to commanders. A generic written format of a running estimate contains six general considerations: situation, mission, course of action, analysis, comparison, and recommendation. *(Fig. 4-2, below)*.

---

1. SITUATION AND CONSIDERATIONS.

a. Area of Interest. Identify and describe those factors of the area of interest that affect functional area considerations.

b. Characteristics of the Area of Operations.

(1) Terrain. State how terrain affects a functional area's capabilities.

(2) Weather. State how weather affects a functional area's capabilities.

(3) Enemy Forces. Describe enemy disposition, composition, strength, and systems in a functional area. Describe enemy capabilities and possible courses of action (COAs) and their effects on a functional area.

(4) Friendly Forces. List current functional area resources in terms of equipment, personnel, and systems. Identify additional resources available for the functional area located at higher, adjacent, or other units. List those capabilities from other military and civilian partners that may be available to provide support in the functional area. Compare requirements to current capabilities and suggest solutions for satisfying discrepancies.

(5) Civilian Considerations. Describe civil considerations that may affect the functional area, including possible support needed by civil authorities from the functional area as well as possible interference from civil aspects.

c. Facts/Assumptions. List all facts and assumptions that affect the functional area.

2. MISSION. Show the restated mission resulting from mission analysis.

3. COURSES OF ACTION.

a. List friendly COAs that were war-gamed.

b. List enemy actions or COAs that were templated that impact the functional area.

c. List the evaluation criteria identified during COA analysis. All staffs use the same criteria.

4. ANALYSIS. Analyze each COA using the evaluation criteria from COA analysis. Review enemy actions that impact the functional area as they relate to COAs. Identify issues, risks, and deficiencies these enemy actions may create with respect to the functional area.

5. COMPARISON. Compare COAs. Rank order COAs for each key consideration. Use a decision matrix to aid the comparison process.

6. RECOMMENDATIONS AND CONCLUSIONS.

a. Recommend the most supportable COAs from the perspective of the functional area.

b. Prioritize and list issues, deficiencies, and risks and make recommendations on how to mitigate them.

Variations on this format, such as the example provided in Figure 4-3 below enable the IO officer to spotlight facts and assumptions, critical planning factors, and available forces. The latter of these requires input from assigned or available IRCs. The graphic format also offers a clear, concise mechanism for the IO officer to articulate recommended high-payoff targets, commander's critical information requirements, and requests for forces. Maintaining both formats simultaneously provides certain benefits: the narrative format enables the IO officer to cut-and-paste sections directly into applicable sections of orders; the graphic format enables the IO officer to brief the commander and staff with a single slide.

Example Graphical IO Running Estimate

Forces or systems available	Facts	Specified tasks	Limitations
• 413 civil affairs BNs • 344 tactical MISO COs • 1-55th Signal CO (-) 3x • 2x EC-130J Commando Solo @ CFACC • OCO available	• Civilian and government-controlled media outlets (radio and television) reach population within AO SWORD • Adversary forces have used civilian radio stations to broadcast coalition forces' troop movements and propaganda in the AO	*Identify key communicators within AO SWORD in order to deliver non-interference*	*MISO messaging and OCO release authority held by CCDR*
Information environment		**Implied tasks**	**HPT nominations**
• Radio is the best medium to reach the civilian population within AO SWORD, followed by social media • Religious leaders within contested areas are key communicators to the population • Displaced civilians in camps along main routes may impede coalition forces' advance		• Deny adversary use of social media messaging during decisive operations • Develop Soldier and leader engagement, and MISO products to support non-interference	• Denial of adversary social media site during decisive operations • Identify tribal leaders
	Assumptions		**CCIR nominations**
	• Civilian population will support HNSF and coalition forces once security is restored • Civilian population will remain in place during attack unless there is a loss of essential services		• Block axis of advance by civilian population during attack • Damage to HN essential services infrastructure and religious structures
			EEFI nominations N/A
Critical planning factors *Air tasking order cycle request 72 hours prior*	**Objectives** 1. Influence civilian population to minimize interference with coalition forces information operations team to prevent civilian casualties 2. Disrupt enemy forces use of media outlets in order to support freedom of movement of coalition forces.		**Request for forces** *Request OCO to deny use of social media site during decisive operations*

AO	area of operations	EEFI	essential element of friendly information
BN	Battalion	HN	host nation
CCDR	combatant commander	HNSF	host-nation security forces
CCIR	commander's critical information requirement	HPT	high-payoff target
CFACC	combined force air component commander	MISO	military information support operations
CO	Company	N/A	not applicable
COMCAM	combat camera	OCO	offensive cyberspace operations

Ref: ATP 3-13.3, fig. 4-3. Example graphical information operations running estimate.

Running estimate development is continuous. The IO officer maintains and updates the running estimate as pertinent information is received. While at home station, the IO officer maintains a running estimate on friendly capabilities. The unit prepares its running estimate based on researching and analyzing the information environment within its region and anticipated mission sets.

See related discussion of the running estimate on p. 4-37.

Information Operations Working Group

The staff has responsibility for conducting IO through synchronizing IRCs. As the staff lead for IO, the IO officer or designated representative develops a range of products and chairs the IO working group.

The IO working group is the primary mechanism for ensuring effects in and through the information environment are planned and synchronized to support the commander's intent and concept of operations.

See pp. 6-1 to 6-4 for in-depth discussion of the information operations working group.

B. Logic of the Effort

An essential part of planning and assessing IO is the need to develop an explicit logic of the effort for each objective or effect. The logic of the effort makes explicit how specific efforts lead to attaining objectives. The value of this logic is that its assumptions are made explicit and can become hypotheses that can then be tested and, if necessary, refined.

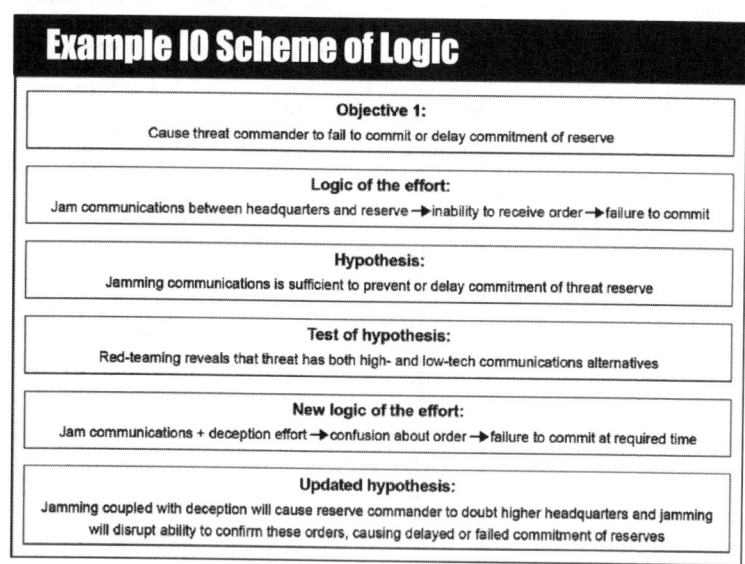

Example IO Scheme of Logic

Objective 1:
Cause threat commander to fail to commit or delay commitment of reserve

Logic of the effort:
Jam communications between headquarters and reserve → inability to receive order → failure to commit

Hypothesis:
Jamming communications is sufficient to prevent or delay commitment of threat reserve

Test of hypothesis:
Red-teaming reveals that threat has both high- and low-tech communications alternatives

New logic of the effort:
Jam communications + deception effort → confusion about order → failure to commit at required time

Updated hypothesis:
Jamming coupled with deception will cause reserve commander to doubt higher headquarters and jamming will disrupt ability to confirm these orders, causing delayed or failed commitment of reserves

ATP 3-13.1, fig. 4-4 provides a simple example of a logic statement and how it evolves when its hypothesis is tested.

C. CCIRs and EEFIs See also pp. 4-7 to 4-8.
Ref: *ATP 3-13.1, The Conduct of Information Operations (Oct '18), pp. 4-7 to 4-8.

Commander's Critical Information Requirements (CCIRs)

Commander's critical information requirements (CCIRs) identify information needed by the commander to visualize an operational environment and make critical decisions. CCIRs also filter information to the commander by defining what is important to mission accomplishment. If the information operation requires the commander to make a timely tactical decision, then staffs include IO input to the CCIRs, with supporting analysis and input to the decision support template produced during war gaming.

CCIRs are derived from information requirements, which are maintained and nominated by each staff element to the intelligence or operations staff officer. From the complete array of these requirements, the staff nominates those critical to the commander's decision making to become CCIRs, using the commander's guidance, higher headquarters' CCIRs, the essential-task list, and the IPB (situation template) to narrow and refine the list. Two types of CCIRs exist: priority intelligence requirements and friendly force information requirements.

Priority Intelligence Requirements

Priority intelligence requirements (PIRs) are information the commander must know about the threat and other aspects of an operational environment. For IO, PIRs focus on conditions in the information environment and adversary actions that affect the information environment. PIRs that may be required for IO include the following questions:

- Hostile forces using or preparing to use a key media outlet to produce or disseminate hostile propaganda.
- Adversary forces preparing to attack friendly information networks (either human or technological).

Friendly Force Information Requirements

Friendly force information requirements (known as FFIRs) are items of information the commander must know about the friendly force. For IO, friendly force information requirements provide information on critical aspects of the command's information system, IRCs, and execution of the information operation. Friendly force information requirements that may be required for IO include the following:

- Death or serious injury of noncombatants by friendly forces.
- Media coverage of alleged friendly force misconduct.

Essential Elements of Friendly Information (EEFIs)

Essential elements of friendly information (EEFIs) are critical aspects of a friendly operation that, if known by the adversary, subsequently lead to compromise, failure, or limited success of an operation, and, therefore, must be protected from detection. In other words, EEFI is a list of information that must be protected from the adversary's intelligence system to prevent the adversary from making timely decisions and allowing friendly forces to retain the initiative. Typically, EEFI include the command intentions, subordinate element status, or the location of critical assets (such as command posts and signal nodes). EEFI should be refined throughout the planning process, as some information may not be identified until COA development. Once EEFI are developed, measures (as tasks to subordinate units) are developed to protect the information (OPSEC process). Two examples of EEFI are:

- Friendly forces' time of departure for an operation.
- Tribal leaders assisting friendly forces.

IV. IO Input to Operation Orders and Plans

Operation orders and plans are products or outputs of planning. They provide a directive for future action. Commanders issue plans and orders to subordinates to communicate their understanding of the situation and their visualization of an operation. Plans and orders direct, coordinate, and synchronize subordinate actions and inform those outside the unit how to cooperate and provide support. As with all other functions and capabilities, IO provides input to these plans and orders.

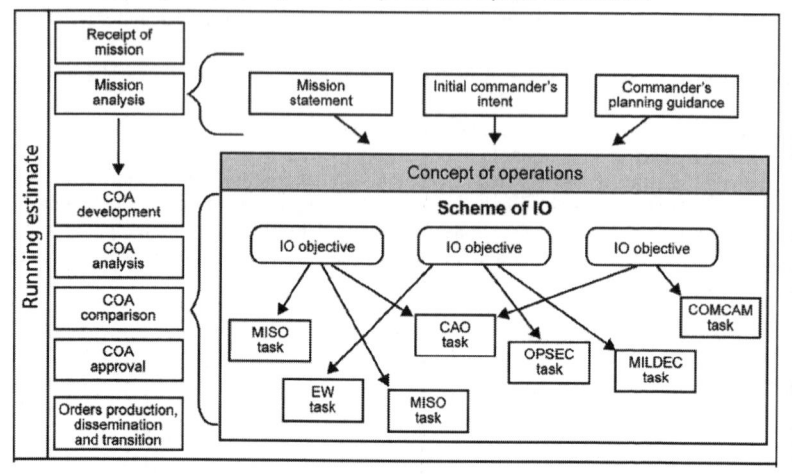

Ref: ATP 3-13.1, fig. 4-7. Relationship of scheme of IO, IO objectives, and IRC tasks.

Base Orders and Plans

While every part of an operation order or plan matters, most personnel read the base order or plan (the initial part of the document before the annexes and appendices) because it contains the most mission-essential information. Usually staff sections or specialists involved with a respective function or capability read only those annexes and appendices. If the base order or plan does not contain that information, it might not get read. Increasingly, some aspect of IO is essential to overall operational success. Sections of the base order or plan in which IO may be found include the following:

- Commander's intent, paragraph 3a.
- Concept of operations, paragraph 3b.
- Scheme of IO, paragraph 3c.x (paragraph number varies)
- Tasks to subordinate units, paragraph 3j.
- Coordinating instructions, paragraph 3k.

Appendix 15 (Information Operations) to Annex C (Operations)

Commanders and staffs use Appendix 15 (Information Operations) to Annex C (Operations) to operation plans and orders to describe how information operations (IO) will support operations described in the base plan or order. The IO officer is the staff officer responsible for this appendix. Products or guidance:

- Combined information overlay. *(See pp. 4-32 to 4-33.)*
- Synchronization matrix. *(See p. 4-16.)*
- Instructions for IRCs not covered by other appendices, such as operations security, visual information, and combat camera.

See pp. 4-61 to 4-64 for a sample annotated format to Appendix 15 (IO) to Annex C (Operations).

A. Mission Statement

Ref: *ATP 3-13.1, The Conduct of Information Operations (Oct '18), p. 4-9.

The IO officer crafts an IO mission statement while preparing or updating the running estimate. They later refine the mission statement to complete Appendix 15 (IO), which occurs with receipt of an order and commencement of mission analysis. FM 6-0 provides a template for attachments, such as annexes and appendixes. For the mission paragraph (paragraph 2), it instructs planners to state the mission of the functional area to support the base plan or order. In the case of Appendix 15, the functional area is IO.

The IO mission statement is a short paragraph or sentence describing what the commander wants IO to accomplish and the purpose for accomplishing it. The IO officer develops the proposed IO mission statement at the end of mission analysis based on the unit's proposed mission statement and IO-related essential tasks. During the mission analysis briefing or shortly thereafter, commanders approve the unit's mission statement and CCIRs. They then develop and issue their commander's intent and planning guidance.

Sample Mission Statement
No later than 130600JAN19, IO supports 1 Stryker Brigade Combat Team's defense of key terrain in AO RAIDER by disrupting Donovian command and control and influencing the population of Erdabil Province to support the Government of Atropia to engage the enemy from a position of advantage.

The IO officer may refine a final IO mission statement based on relevant input from the commander's intent and planning guidance and get it approved by the operations officer. The final IO mission statement includes IO effects and most significant IO-related target categories identified in the information environment during mission analysis.

The mission statement differs from the scheme of IO in its level of detail. The mission statement describes IO in the aggregate. The scheme of IO addresses how IRCs contribute to the scheme and, as a result, accomplish the mission.

Note. There is legitimate debate about whether more than one mission statement can or should exist for a given operation. Some commanders may direct that all attachments reiterate the restated mission in the base order. Functional mission statements are not intended as replacements for the base order mission but, instead, to support it. They are doctrinally justified per FM 6-0

B. Scheme of Information Operations

Ref: *ATP 3-13.1, The Conduct of Information Operations (Oct '18), pp. 4-9 to 4-11.

The scheme of IO begins with a clear, concise statement of where, when, and how the commander intends to employ synchronized IRCs to create effects in and through the information environment to support the overall operation and accomplish the mission. Based on the commander's planning guidance, the IO officer develops a separate scheme of IO for each COA the staff develops during COA development. IO schemes of support are expressed both narratively and graphically, in terms of IO objectives and IRC tasks required to achieve these objectives.

Figure 4-5 provides a sample scheme of an IO statement. Figure 4-6 illustrates a supporting sketch with articulated objectives and IRCs.

---

1 SBCT coordinates, deconflicts, and synchronizes IRCs in support of Phase III (Defense) in AO RAIDER. CO collects against Donovian frequencies and communications east of PL MAINE. EW conducts jamming of Donovian armor mission command systems in EAs THOMPSON, UZI, and RUGER. CMOC informs IDPs of collection instructions and safe rally points. MISO influences IDPs to not interfere with military movements and counters Donovian propaganda. The goal of all IRCs is to elicit the surrender or desertion of enemy forces, reduce CIVCAS, and prevent massing of enemy armor and indirect fires. PA controls release of operational information in order to bolster OPSEC and facilitates media engagement strategy to highlight operational successes. Maneuver, CAO, and MISO will conduct SLEs to enable 1 SBCT elements freedom of maneuver throughout AO RAIDER. Finally, 1 SBCT will capture operational successes through COMCAM and other visual information capabilities while OPSEC will protect EEFIs.

AO	area of operations	IDP	internally displaced person
CAO	civil affairs operations	IRC	Information-related capability
CIVCAS	civilian casualty	MISO	military information support operations
CMOC	civil-military operations center	OPSEC	operations security
CO	cyberspace operations	PA	public affairs
COMCAM	combat camera	PL	phase line
EA	engagement area	SBCT	Stryker brigade combat team
EEFI	essential elements of friendly information	SLE	Soldier and leader engagement
EW	electronic warfare		

Ref: ATP 3-13.1, fig. 4-5. Sample scheme of information operations statement.

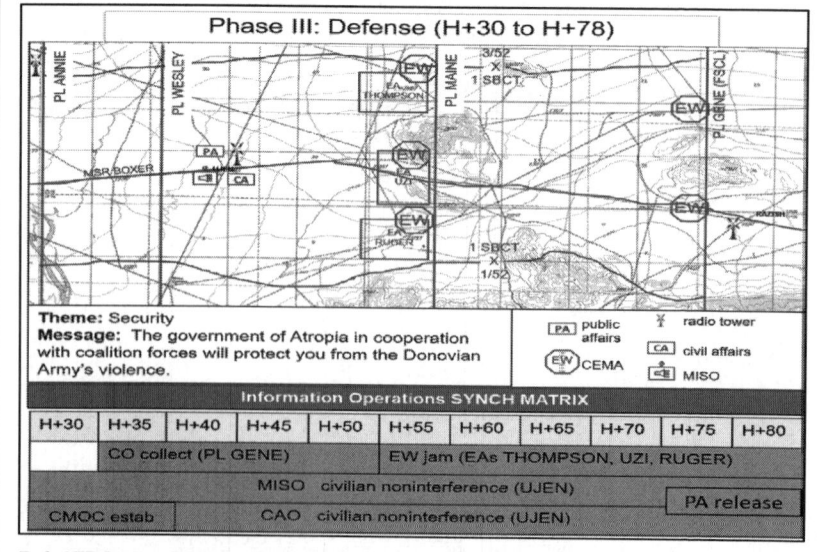

Phase III: Defense (H+30 to H+78)

Theme: Security
Message: The government of Atropia in cooperation with coalition forces will protect you from the Donovian Army's violence.

Legend: PA public affairs, EW CEMA, CA civil affairs, MISO, radio tower

Information Operations SYNCH MATRIX

H+30	H+35	H+40	H+45	H+50	H+55	H+60	H+65	H+70	H+75	H+80
	CO collect (PL GENE)				EW jam (EAs THOMPSON, UZI, RUGER)					
			MISO civilian noninterference (UJEN)							
									PA release	
CMOC estab			CAO civilian noninterference (UJEN)							

Ref: ATP 3-13.1, fig. 4-6. Example scheme of information operations sketch.

IO Mission Statement. NLT 130600JAN19, IO supports 1 SBCT's defense of key terrain in AO RAIDER by disrupting Donovian C2 and influencing the population of Erdabil Province to support the government of Atropia IOT engage the enemy from a position of advantage.

IO OBJ 1: Influence populace in UJEN to not interfere with 1 SBCT combat operations IOT limit CIVCASs. IO

OBJ 2: Disrupt enemy communications in EAs THOMPSON, RUGER, and UZI to degrade C2 IOT prevent massing of combat power.

Key Tasks
CO/EW
T: Jam Donovian armor mission command systems in EAs THOMPSON, UZI, and RUGER.
P: Degrade C2 capability IOT prevent massing of combat power. M/MOP: Three precision jamming delivered by tech ops to Donovian C2 systems in EAs THOMPSON, UZI, and RUGER.
M/MOP: Three precision jamming delivered by tech ops to Donovian C2 systems in EAs THOMPSON, UZI, and RUGER.

MISO
T: Persuade populace in UJEN to not interfere with 1 SBCT movements.
P: Counter Donovian propaganda that misdirects Atropians IOT prevent CIVCASs.
M/MOP: Eight broadcasts by loudspeaker to local nationals vic UJEN.

CAO
T: Inform IDPs of safe rally points.
P: Prevent CIVCASs IOT allow 1 SBCT freedom of movement.
M/MOP: One CMOC established to communicate civil control information with local nationals vic UJEN NLT H+35.

PA
T: Publicize Donovian battle losses to key audiences.
P: Facilitate media engagement strategy IOT highlight operational successes.
M/MOP: Three releases accessible via public sites to key audiences.

MOE 1: Decrease in daily observed number of civilian vehicles or foot traffic on MSR BOXER by 25% from baseline at H+6.

MOE 2: Increase in numbers of tips providing enemy locations and activity by 50% compared to those received at H+6.

AO	area of operations	MISO	military information support operations
C2	command and control	MOE	measure of effectiveness
CAO	civil affairs operations	M/MOP	method/measure of performance
CEMA	cyberspace electromagnetic activities	MSR	main supply route
CIVCAS	civilian casualty	NLT	no later than
CMOC	civil-military operations center	OBJ	objective
CO	cyberspace operations	P	purpose
EA	engagement area	PA	public affairs
estab	establishment	PL	phase line
EW	electronic warfare	SBCT	Stryker brigade combat team
H	hour	synch	synchronization
IDP	internally displaced person	T	task
IO	information operations	tech ops	technical operations
IOT	in order to vic vicinity	vic	vicinity

Ref: ATP 3-13.1, fig. 4-6. Example scheme of information operations sketch (continued).

C. IO Objectives & IRC Tasks

Ref: *ATP 3-13.1, The Conduct of Information Operations (Oct '18), pp. 4-11 to 4-13.

Information Operations Objectives

IO objectives express specific and obtainable outcomes or effects that commanders intend to achieve in and through the information environment. In addition to being specific, these objectives enable measurable, achievable, realistic, and time-bounded (known as SMART) measures of effectiveness and performance, which facilitate attaining and assessing established objectives (see Chapter 6 for more details on measures of effectiveness and performance). IO objectives do not stand alone but support the commander's operational intent. Based on the definition of IO, objectives are framed to accomplish the following:

- Attack enemy or adversary decision making and the capabilities or conditions that facilitate that decision making.
- Preserve friendly decision making and the capabilities or conditions that facilitate it.
- Otherwise shape the information environment to provide operational advantage to friendly forces, including freedom of maneuver in this environment.

For example, if an operational objective is to prevent an enemy force or weapon system from moving from Objective Black before attack, then possible associated IO objectives could be to—

- Disrupt adversary communications within AO Blue to prevent early warning.
- Deceive adversary decision makers on Objective Black to prevent relocation of command and control.
- Influence local populace in Operational Area Blue to support friendly force operations and prevent populace reporting on friendly force activities.

For each mission or COA considered, IO planners develop IO objectives based on the tasks for IO identified during mission analysis. Depending upon the complexity or duration of the mission (for example, a tactical direct-action mission versus a long-term foreign internal defense mission), there may be only one or numerous IO objectives developed for each phase of the overall operation. Generally, regardless of the mission, no more than five objectives are planned for execution at any one time in the operation.

Accurate situational understanding is key to establishing IO objectives. Operational- and tactical-level IO objectives must nest with strategic theater objectives. IO objectives further help the staff determine tasks to subordinate units during COA development and analysis.

No prescriptive format exists for an IO objective. One possible format uses effect, target or target audience, action, and purpose (known as ETAP):

- **Effect** describes the outcome (i.e., influence, destroy, degrade, disrupt, or deceive).
- **Target or target audience** describes the object of the desired effect.
- **Action** describes the behavior expected of the recipient.
- **Purpose** describes what will be accomplished for the friendly force.

Note. Around 2010, the definition of "target" was revised to specify that a target is an entity or object that performs a function for the adversary. However, the definition of "target audience" was not similarly adjusted. Per the DoD Dictionary of Military and Associated Terms, a target audience is an individual or group selected for influence.

IO objectives are written in terms of effects, because the desired effect focuses the activities (tasks) of IRCs. For IO, a proper effect falls into one of three categories:

- **Effects against the enemy or an adversary.** IO effects against the enemy or an adversary focus on the threat's ability to collect, protect, and project information. For example, an IO objective might disrupt (effect) an enemy formation's (target) ability to conduct command and control (action) to surprise adversary forces in and around Objective X (purpose).

- **Effects to defend friendly forces.** IO effects regarding friendly forces seek to prevent enemy or adversary interference with friendly abilities to collect, protect, and project information. For example, an IO objective might deny (effect) enemy IRCs (target) the ability to exploit negative effects of friendly force operations (action) to prevent attrition of local populace support away from coalition forces to the enemy (purpose).
- **Effects to shape the information environment.** IO effects shape information content and flow in the operational area's information environment. For example, an IO objective might influence (effect) local populace (target audience) perception of the enemy (action) to increase reporting of enemy activity and locations to coalition forces (purpose).

Because it is impossible to anticipate all possible effects, terms other than those presented in this publication may be used to describe the desired effects for IO. Effects terms should describe a condition— not a task. Definitions for the same effect may vary based on the physical, informational, and cognitive nature of the effect and the target of the effect.

As IO officers develop IO objectives, they establish the criteria—measures of effectiveness (MOEs)— and methods to collect the indicators. If planners cannot identify adequate indications and collection means, then they may need to refine the objective to produce measurable and detectable results. If an objective's MOE is focused on behavior or beliefs, planners must consider physical actions that result from the desired behavior or belief as an indicator.

Information-Related Capability Tasks

Once IO officers write IO objectives, they develop tasks to subordinate units and staff elements that possess the IRCs needed to accomplish these objectives. These tasks are conveyed through the various types of orders dictated by the MDMP. IRC tasks to subordinate units translate the broad concepts of the objectives into discreet actions. Tasks are often written as **task, purpose, method.**

Units take care to ensure that developed tasks do not cause IRCs to violate relevant authorities. For example, MISO tasks are tied directly to Office of the Secretary of Defense-approved MISO objectives (joint) or psychological objectives (Army), which are provided in a MISO program or applicable order. Through direct coordination or the IO working group, IO officers synchronize MISO objectives with IO objectives and align tasks so they support MISO, psychological, and IO objectives simultaneously. Alternatively, if an IO objective requires a MISO task not currently approved, then the IO working group seeks approval through MISO channels, reinforcing the need to plan selected IO objectives well in advance.

Similar to effects, tasks can be organized into three categories:

- **Tasks against the adversary.** These tasks target threat capabilities and vulnerabilities to collect, protect, and project information (as identified during the COG analysis). An example task might counter enemy propaganda to maintain populace support for capture or kill missions.
- **Tasks to protect friendly forces.** These tasks seek to protect friendly force vulnerabilities in the information environment from threat capabilities to collect and project information. An example task might detect intrusions into friendly force information systems to prevent enemy or adversary collection of critical information.
- **Tasks to shape the information environment.** These tasks shape information content and movement by impacting the key nodes in each sub information environment to influence local populace perceptions and behavior. An example task might engage religious leaders to bolster friendly credibility and legitimacy.

D. IO Synchronization Matrix

Ref: *ATP 3-13.1, The Conduct of Information Operations (Oct '18), pp. 4-14 to 4-16.

The synchronization matrix is used to monitor progress and results of IO objectives and IRC tasks as well as to keep IO execution focused on contributing to the overall operation. It is one of the IO working group's primary tools for monitoring and evaluating progress and assessing whether planned effects have been achieved.

Tasked unit or system	IO task	Time on target or time of effect	Location	Remarks
EA-6B	EW-01	H-1 through H-hour	TAI 002 and 003	Successful if enemy is unable to send early warning
Tactical PSYOP team	MISO-01	H-24 and continue	Objective SPRUCE	Successful if no civilian interference
Civil affairs team	CAO-01	H-24 through H-hour	Objective PINE	N/A

Ref: ATP 3-13.1, table 4-2. Example 2 – Information operations synchronization matrix.

IRC	Phase I	Phase II	Phase III	Phase IV
EW	Monitor signals of interest. Electronic protection for personnel and equipment.	Electronic attack to disrupt enemy communications. Electronic protection for personnel and equipment.	N/A	N/A
MISO	Broadcast harassment messages against enemy. Broadcast noninterference messages for local populace.	N/A	Broadcast via mobile radio to keep population informed on mission.	Broadcast on mission success. Coordinate with COMCAM for post-mission messaging and countering the effect of adversary information activities.
OPSEC	Determine essential elements of friendly information for mission.	Implement measures to protect essential elements of friendly information to protect movement routes, mission command, and objective.	N/A	N/A
MILDEC	N/A	N/A	N/A	N/A
CAO	Prepare Commander's Emergency Response Program paperwork for funds disbursement. Coordinate with Provincial reconstruction team.	N/A	N/A	Assist personnel returning to villages. Assess small-scale immediate projects.
PA	Prepare press releases. Embed media.	N/A	N/A	Distribute press releases. Conduct press conference and set up interviews with subject matter experts.
COMCAM	Document operation.	Document operation.	Document operation.	Document operation.

Ref: ATP 3-13.1, table 4-1. Example 1 – Information operations synchronization matrix.

II. Information Environment Analysis

Ref: *ATP 3-13.1, Conduct of Information Operations (Oct '18), chap. 2. (*See note p. 1-2.)

IO and Intelligence Preparation of the Battlefield (IPB)

The mechanics of analyzing the information environment and enemy or adversary operations in the information environment are generally the same as those established to support intelligence preparation of the battlefield (IPB) for other military planning. IPB is a critical component of the military decisionmaking process (MDMP). It provides a systematic approach to evaluating the effects of significant characteristics of an operational environment for missions.

IPB to support IO refines traditional IPB to focus on the information environment. Its purpose is to gain an understanding of the information environment in a geographic area and determine how the enemy or adversary will operate in this environment. The focus is on analyzing the enemy's or adversary's use of information to gain positions of relative advantage. The end state is the identification of threat information capabilities in the information environment against which friendly forces must contend and threat vulnerabilities that friendly forces can exploit with IO.

Analyze and Depict the Information Environment

To achieve advantage in the information environment, commanders, with specialized advice and support from the IO officer, ensure that IO planning is fully integrated into the operations process. This begins with analysis to understand, visualize, and describe the information environment.

A significant part of what makes the operational environment complex is the information environment because it includes such components as cyberspace, the electromagnetic spectrum, data flow, encryption and decryption, the media, biases, perceptions, decisions, key leaders and decision makers, among many others. What occurs in the physical dimension of the information environment and, more broadly, the operational environment, always has second- and third-order effects in the informational and cognitive dimensions of the information environment. Thus, there must be holistic and nuanced understanding of how these various components and dimensions interrelate and the whole operates.

This understanding is depicted through a series of information overlays and comprehensive combined information overlays, which vary depending on commanders' priorities, the nature of the operation, and the type of analysis being conducted. Modeling or mapping social or human networks also enhances this understanding. While complex, the information environment still needs to be captured in a way that the commander can visualize and understand it, draw necessary insights and conclusions, and make informed decisions. The IO officer should not be locked into any specific method for analyzing and depicting the information environment but develop a process and overlays that best serve the commander and, as appropriate, follow unit standard operating procedures. As new technologies and interactive capabilities emerge, they should be incorporated as tools to facilitate the visualization and understanding processes.

Information Environment Analysis

Ref: *ATP 3-13.1, The Conduct of Information Operations (Oct '18), pp. 2-1 to 2-2.

The information environment is the aggregate of three components—individuals, organizations, and systems—that collect, process, disseminate, or act on information. Understanding this environment requires an analyst—chiefly the IO officer or designated representative—to analyze each component of the environment as well as their aggregate. The analyst determines how the components interrelate.

The information environment also has three dimensions: physical, informational, and cognitive. All are important. The physical dimension consists of what users see—the physical content of the environment. This dimension contains observable behavior. This behavior enables the commander and staff to measure the effectiveness of their efforts to influence enemy and adversary decision making and the attendant actions that must occur across all audiences in the area of operations (AO). The informational dimension is the code that captures and organizes information that occurs in the physical dimension so that it can be stored, transmitted, processed, and protected. This dimension links the physical and cognitive dimensions. The cognitive dimension consists of the perspective of those who inhabit the environment; their individual and collective efforts to give context to what is happening or has happened and make sense of it. In this dimension, sense making occurs. If conflict is ultimately a contest of wills and victory is achieved by defeating the enemy or adversary psychologically, then achieving effects in the cognitive dimension can be decisive. The cognitive dimension is the hardest to understand. Therefore, the better that units operate in and exploit the physical and informational dimensions, the more they can overcome the challenges associated with the cognitive dimension. Table 2-1 explores the three dimensions:

Types	Affects	Examples
Physical	Content	• The physical world and its content, particularly that which enables and supports exchanging ideas, information, and messages. • Information systems and physical networks. • Communications systems and networks. • People and human networks. • Personal devices, handheld devices, and social media graphical user interface. • Mobile phones, personal digital assistants, and social media graphical user interfaces.
Informational	Code	• Collected, coded, processed, stored, disseminated, displayed, and protected information. • Information metadata, flow, and quality. • Social media application software, information exchange, and search engine optimization. • The code itself. • Any automated decision making.
Cognitive	Context	• The impact of information on the human will. • The contextualized information and human decision making. • Intangibles, such as morale, values, worldviews, situational awareness, perceptions, and public opinions. • Mental calculations in response to stimuli, such as liking something on a social media application.

Table 2-1. Information environment dimensions. See pp. 1-7 and 2-10 for further discussion of the three dimensions.

One purpose of IO involves affecting an adversary's ability to make sense of unfolding events. Affecting the adversary's perception of an event can indirectly impair, disrupt, or disable the adversary's ability to lead and direct operations. At the same time IO affects those perceptions, it attempts to preserve friendly commanders' ability to lead their forces and understand, visualize, describe, and direct operations. IO uses social media—a dominant aspect of the information environment—across and among all three dimensions. Messages, images, graphics, and sounds transmitted via social media affect perceptions and behaviors in real time and with profound impact.

Actions that occur in an operational environment almost always create effects in all three dimensions of the information environment. Through effective, proactive planning, units account for intended primary, secondary, and tertiary effects to support the commander's intent and concept of operations, while mitigating unintended effects. Precise effects across all three dimensions are only possible if the unit commander analyzes, understands, and visualizes the information environment and operational environment as a whole. Even the most prepared staff cannot anticipate all potential effects; however, understanding the information environment enables the staff to prepare for and react to unintended effects and determine why they occurred.

The mechanics of analyzing the information environment and enemy or adversary operations in the information environment are generally the same as those established to support intelligence preparation of the battlefield (IPB) for other military planning. IPB is a critical component of the military decisionmaking process (MDMP). It provides a systematic approach to evaluating the effects of significant characteristics of an operational environment for missions (for a full discussion of IO and the MDMP, see FM 3-13). IPB to support IO refines traditional IPB to focus on the information environment. Its purpose is to gain an understanding of the information environment in a geographic area and determine how the enemy or adversary will operate in this environment. The focus is on analyzing the enemy's or adversary's use of information to gain positions of relative advantage. The end state is the identification of threat information capabilities in the information environment against which friendly forces must contend and threat vulnerabilities that friendly forces can exploit with IO.

In addition to the running estimate, IPB to support IO results in producing a graphic or visualization product known as the combined information overlay. This overlay results from a series of overlays that depict where and how information aspects such as infrastructure, content, and flow potentially affect military operations. In certain instances, staffs may need more than one combined information overlay to capture the full complexity of the information environment (see paragraph 2-50 for a discussion on combined information overlay).

During mission analysis, the IO officer or representative ensures that IPB addresses the information environment and supports the planning and execution of operations. The intent is to better visualize the impact of the information environment on unit operations and to identify potential threat capabilities and vulnerabilities that the unit can protect against or exploit.

This analysis involves four substeps that mirror the steps discussed in ATP 2-01.3 (IPB):

- Define the information environment (*See pp. 4-22 to 4-23.*)
- Describe the information environment's effects. (*See pp. 4-24 to 4-27.*)
- Evaluate the threat's information situation. (*See pp. 4-28 to 4-33.*)
- Determine threat courses of action in the information environment. (*See p. 4-34.*)

IPB Considerations for the Information Environment

Ref: ATP 2-01.3 (w/Chg 1), Intelligence Preparation of the Battlefield (Jan '21), pp. 8-6 to 8-8.

The information environment is the aggregate of individuals, organizations, and systems that collect, process, disseminate, or act on information (JP 3-13). Although defined separately, the information environment and OE are interdependent and integral to each other. A unit's information operations officer or designated representative supports IPB with a specific focus on the information environment. (See FM 3-13 and ATP 3-13.1 for a detailed discussion on information operations considerations for IPB.)

The information environment consists of three interrelated dimensions—physical, informational, and cognitive. Cyberspace, a significant component of the information environment, overlaps the physical and informational dimensions. The IPB process must determine a threat's capabilities to operate in each of these dimensions of the command's battlefield:

- **Physical dimension** comprises C2 systems, key decision makers, and supporting infrastructure that enable individuals and organizations to create effects. It includes but is not limited to people, computers, smart phones, and newspapers.

- **Informational dimension** encompasses where and how information is collected, processed, stored, disseminated, and protected. It includes but is not limited to C2 systems, knowledge management TTP, and physical and operational security policies. A key aspect of this dimension is determining where, how, and when friendly, neutral, and threat information and information systems will be vulnerable to exploitation and attack.

- **Cognitive dimension** encompasses the minds of those who transmit, receive, and respond to or act on information. It includes but is not limited to cultural norms, perspectives, beliefs, and ideologies.

Table 8-1 provides IPB considerations for the information environment. It assists in organizing where and how information and information capabilities reside, are employed, and disseminated.

Informational aspects relevant to friendly forces
• Those that enable friendly capabilities, including each warfighting function.
• Where, how, and when information can be employed to support operations. **Note.** This is not just outward application of capabilities; it includes knowledge management, information assurance, information security, operations security, as well as how the command leverages information to enable the staff and maneuver units to achieve the mission.
• Where, how, and when friendly information and information systems will be vulnerable to exploitation and attack by others.
• Identifying components of information infrastructure or nodes of information systems that must be destroyed, disabled, or left in place.
Informational aspects relevant to neutral forces
• Where, how, and when information can be employed to support operations.
• Those that enable neutral capabilities, including each warfighting function or system. **Note.** Analysts should avoid putting U.S. architectures (warfighting functions) on other force's constructs as this may create a gap in analysis.
• Where, how, and when neutral information and information systems will be vulnerable to exploitation and attack by others.
• Identifying components of information infrastructure or nodes of information systems that neutral forces will consider for destruction and disablement, or will leave in place.
Informational aspects relevant to threat forces
• Where, how, and when information can be employed to support threat operations.
• Those that enable threat capabilities, including each warfighting function or system.
• Where, how, and when threat information and information systems will be vulnerable to exploitation and attack by others.
• Identifying components of information infrastructure or nodes of information systems that threat forces will consider for destruction and/or disablement, or will leave in place.
Informational aspects relevant to populations
• Information capabilities that enable population support systems.
• Sources of information that inform and influence decisions.

Ref: ATP 2-01.3 (w/Chg 1), table 8-1. IPB considerations for the information environment.

The Electromagnetic Spectrum (EMS)

The electromagnetic spectrum is the entire range of electromagnetic radiation from zero to infinity. It is divided into 26 alphabetically designated bands (JP 3-13.1). The EMS is a continuum of all electromagnetic waves arranged according to frequency and wavelength. The frequency range suitable for radio transmission(the radio spectrum) extends from 10 kilohertz to 300,000 megahertz, which is divided into a number of bands. Below the radio frequency spectrum, and overlapping it, is the audio frequency band, extending from 20 to 20,000 hertz. Above the radio frequency spectrum are heat and infrared, the optical (visible) spectrum(light in its various colors), ultra-violet rays, x-rays, and gamma rays. Within the radio frequency range, from 1 to 40 gigahertz (1,000 to 40,000 megahertz), between the ultrahigh frequency and extremely high frequency are additional bands, defined as follows:

- L band: 1 to 2 gigahertz.
- S band: 2 to 4 gigahertz.
- C band: 4 to 8 gigahertz.
- X band: 8 to 12 gigahertz
- Ku band: 12 to 18 gigahertz.
- K band: 18 to 27 gigahertz.
- Ka band: 27 to 40 gigahertz

Maritime radar systems commonly operate in the S and X bands, while satellite navigation system signals are found in the L band. The break of the K band into lower and upper ranges is necessary because the resonant frequency of water vapor occurs in the middle region of this band, and severe absorption of radio waves occurs in this part of the spectrum.

EMS-based operations must be understood to accurately depict possible threat COAs and how these COAs may impact friendly operations. EMS effects and the systems that use the EMS are critical IPB considerations that highlight the multi-domain nature of friendly and threat operations. Although interrelated by the EMS, each domain has different functions and objectives.

SIGINT, cyberspace operations, EW, and spectrum management operations all operate within the EMS. When performing IPB, considering the EMS maximizes the employment of friendly SIGINT and EW assets by providing direction to the collection management effort, electronic node analysis, and decisionmaking, as well as a thorough understanding of the threat's communications and SIGINT and EW asset noncommunications, electronic surveillance, and electronic countermeasure capabilities.

SIGINT is the interception and collection of signals in the EMS. Electronic warfare is military action involving the use of electromagnetic and directed energy to control the electromagnetic spectrum or to attack the enemy (JP 3-13.1). SIGINT and EW information is integrated into the threat, situation, and event templates developed during the IPB process and the DST developed during the MDMP. These templates graphically portray threat dispositions, vulnerabilities, and capabilities, including capabilities to employ electronic systems throughout the AO for electronic countermeasures, C2, target acquisition, maneuver, and CAS and airspace management.

EMS considerations are integrated into DSTs during the MDMP to assist commanders and staffs in the decision-making process by depicting critical points on the battlefield and identifying HPTs, which, when exploited, provide friendly forces with an advantage. The intelligence staff integrates electronic data into the IPB process to verify threat unit identification, locations, and types and sizes.

The intelligence staff conducts the initial electronic data assessment, to include developing the intelligence portion of the DST. The EW officer integrates SIGINT and EW information into the DST with enough detail to satisfy targeting priorities and determine the effectiveness of threat EW systems. The DST,together with the MCOO, is forwarded to subordinate elements, where they are further refined to meet functional mission requirements.

See p. 3-60 for related discussion of the electromagnetic spectrum (EMS).

In addition to the **running estimate**, IPB to support IO results in producing a graphic or visualization product known as the **combined information overlay**. This overlay results from a series of overlays that depict where and how information aspects such as infrastructure, content, and flow potentially affect military operations. In certain instances, staffs may need more than one combined information overlay to capture the full complexity of the information environment.

See pp. 4-6 to 4-7 for discussion of running estimates and pp. 4-32 to 4-33 for discussion of the combined information overlay.

During mission analysis, the IO officer or representative ensures that IPB addresses the information environment and supports the planning and execution of operations. The intent is to better visualize the impact of the information environment on unit operations and to identify potential threat capabilities and vulnerabilities that the unit can protect against or exploit. This analysis involves four substeps that mirror the steps discussed in ATP 2-01.3 (IPB):

Step 1: Define the Information Environment

During the first step of mission analysis, the IO officer or representative coordinates with other staff officers and elements, particularly the intelligence staff section. Defining the information environment begins by clearly delineating the AO, as well as areas of interest, including contiguous areas to the AO that may affect information flow and decision making. Once delineated, the IO officer identifies the significant characteristics of the information environment within this defined area in all three dimensions (physical, informational, and cognitive) that can affect friendly and threat operations, as well as influence friendly courses of action and command decisions. These significant characteristics can include, but are not limited to, the following:

- Terrain (and weather).
- Populace.
- Societal structures.
- Military or government information and communications infrastructure.
- Civilian information and communications infrastructure.
- Media.
- Third party organizations.

Terrain (and Weather)

One characteristic that the IO officer identifies is the terrain (and weather). The IO officer looks at the various ways physical, geographical, and atmospheric aspects of the AO impact information content and flow. These aspects can include compartmentalization, canalization, signal attenuation, radio wave propagation, and atmospheric and environmental limits on employing information systems.

Populace

Populace is another characteristic that the IO officer identifies. This characteristic involves identifying the human composition of the AO or area of interest in all its diversity to determine factors that impact information flow, receipt, and understanding. These factors tend to be static and non-voluntary; they are enduring traits or patterns of behavior that are innate or culturally ingrained to the point they are habitual and non-reflexive. Often IO officers study demographic and linguistic factors such as age, gender, education level, literacy, birth rate, ethnic composition, family structure, employment or unemployment rates, and languages.

Societal Structures

Societal structures affect friendly and threat operations. IO officers identify human networks, groups, and subgroups that affiliate along religious, political, or cultural lines, including commonly held beliefs and local narratives. These affiliations are voluntary and varied—over time, over space, and among individuals. IO officers focus their analysis on preferred means, methods, and venues that each social affiliation uses to interact and communicate and the ways each collectively constructs reality. Analysis examines biases, pressure points, general leanings, and proclivities, especially as they pertain to support or opposition of friendly and adversarial forces. Analysis also explores how these networks, groups, and subgroups express themselves and their commonly held beliefs through written and spoken narratives, stories, and messages.

Military or Government Information and Communications Infrastructure

The IO officer identifies another characteristic: the military or government information and communications infrastructure. Details of this characteristic involve understanding informational networks and communications systems that move information through the information environment to support military and governmental activities and facilitate decision making. Key networks and systems include special or enclave telecommunications means, methods, and capacity, such as telecommunications towers, fiber-optic networks, telephone networks (wired or wireless), microwave, satellite, and internet. IO officers understand the type and volume of information passed over or through these systems. These officers also benefit from knowing military or governmental authorities (leaders, decision makers, and military and civilian workforce) who use, manage, and control these systems.

Civilian Information And Communications Infrastructure

A related characteristic that the IO officer identifies is civilian information and communications infrastructure. This characteristic involves understanding informational networks and communications systems servicing the general population that move information throughout the information environment. Key systems include telecommunications towers, fiber-optic networks, telephone networks (wired or wireless), microwave, satellite, internet, and cellular networks. IO officers identify these systems and the informational content moved on them before understanding ways that forces—friendly or adversarial—can exploit these systems to influence indigenous populations.

Media

Characteristics of media can affect friendly and threat operations. Media includes physical, informational, and cognitive means by which local populations (including the adversary) receive and have their thinking shaped by information. Examples of the media's characteristics include radio and television broadcast facilities; print production facilities; news reporting, production, and dissemination sources; and outlets servicing the AO and areas of interest. Other examples include the primary and backup information systems used to move information from point to point, the information reported in terms of volume and content, the media's range and distribution capabilities, the audiences being marketed to and affected by the media, and the observed bias of the media and its cognitive effect on government, military, and civilian leaders, decision makers, and the general population.

Third-Party Organizations

Third-party organizations also have characteristics that affect operations. These organizations that simultaneously message in the information environment vary from nongovernmental and private organizations to other government agencies and international organizations. IO officers place analytical emphasis on identifying these organizations, determining their audiences, discerning their agendas, and estimating their impact on friendly operations.

Identifying and defining the significant aspects of the information environment helps to focus the IPB to support IO on those characteristics that will influence friendly courses of action (COAs) and command decisions. This focus thereby prevents unnecessary analysis and wasted effort. The initial analysis in this step determines the resources and time the IO officer or element commits to the detailed analysis that occurs in Step 2.

Step 2: Describe the Information Environment Effects

In this step, the IO officer examines the significant characteristics or features of the information environment identified in Step 1 and determines their potential effects or impacts on friendly and threat operations in each dimension. As with IPB in general, this step focuses on how the threat, terrain and weather, and civil considerations can affect operations.

A. Describe How the Threat Can Affect Friendly Operations

The enemy or adversary is part of an operational environment and information environment. The threat's physical posture alone influences friendly decisions and operations, as well as the decisions and actions of the populace in the AO in ways that benefit the threat commander's intent.

For many adversaries, the information environment is decisive terrain. Adversaries actively seek to shape it to their advantage, often well before hostilities begin. Although a detailed analysis of enemy forces occurs during Step 3 and Step 4 of the IPB process to support IO, Step 2 defines the type of enemy forces and their general information capabilities. This is done to place the existence of these forces and their capabilities in context with other variables to understand their relative importance to the information environment. Sometimes the mere presence of a threat force is the most important characteristic in the information (as well as operational) environment. This force presence is the chief locus of influence. In other instances, the presence or absence of communications infrastructure or other feature will be the predominating characteristic.

The results of this substep are typically reflected in threat and situational overlays, which are visual depictions of doctrinal and current physical dispositions of all potential threat information forces or capabilities in the AO and area of interest. In addition to locations, these graphics include the identity, size, strength, AO, and coverage or reach for each potential threat information unit or capability. The IO officer often supplements these overlays with a threat description table that describes the threat's broad information capabilities (see Steps 3 and 4 of the IPB process for additional information about these overlays).

B. Describe How Terrain and Weather Can Affect Friendly and Threat Operations

Terrain analysis is the collection, analysis, evaluation, and interpretation of geographic information on the natural and man-made features of the terrain, combined with other relevant factors, to predict the effect of the terrain on military operations (JP 2-03). Just as terrain can canalize friendly or threat movement and maneuver, it can canalize the flow of information, thereby affecting the timeliness and effectiveness of decision making. Weather analysis is the evaluation of the direct and indirect effects of weather and climate on operations in the information environment. These effects can be as simple as directly affecting the employment of capabilities, such as EC-130J Commando Solo, or as complex as indirectly affecting AO-wide efforts to inoculate the local populace against enemy propaganda.

Terrain analysis involves identifying obstacles and key terrain, but IPB to support IO analyzes these features in terms of how they will affect the employment of IRCs, the flow of information, and decision making. Similarly, IPB to support IO analyzes weather patterns, forecasts, and climate data to determine their impact on IRC's employment, information flow, and decision making.

C. Describe How Civil Considerations Can Affect Friendly and Threat Operations

An understanding of civil considerations enhances the selection or formulation of IO objectives, the weighting of IO efforts (attack, defend, or stabilize), the appropriate mix of IRCs, and their employment, among other aspects. Such understanding begins even before deployment and leverages the entire staff, as well as outside agencies and unified action partners, who has relevant regional knowledge and expertise in civil considerations.

One method to discern significant civil consideration characteristics is depicted in table 2-2 on the following pages, which crosswalks civil considerations with operational variables. Operational variables are known by the acronym PMESII-PT (political, military, economic, social, information, infrastructure, physical environment, and time). Civil considerations, a subset of mission variables—mission, enemy, terrain and weather, troops and support available, time available, and civil considerations (known as METT-TC)— comprise areas, structures, capabilities, organizations, people, and events (known as ASCOPE). Staffs use operational variables to develop a comprehensive understanding of operational and information environments. Civil considerations refine this understanding so that staffs can visualize and describe operational and information environments in a manner that fosters shared understanding. This crosswalk helps the IO staff refine its understanding of what is relevant to missions and operations from its perspective. The staff can complete it with any single mission variable to the operational variables.

See following pages (pp. 4-26 to 4-27) for examples of operational variables crosswalked with civil considerations and overlays. Refer to appendix A of FM 6-0 for information about operational variables; refer to chapter 4 ATP 2-01.3 for information on specific civil considerations in the IPB process.

Next, the IO officer or planner refines the information overlays to produce a "so what" statement for each. Put another way, IO officers iteratively refine information overlays to capture and display those features and impacts that most affect mission accomplishment. The information environment is complex. While oversimplifying it can lead to faulty conclusions and decisions, staffs must competently represent it in a few products that enable commanders to visualize and understand it sufficiently to make informed decisions.

Once IO planners have generated an information overlay for each significant characteristic, they determine the aggregate impacts across all significant characteristics, mindful of these questions, among others:

- How will each significant characteristic impact the others?
- How does the interaction among significant characteristics impact employing IRCs and the content and flow of information?
- What slow-go or no-go areas in the information environment constrict, restrict, or prevent information flow; what areas facilitate or hasten its flow?

Examples of Operational Variables Crosswalked with Civil Considerations

Ref: *ATP 3-13.1, The Conduct of Information Operations (Oct '18), pp. 2-5 to 2-6.

Due to the complexity and volume of data involving civil considerations, no simple or single model exists for presenting this analysis. It typically comprises a series of products, such as data files, overlays, and assessments.

	Political	Military	Economic	Social	Information	Infrastructure
Areas	• Enclave, province, district • National boundaries • Shadow government influence area	• Areas of influence and interest • Area of operations • Safe haven • Local nation base or training area	• Commercial • Fishery • Industrial • Markets • Mining • Smuggling routes • E-commerce	• Refugee camp • Ethnic, social, tribal enclave • School district • Online group	• Broadcast coverage area • Social media reach or penetration • Word of mouth • Graffiti	• Road system • City limit • Power grid • Irrigation network • Suburb, exurb, urban core
Structures	• Court house • Government center • Capitol building • Meeting hall	• Base and base buildings • Training facility • Known leader house	• Banking • Fuel • Factory • Warehousing • Online store • "Wall Street" versus "Main Street"	• Club • Jail • Library • Religious building • Restaurant • Social media platform	• Cell tower • Broadcast facility • Physical internet structure • Postal service • Print shop	• Emergency shelter • Public building • Airfield, bridge, railroad • Construction sites • Electric station
Capabilities	• Civil authority, practices and rights • Executive, legislative, and judicial functions • Dispute resolution	• Doctrine • Organization • Training • Materiel • Leadership • Personnel • Facilities • Civil-military relationship	• Currency • Food security • Market or black market • Raw material • Tariff • BITCOIN • Imports or exports	• Social network • Nonprofit support to disasters • Social services	• News operation • Newspaper • Social media platform • Literacy rate • Intelligence service • Internet access	• Law enforcement • Fire fighting • Maintenance • Transportation • HVAC (heating, ventilation, and air conditioning)
Organizations	• Major political party • Nongovernmental organization • Host government • Court system • Insurgent group affiliation	• Host-nation forces • Insurgent group or network • Terrorist • Military lobbying group	• Bank • Business organization • Guild • Labor union • Landowner • Cooperative	• Clan • Online or in-person affinity group • Patriotic or service organization • Familial	• Media group • Public relations firm • Social media information group • News organization	• Construction company • Trade union • Cooperative
People	• United Nations representative • Political leader • Governor • Elder • Legislator, judge, and prosecutor	• Key leader • Thought leader	• Banker • Employer or employee • Employment rate • Merchant • Smuggler	• Community leader • Teacher • Entertainer • Criminal • Migration patterns	• Decision maker • Elder • Religious leader • Internet personality	• Builders • Local development council • Road repairers • Police, fire fighter
Events	• Election • Council meeting • Treaty signing • National parade • Speech • Significant legal trial	• Combat • Military parade • Unit relief • Loss of leadership	• Drought, yield • Labor migration • Market day • Payday • Business opening	• Celebration • Civil disturbance • Funeral • Online forum • Social media livestream	• Censorship • Publishing dates • Online launch • Press briefing • Interview • Disruption of service	• Scheduled maintenance • School construction • New bridge opening • Disaster, man-made or natural

Table 2-2. Examples of operational variables crosswalked with civil considerations.

Example Overlay

Ref: ATP 3-13.1, The Conduct of Information Operations (Oct '18), pp. 2-7 to 2-11.

IO officers and planners often use one common technique to present analysis. They prepare an overlay (graphical depiction) for each significant characteristic that visually displays its salient features and identifies gaps in intelligence or information that are subsequently refined into requirements for collection (requests for information, requests for collection).

The following figures provide example overlays. The first focuses on population centers and the second focuses on communications infrastructure. Both examples are based on the Decision Action Training Environment or DATE scenario as employed at the Joint Readiness Training Center.

Note. These overlays depict "a" way, not "the" way. IO officers or representatives must adapt their products to the situation at hand, their units' standard operating procedures, and commander's preference.

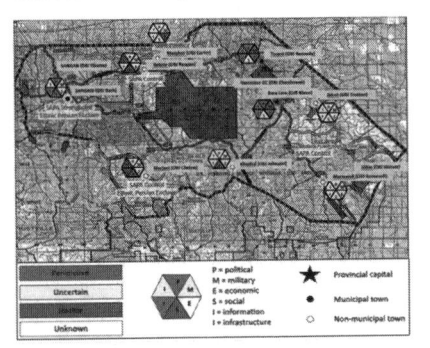

Figure 2-1. Example overlay that depicts relevant information about the populace in the area of operations.

Figure 2-2. Example overlay that depicts relevant information about communications infrastructure in the area of operations.

Sangari:	Turani:	Janan:
• 2nd largest town in Kirsham	• Joint municipality with Dara Lam	• Small rural village
• Strong allegiance to ROA (pre-SAPA)	• Strong allegiance to ROA	• Dependent on NGO/IGO for essential services
• Active municipal gov. (pre-SAPA)	• Strong economic growth	• Agricultural economy; minimal growth
• ROA/U.S. built 'Model City'	• Majority ethnic Atropian	• Majority ethnic Atropian; dislike SAPA/likely support anti-SAPA activity
• Regular access to school, medical facility, and emergency services	• Moderate inter-ethnic friction	
• Many businesses	• Adequate transportation	
• Majority ethnic Persian; minimal ethnic tension pre-SAPA	• USAID and NGO activity	• Ethnic unrest; Persian residents likely support insurgents/resent U.S. presence
• SAPA restricts information flow	• Clinic funded and operated by town (USAID rehabilitation project)	• Inadequate transportation

AO	area of operations	OA	operational area	
ASR	alternate supply route	ROA	Republic of Atropia	
MSR	main supply route	SAPA	South Atropian People's Army	
IGO	intergovernmental organization	USAID	United States Agency for International Development	
NGO	nongovernmental organization			

Figure 2-1. Example overlay that depicts relevant information about the populace in the area of operations (continued).

Step 3: Evaluate the Threat's Information Situation

Opposing forces use the information environment just as they use the physical domains of air, land, maritime, and space. They aim to gain positions of relative advance, place their enemy at a disadvantage, dominate the information environment, and achieve their objectives. In this step, the staff determines threat capabilities; doctrinal principles; and tactics, techniques, and procedures that threat forces prefer to employ. The IO staff identifies how enemies or adversaries view and use the information environment, including how they array their forces and employ capabilities to create effects in this environment.

The IO staff applies critical thinking to avoid confirmation bias, groupthink, and other biases. A common mistake is presuming that the adversary views and, therefore, uses the information environment in the same way as U.S. forces—that they are bound by the same constraints or limited by the same means. To avoid mirror-imaging the friendly concept of IO upon the enemy or adversary and prevent mismatching U.S. capabilities and vulnerabilities, the IO staff views adversary operations in the information environment in terms of activities to collect, protect, and project information. These three functions are universal to any armed force's ability to use information as combat power regardless of its organization, capabilities, and mission. As such, these functions form the basis of a threat's capabilities (and vulnerabilities) in the information environment.

Term	Definition
Collect	To plan and execute operations, the adversary must collect accurate and timely information
Protect	To ensure its ability to make timely and informed decisions, the adversary must protect its critical information from collection and maintain its means of communication
Project	To further its goals and objectives, the adversary must project the information into the information environment to influence the perceptions of its target audiences

Table 2-3. Adversary functions.

Depending on the threat, the means used can be as simple as direct human observation and open sources (collect); couriers and intimidation (protect); and public broadcasts, printed materials, graffiti, or lethal action. When taken together, these means create a cohesive narrative (project). Ideally, analysis of how the adversary operates in the information environment is based on modeling or templating. Two common tools to conduct this analysis are threat templates and center of gravity (COG) analysis.

The resulting analysis is an understanding of threat capabilities and vulnerabilities under unconstrained conditions in the information environment. IO planners then refine this understanding using the actual, constrained conditions identified in the information environment analysis and depicted in information overlays and the combined information overlay.

This step consists of two substeps:

    A. Threat templates *(See facing page)*

    B. Threat Center of Gravity Analysis *(See following pages pp. 4-30 to 4-31)*

For more information on threat templates and center of gravity analysis, refer to JP 2-01.3 and ATP 5-0.1, respectively.

A. Threat Templates

Ref: *ATP 3-13.1, The Conduct of Information Operations (Oct '18), pp. 2-12 to 2-13.

Threat templates graphically portray how the threat might use its capabilities to perform the functions required to accomplish its objectives when not constrained by the effects of an operational environment. Threat templates are scaled to depict the threat's disposition and actions for a particular type of operation (for example, offense, defense, insurgent ambush, or terrorist kidnapping). Threat templates are the result of careful analysis of a threat's capability, vulnerabilities, doctrinal principles, and preferred tactics, techniques, and procedures that, in turn, lead to developing threat models and situation templates (refer to ATP 2-01.3). When possible, IO planners place these threat templates on a terrain product (such as a paper or digital map), adjusting time and distance relationships as necessary, but without violating the threat's fundamental doctrinal precepts. When not practical to overlay these templates on a terrain product, templates nonetheless depict doctrinal interrelationships of threat information warfare forces, key personnel, capabilities, and assets.

In terms of threat information warfare, threat templates seek to depict doctrinal information usage and flow, decision-making nodes, and locating IRCs, informational systems, sub systems, and associated assets. IO planners typically use three templates:

- Decision-making or information exchange template.
- Information infrastructure template.
- Information tactics template.

IO planners coordinate with the intelligence staff officer to incorporate information-related threat templates into the threat model. This coordination creates accurate situation templates and subsequent COAs in Step 4 of IPB. Threat templates allow the staff to fuse all relevant combat information and identify intelligence gaps. Further, they enable the staff to predict threat activities—in this case, in the information environment—and adopt COAs, as well as synchronize information collection.

Decision-Making Template

Also termed an information exchange template, this model considers and then depicts who makes or supports decisions and how they exchange information to support their decision making. It reveals human nodes and links that a threat organization uses to exchange information, with particular emphasis on ways the threat commander receives and disseminates information. Developing this template requires an understanding of threat organizational structures, critical links and interrelationships, and key personnel affecting the decision-making process.

Information Infrastructure Template

This template considers and then depicts the assets and means the threat employs to exchange information. If the decision-making template focuses on who is involved with information exchange, the infrastructure template focuses on what enables them to exchange that information. It depicts known infrastructure to exchange information internally and externally. Examples include satellite uplinks or downlinks, radio antennas, cell towers, couriers, and face-to-face interactions.

Information Tactics Template

The tactics template models how the threat arrays or employs its information assets and capabilities. While the first two templates do not necessarily have to be overlaid on terrain, the tactics template works best depicted as an overlay, so that staffs can clearly see and understand time and distance relationships. Not every adversary will have formal organizations or doctrine for employing information assets and capabilities; thus, the IO officer carefully avoids mirroring U.S. doctrine, capabilities, and methods onto the threat.

B. Threat Center of Gravity Analysis

Ref: *ATP 3-13.1, The Conduct of Information Operations (Oct '18), pp. 2-13 to 2-15.

An IO planner uses a COG analysis to identify threat capabilities, requirements, and vulnerabilities. The IO officer does not conduct a separate COG analysis but participates in and contributes to the staff COG effort, led by the intelligence staff officer. The IO officer brings to this effort expertise in the information environment.

COG analysis, with an emphasis on the information environment, is used to—

- Identify potential threat COGs.
- Identify critical capabilities.
- Identify critical requirements for each critical capability.
- Identify critical vulnerabilities for each critical requirement.
- Prioritize critical vulnerabilities.

Identify Potential Threat Centers of Gravity

In this step, the staff visualizes the threat as a system of functional components. Based upon how the threat organizes, fights, makes decisions, and uses its physical and psychological strengths and weaknesses, the staff selects the threat's primary source of moral or physical strength, power, and resistance. Depending on the level (strategic, operational, and tactical), COGs may be tangible entities or intangible concepts. To test the validity of the COG, the staff asks: "Is the COG capable of achieving the threat's objective?" The COG is supported, not supporting; if something provides support or contributes to a function that ultimately achieves the threat's objective, then it is a capability or a requirement, not a COG. Typically, a threat COG in the information environment is the threat's information position, which is a way of describing the quality of information the threat possesses and its ability to use that information.

Identify Critical Capabilities

The IO planner analyzes each COG to determine what primary abilities (functions) the threat possesses in the context of the operational area and friendly mission that can prevent friendly forces from accomplishing the mission. Critical capabilities are not tangible objects; rather, they are threat functions. To test the validity of a critical capability, the staff asks: "Is the identified critical capability a primary ability in context with the given missions of both threat and friendly forces? Is the identified critical capability directly related to the COG?" A critical capability is a crucial enabler for a COG to function and, as such, is essential to accomplishing the adversary's specified or assumed objectives.

Note. The threat's critical capabilities relate to the functions in the information environment— collect, protect, and project.

Identify Critical Requirements for Each Critical Capability

The IO planner analyzes each critical capability to determine what conditions, resources, or means enable threat functions or mission. To test validity of a critical requirement, the staff asks: "Will an exploitation of the critical vulnerability disable the associated critical requirement? Does the friendly force have the resources to affect the identified critical vulnerability?" If either answer is no, then the IO planner must review the threat's identified critical factors for other critical vulnerabilities or reassess how to attack the previously identified critical vulnerabilities with additional resources.

Note. Critical requirements usually are tangible elements such as communications means, nodes, or key communicators.

Identify Critical Vulnerabilities for Each Critical Requirement

The IO planner analyzes each critical capability to determine which critical requirements (or components thereof) are vulnerable to neutralization, interdiction, or attack. As a planner develops the hierarchy of critical requirements and critical vulnerabilities, the staff seeks interrelationships and overlapping between the factors to identify critical requirements and critical vulnerabilities that support more than one critical capability. When selecting critical vulnerabilities, a critical-vulnerability analysis is conducted to pair critical vulnerabilities against friendly capabilities.

Note. Critical vulnerabilities may be tangible structures or equipment, or intangible perception, populace belief, or susceptibility.

Prioritize Critical Vulnerabilities

A tool for prioritizing critical vulnerabilities is CARVER, which stands for criticality, accessibility, recuperability, vulnerability, effect, and recognizability. As a methodology or process, CARVER weighs and ranks six target criteria for targeting and planning decisions. The IO planner applies the six criteria against the critical vulnerability to determine impact on the threat organization as follows:

- **Criticality** is estimating the critical vulnerability's or target's importance to the enemy. Vulnerability will significantly influence the enemy's ability to conduct or support operations. As applied to targeting, criticality means target value and relates to how much a target's destruction, denial, disruption, and damage will impair the enemy or adversary's political, economic, or military operations or how much a target component will disrupt the function of a target complex.
- **Accessibility** is determining whether the critical vulnerability or target is accessible to the friendly force; it is the ease with which a target can be reached.
- **Recuperability** is evaluating how much effort, time, and resources the enemy or adversary must expend if the critical vulnerability or target is successfully affected.
- **Vulnerability** is determining whether the friendly force has the means or capability to affect the critical vulnerability or target using available assets. A target is vulnerable if friendly forces can attack it.
- **Effect** is determining the extent of the effect achieved if the critical vulnerability is successfully exploited. Effect means the impact on the enemy or adversary decision maker or makers. A target should not be attacked unless it can achieve the desired military effect.
- **Recognizability** is determining if the critical vulnerability or target, once selected for an exploitation, can be identified during the operation by the friendly force, and can be assessed for the impact of the exploitation.

The resulting analysis provides a prioritized list of objectives or targets that can then be discussed in context of each possible COA, aiding COA analysis. Each COA will dictate the capability to be employed.

Refer to ATP 3-05.20 for an overview of Army special operations forces targeting methodology that includes COG analysis and CARVER criteria; see ATP 2-33.4 and ATP 3-60 for the Army use of CARVER as a target value analysis tool).

Note. Planners also use COG analysis to identify friendly COGs, capabilities, requirements, and vulnerabilities and CARVER to identify friendly targets that are vulnerable to attack and for defensive purposes.

Combined Information Overlay (CIO)

Ref: *ATP 3-13.1, The Conduct of Information Operations (Oct '18), pp. 2-16 to 2-17.

In addition to the running estimate, IPB to support IO results in producing a graphic visualization product known as the combined information overlay (CIO). The CIO results from the prior analysis conducted in Steps 1 through 4, aggregating the information, threat, and situation templates (or overlays) to depict where and how aspects—such as infrastructure, terrain, and populace—can affect military operations. In certain instances, the IPB may require more than one CIO to capture the full complexity of the information environment.

The CIO gives the commander and the staff a visual depiction of the ways in which information affects the AO. Similar to the modified combined obstacle overlay, which the intelligence staff officer develops during the IPB, the CIO is a simplified depiction of numerous interconnected variables. The CIO is a tool to visualize a collection of inputs that can never be completely synthesized. As such, it never becomes a final product; it is continually updated as new information arises and as time and staffing permits.

Reachback capabilities, such as provided by the 1st IO Command, sometimes provide a starting point for a CIO, but the IO working group must verify and refine these products with more localized analysis. The IO officer, aided by the IO working group, is ultimately responsible for the product. Although the CIO may include classified information, particularly when dealing with technical or military aspects of an operational environment or intelligence products, it primarily consists of open-source and publically available information that is useful once validated. With a request for information, the IO officer can obtain additional information about the threat from the intelligence staff.

Note. Using open-source and publically available information for other than intelligence purposes should not be confused with open-source intelligence (known as OSINT). Only intelligence personnel conduct open-source intelligence (refer to ATP 2-22.9 for more on this topic).

A thorough understanding of the current state of the information environment, local communications means, methods, trusted sources, key influencers, established cognitive patterns, cultural norms, perspectives, historical narrative, system of opposition, and adversary and HN IRCs is critical to the development of the commander's communication synchronization effort.

Significant characteristics, further analyzed within the physical, informational, and cognitive dimensions, can be graphically represented on a **combined information overlay.** The analyst can use this overlay to identify strengths and/or vulnerabilities of the information environment that can be exploited by friendly or adversary forces. The adversary mindset should be evaluated to determine the probable state of morale in both the civil and military population. Morale is a significant factor not only in assessing the overall capability of a military force, but also in evaluating the extent to which the civil populace will support military operations. The degree of regime loyalty should be assessed not only for the populace but also, if possible, for individual leaders. Depending on the situation, factors such as ethnic, religious, political, or class grievances or differences may be exploitable for military information support operations (MISO) purposes. Psychological profiles on military and political leaders may facilitate understanding an adversary's behavior, evaluating an adversary's vulnerability to deception, and assessing the relative probability of an adversary's adopting various COAs.

- JP 2-01.3, Joint Intelligence Preparation of the Operational Environment (May '14), pp. III-23 to III-24.

Figure 2-5 below illustrates a sample CIO. What appears in or on the CIO depends on the situation, mission, commander preferences, and the resulting analysis. Templates include a combination of narrative (descriptive) elements, pictorial elements, and graphical elements. Whether the "so what" statement appears on the template itself or in accompanying notes, it needs to be conveyed concisely to the commander. The proportion of one element to the others depends on the conclusions the IO officer reaches and a judgement call on the best way to convey these conclusions.

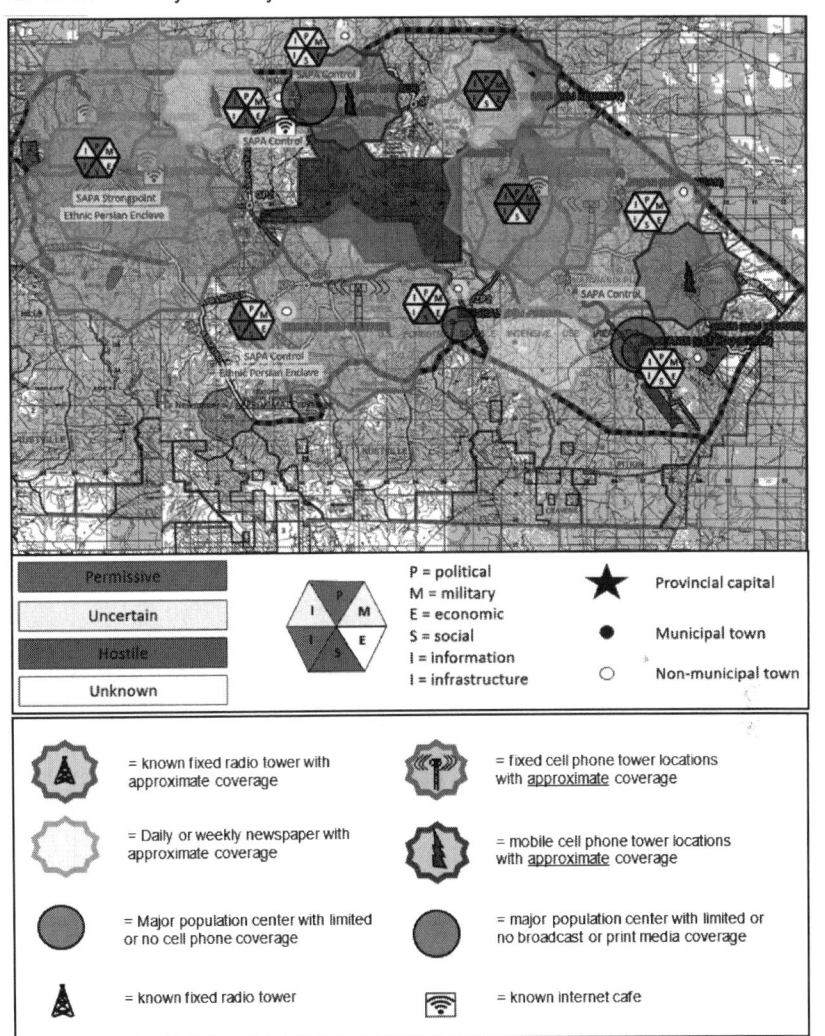

Figure 2-5. Example of combined information overlay.

Step 4: Determine Threat Courses of Action

Developing a threat COA is a six-step process that requires an understanding of the threat characteristics and the effects of terrain, weather, and civil considerations on operations (refer to ATP 2-01.3 for a detailed discussion on the threat). These steps include:

- Identify likely objectives and end state.
- Identify the full set of COAs available to the threat.
- Evaluate and prioritize each threat COA.
- Develop each COA in the detail that time allows.
- Identify high-value targets for each COA.
- Identify initial collection requirements for each COA.

The IO officer or planner filters each step of the process through an information lens, determining possible COAs that rely on the information environment to achieve an advantage. When developing each COA, the IO officer or planner coordinates closely with the intelligence staff officer to ensure each situation template depicts where, when, and why the threat employs its information systems and capabilities. The IO officer or planner develops IO-specific situation templates as a first step in the coordination process. These templates do not stand alone; instead, they contribute to the intelligence staff officer's situation templates. Continual coordination during IPB ensures that the staff develops the most accurate threat COAs.

While threat templates reflect how the threat should operate based on doctrine or preferred methods, the situation template conveys how the threat actually operates and employs its forces and capabilities based on an operational environment.

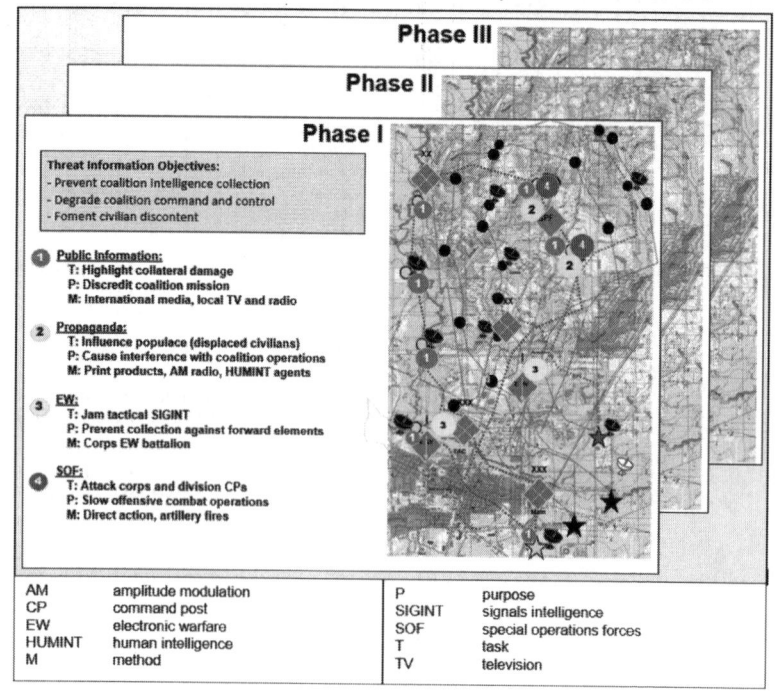

Figure 2-4. Example information situation template.

III. IO & the MDMP

Chap 4

Ref: *FM 3-13, Information Operations (Dec '16), pp. 4-2 to 4-29. (*See note p. 1-2.)

Commanders use the MDMP to understand the situation and mission confronting them and make informed decisions resulting in an operations plan or order for execution. Their personal interest and involvement is essential to ensuring that IO planning is integrated into MDMP from the beginning and effectively supports mission accomplishment.

See pp. 2-22 to 2-25 for related discussion of IO planning as related to the joint planning process (JPP).

IO planning is integral to several other processes, to include intelligence preparation of the battlefield (IPB) and targeting. The G-2 (S-2) and fire support representatives participate in the IO working group and coordinate with the IO officer to integrate IO with their activities and the overall operation.

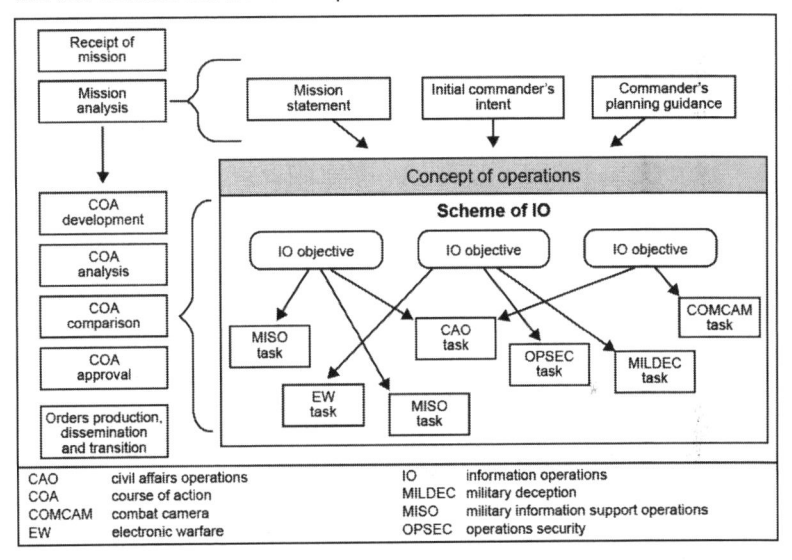

CAO	civil affairs operations	IO	information operations
COA	course of action	MILDEC	military deception
COMCAM	combat camera	MISO	military information support operations
EW	electronic warfare	OPSEC	operations security

Figure 4-1. Relationship among the scheme of IO, IO objectives, and IRC tasks.

Commanders use their mission statement for the overall operation, the IO mission statement, scheme of IO, IO objectives, and IRC tasks to describe and direct IO, as seen in fig. 4-1. *See pp. 4-3 to 4-16 for in-depth discussion IO mission statement, scheme of IO, IO objectives, and IRC tasks (synchronization of IRCs).*

Refer to BSS7: The Battle Staff SMARTbook, 7th Ed., updated for 2023 to include FM 5-0 w/C1 (2022), FM 6-0 (2022), FMs 1-02.1/.2 (2022), and more. Focusing on planning & conducting multidomain operations (FM 3-0), BSS7 covers the operations process; commander/ staff activities; the five Army planning methodologies; integrating processes (IPB, information collection, targeting, risk management, and knowledge management); plans and orders; mission command, command posts, liaison; rehearsals & after action reviews; operational terms & military symbols.

(INFO Planning) III. IO & the MDMP 4-35

Scheme of IO *See pp. 4-12 to 4-13.*

The scheme of IO is a clear, concise statement of where, when, and how the commander intends to employ and synchronize IRCs, to create effects in and through the information environment to support overall operations and achieve the mission. Based on the commander's planning guidance, to include IO weighted efforts, the IO officer develops a separate scheme of IO for each course of action (COA) the staff develops. IO schemes of support are written in terms of IO objectives—and their associated weighted efforts—and IRC tasks required to achieve these objectives. For example, the overall scheme may be oriented primarily on defending friendly information but also include attack and stabilize objectives.

IO Objectives *See pp. 4-14 to 4-15.*

IO objectives express specific and obtainable outcomes or effects that commanders intend to achieve in and through the information environment. In addition to be being specific, these objectives are measurable, achievable, relevant, and time-bounded (or SMART), which facilitates their attainment and assessment (see chapter 8).

IO objectives serve a function similar to that of terrain or force-oriented objectives in maneuver operations. They focus the IO effort on achieving synchronized IRC effects, at the right time and place, to accomplish the unit's mission and support the commanders' intent and concept of the operation.

Accurate situational understanding is key to establishing IO objectives. Operational- and tactical-level IO objectives must nest with strategic theater objectives. Joint and component staffs develop IO objectives to help integrate and synchronize their campaigns and major operations.

The IO officer develops objectives as part of developing the scheme of IO during COA development. These objectives help the staff determine tasks to subordinate units during COA development and analysis.

IRC Tasks *See pp. 4-14 to 4-15.*

Tasks are developed to support accomplishment of one or more IO objectives. These tasks are developed specifically for a given IRC. In concert with IRC representatives, the IO officer develops tasks during COA development and finalizes them during COA analysis. During COA development and COA analysis, tasks are discussed in general terms but not assigned to a subordinate unit. During orders production, these tasks are assigned to IRC units.

Step I. Receipt of Mission

Upon receipt of a mission, the commander and staff perform an initial assessment. Based on this assessment, the commander issues initial guidance and the staff prepares and issues a warning order (WARNORD). Between receiving the commander's initial guidance and issuing the WARNORD, the staff performs receipt of mission actions.

See pp. 4-48 to 4-49 for a summary of the inputs, actions and outputs required of the IO officer during mission analysis.

During receipt of mission, the IO officer—

- Reviews and updates the running estimate.
- Participates in the initial assessment.
- Provides input to the commander's initial guidance.
- Provides input to the warning order.
- Prepares for subsequent planning.

See facing page for an overview of the running estimate. See also pp. 4-6 to 4-7 for more in-depth discussion.

A. Review and Update the Running Estimate

Ref: *FM 3-13, Information Operations (Dec '16), p. 4-4. (See also pp. 4-6 to 4-7.)

Running estimates are integral to IO planning. A running estimate is the continuous assessment of the current situation, and is used to determine if the current operation is proceeding according to the commander's intent and if planned future operations are supportable (ADP 5-0). Running estimates help the IO officer record and track pertinent information about the information environment leading to a basis for recommendations to the commander.

The IO officer uses the running estimate to assist with completion of each step of the MDMP. An effective running estimate is as comprehensive as possible within the time available but also organized so that the information is easily communicated and processed. Normally, the running estimate provides enough information to draft the applicable IO sections of WARNORDs as required during planning and ultimately to draft applicable IO sections of the operation order (OPORD) or operation plan (OPLAN).

Running estimate

Forces/systems available
- 413 civil affairs BN
- 344 tactical MISO company
- 1/55 SIGNAL CO (-) 3x COMCAM teams
- 2x EC-130J Commando Solo @ CFACC
- OCO available

Information environment
- Radio is the best medium to reach the civilian population within AO SWORD, followed by social media
- Religious leaders within contested areas are key communicators to the population
- Displaced civilians in camps along main routes may impede coalition forces advance

Critical planning factors
- ATO cycle request 72 hours prior

Facts
- Civilian and government-controlled media outlets (radio, television) reach population within AO SWORD
- Adversary forces have used civilian radio stations to broadcast coalition forces troop movements and propaganda in the AO

Assumptions
- Civilian population will support HNSF and coalition forces once security is restored
- Civilian population will remain in place during attack unless there is a loss of essential services

Objectives
1. Influence civilian population to minimize interference with coalition forces information operations team to prevent civilian casualties
2. Disrupt enemy forces use of media outlets in order to support freedom of movement of coalition forces.

Specified tasks
- Identify key communicators within AO SWORD in order to deliver non-interference

Implied tasks
- Deny adversary use of social media messaging during decisive operations
- Develop Soldier and leader engagement, and MISO products to support non-interference

Limitations
- MISO messaging and OCO release authority held at CCDR

HPT nominations
- Denial of adversary social media site during decisive operations
- Identify tribal leaders

CCIR nominations
- Block axis of advance by civilian population during attack
- Damage to HN essential services infrastructure and religious structures

EEFI nominations
- N/A

Request for forces
- Request OCO to deny use of social media site during decisive operations

AO	area of operations		COMCAM	combat camera
ATO	air tasking order		EEFI	essential elements of friendly information
BN	battalion		HN	host nation
CCDR	combatant commander		HNSF	host-nation security forces
CCIR	commander's critical information requirement		HPT	high-payoff target
CFACC	combined force air component commander		MISO	military information support operations
CO	company		OCO	offensive cyberspace operations

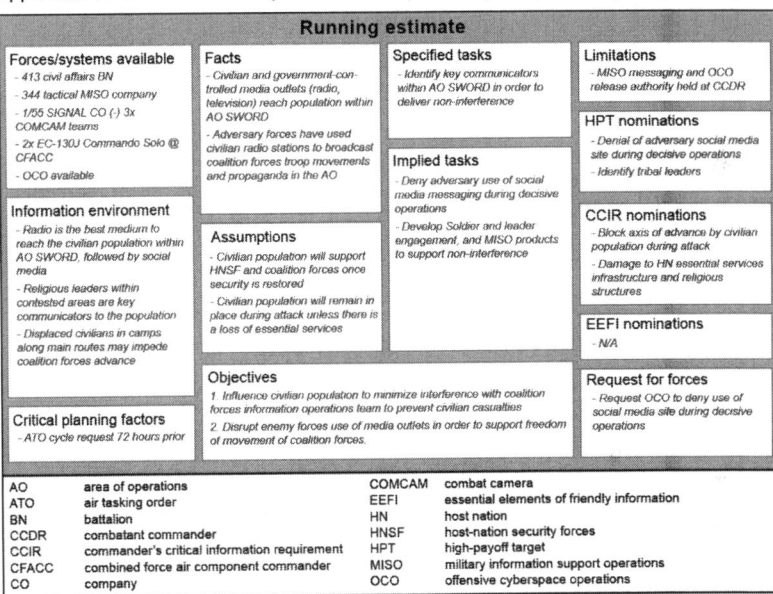

Figure 4-2. Example graphical IO running estimate.

Variations on the standard, narrative format, such as the example provided in figure 4-2, enable the IO officer to spotlight facts and assumptions, critical planning factors, and available forces. The latter of these requires input from assigned or available IRCs. The graphical format also offers a clear, concise mechanism for the IO officer to articulate recommended high-payoff targets, commander's critical information requirements, and requests for forces. Maintaining both formats simultaneously provides certain benefits: the narrative format enables the IO officer to cut-and-paste sections directly into applicable sections of orders; the graphical format enables the element to brief the commander and staff with a single slide.

Running estimate development never stops. The IO officer continuously maintains and updates the running estimate as pertinent information is received. While at home station, the IO officer maintains a running estimate on friendly capabilities. If regionally aligned, the unit prepares its estimate based on research and analysis of the information environment within its region and anticipated mission sets.

B. Participate in Commander's Initial Assessment

Initial assessment primarily focuses on time and resources available to plan, prepare and begin execution of an operation. The IO officer assesses readiness to participate in ADM and MDMP, as well as what external support might be necessary to ensure effective IO planning.

During the initial assessment, the IO officer establishes a battle rhythm, including locations, times, preparation requirements, and the anticipated schedule. Upon receiving a new mission, the IO officer begins gathering planning tools, including a copy of the higher command OPLAN or OPORD, maps of the area of operations, appropriate references, and the running estimate. During initial assessment, the IO officer also coordinates with organic, assigned, and available IRCs and subordinate units to gauge their planning readiness.

Initial time allocation is important to IO because some operations and activities require significant time to produce effects or for assessment. The time available may be a limiting factor for some IRCs. The IO officer identifies activities for which this is the case and includes these limitations in estimates and recommendations.

The commander determines when to execute time-constrained MDMP. Under time-constrained conditions, the IO officer relies on existing tools and products, either his or her own or those of higher headquarters. The lack of time to conduct reconnaissance requires planners to rely more heavily on assumptions and increases the importance of routing combat information and intelligence to the people who need it. A current running estimate is essential to planning in time-constrained conditions.

C. Provide Input to Commander's Initial Guidance

Commanders include IO-specific guidance in their initial guidance, as required. Examples include authorized movements of IRCs, initiation of information collection necessary to support IO, and delineation of IRs.

D. Provide Input to the Initial Warning Order

A WARNORD is issued after the commander and staff have completed their initial assessment and before mission analysis begins. It includes, at a minimum, the type and general location of the operation, initial timeline, and any movements or reconnaissance that need to be initiated. When they receive the initial WARNORD, subordinate units begin parallel planning.

Parallel planning and collaborative planning are routine MDMP techniques. The time needed to achieve and assess effects in the information environment makes it especially important to successful IO. Effective parallel or collaborative planning requires all echelons to share information fully as soon as it is available. Information sharing includes providing higher headquarters plans, orders, and guidance to subordinate IO officers or representatives.

Because some IRCs require a long time to plan or must begin execution early in an operation, follow-on WARNORDs may include detailed IO information. Although the MDMP includes three points at which commanders issue WARNORDs, the number of WARNORDs is not fixed. WARNORDs serve a purpose in planning similar to that of a fragmentary order (FRAGORD) during execution. Commanders issue both, as the situation requires. Possible IO officer input to the initial WARNORD includes:

- Tasks to subordinate units and IRCs for early initiation of approved IO actions, particularly for military deception operations and MISO.
- Essential elements of friendly information (EEFIs) to facilitate defend weighted efforts and begin the OPSEC process.
- Known hazards and risk guidance.
- Military deception guidance and priorities.

Step II. Mission Analysis

Commanders and their staff conduct mission analysis to better understand the situation and problem, and to identify the purpose of the operation. It is the most important step in MDMP and consists of 18 sub-steps, many of which are performed concurrently. (See FM 6-0, Chapter 9) The IO officer ensures each output or product from this step includes relevant factors or tie-ins. The IO officer also participates in other staff processes (such as IPB and targeting) to ensure IO is properly integrated. For the IO officer, mission analysis focuses on developing information and products that will be used during the rest of the operations process.

A. Analyze Higher Headquarters' Plan Or Order

Mission analysis begins with a thorough examination of the higher headquarters OPLAN/OPORD in terms of the commander's initial guidance. By examining higher echelon plans, commanders and staffs learn how higher headquarters plan to conduct IO and which resources and higher headquarters assets are available. The IO officer researches these plans and orders to understand the—

- Higher commander's intent and concept of operations.
- Higher headquarters area of operations and interest, mission and task constraints, acceptable risk,and available assets.
- Higher headquarters schedule for conducting the operation.
- Missions of adjacent units.

Planning to conduct IO without considering these factors may result in an uncoordinated operation, which will hamper overall mission effectiveness. A thorough analysis also helps to determine if additional, external IO support is necessary.

B. Perform Initial Intelligence Preparation of the Battlefield (IPB)

During mission analysis, the G-2 (S-2) prepares IPB products or updates existing products and the initial IPB is performed upon receipt of the mission. The G-2 (S-2), with assistance and input from other staff elements, uses IPB to define the area of operations/interest, describe its effects, evaluate the threat, and determine threat courses of action.

See pp. 4-40 to 4-41 for an overview and listing of possible IO-related factors to consider during each IPB step. During IPB, the IO officer works with the G-2 (S-2) to determine threat capabilities and vulnerabilities in the information environment regarding both the threat and other relevant targets and audiences in the area of operations.

See also pp. 4-17 to 4-34 for an in-depth discussion of information environment analysis (IO & IPB) from ATP 3-13.3.

C. Determine Specified, Implied, & Essential Tasks

While the staff determines specified, implied, and essential tasks the unit must perform, the IO officer identifies specified IO tasks in the higher headquarters OPLAN or OPORD. The IO officer also develops IO-related implied tasks that support accomplishing identified specified tasks. These identified tasks are the basis of the initial scheme of IO developed during COA development.

IO officers look for specified tasks that may involve IO in the higher headquarters OPLAN or OPORD, paying particular attention to:

- Paragraph 1, Situation.
- Paragraph 2, Mission.

Perform Initial IPB (Overview)

Ref: *FM 3-13, Information Operations (Dec '16), p. 4-4.

During mission analysis, the G-2 (S-2) prepares IPB products or updates existing products and the initial IPB is performed upon receipt of the mission. The G-2 (S-2), with assistance and input from other staff elements, uses IPB to define the area of operations/interest, describe its effects, evaluate the threat, and determine threat courses of action.

Figure 4-3 below provides an overview and listing of possible IO-related factors to consider during each IPB step. During IPB, the IO officer works with the G-2 (S-2) to determine threat capabilities and vulnerabilities in the information environment regarding both the threat and other relevant targets and audiences in the area of operations.

See pp. 4-17 to 4-34 for an in-depth discussion of information environment analysis (IO & IPB) from ATP 3-13.3.

Define the Operational Environment	Describe the Environmental Effects on Operations	Evaluate the Threat	Determine Threat COAs
Portions or aspects of the information environment that can effect friendly operations.	IE effects on decisionmakers, C2 or mission command systems, and decision-making processes.	Adversary and other group C2 systems, including functions, assets, capabilities, and vulnerabilities (both offensive and defensive).	How threats and other groups pursue operational or decisive advantage in the IE.
Features/activities that can influence information and threat command and control (C2) or friendly mission command systems.	How the IE relates to the area of operations.	Assets and functions (such as decisionmakers, C2 systems, and decision-making processes) that adversaries and others require to operate effectively.	How, when, where, and why (to what purpose) threats and other groups will use information-related capabilities to achieve their likely objectives.
Political and governmental structures and population demographics.	IE effects on friendly, threat, and other operations.		
Major cultures, languages, religions, and ethnic groups.	Combined effects of friendly, threat, and other information, and C2 or mission command systems on the information environment.	Adversary capabilities to attack friendly information systems and defend their own.	
Civilian communication and power infrastructures (both physical and informational).	Effects of terrain, weather, and other characteristics of the area of operations on friendly and enemy information and C2 or mission command systems.	Models of threat and other group C2 systems.	
Non-state actors, non-governmental organizations and significant non-threat groups.	Effect of public media or press on friendly and threat operations.	IO or information-related strength, vulnerabilities, and susceptibilities of adversaries and other groups.	
Types of and public access to media or press outlets.			
C2 command and control	**COA** Course of Action	**IE** Information environment	**IO** Information operations

Figure 4-3. IO-related factors to consider during IPB

Define the Information Environment

The information environment has always affected military operations. IO officers, working with the G-2 (S-2), use available intelligence to analyze the information environment and the threat's use of information. This information is submitted to the G-2 (S-2) to answer intelligence gaps that address how information environment factors affect operations. The G-2 (S-2) obtains the information from strategic and national-level databases, country studies, collection assets and, when necessary, other intelligence agencies.

As part of defining the battlefield environment, the G-2 (S-2) establishes the limits of the area of interest. The area of interest includes areas outside the area of operations that are occupied by threat or other forces/groups that can affect mission accomplishment. This fact is particularly true from an information environment perspective. The ability to obtain and pass information has vastly expanded the capacity of actors to affect areas of operations from anywhere. The IO officer ensures that the G-2 (S-2) considers this factor of the information environment in defining the area of interest for IPB.

One of the enabling activities of IO is analyzing and understanding the information environment in all its complexity. Using the IPB process to accomplish this task, the IO officer develops a series of information overlays, as well as combined information overlays, to depict the information aspects of the operational environment.

The IO officer provides input to help the G-2 (S-2) develop IPB templates, databases, social network diagrams, and other products that portray information about threats and other key groups or audiences in the areas of operation and interest. These products contain information about each group's leaders and decision makers. Information relevant to conducting IO includes, but is not limited to:
- Religion, language, culture, and internet activities of key groups and decision makers.
- Agendas of non-governmental organizations.
- Military and civilian communication infrastructures and connectivity.
- Population demographics, linkages, and related information.
- Location and types of radars, jammers, and other non-communication information systems.
- Audio, video, and print media outlets and centers; the populations they serve; and their dissemination characteristics, such as frequency, range, language, etc.
- Command and control or mission command vulnerabilities of friendly, adversary, and other forces or groups.
- Conduit analysis describing how threat decision makers receive information.

Threat templates portray how adversaries use forces and assets unopposed by friendly forces and capabilities. Threat templates are often developed before deployment. The G-2 (S-2) and IO officer may add factors from the information environment to a maneuver-based threat template, or they may prepare a separate IO threat template. The situation, available information, and type of threat affect the approach taken. IO-related portions of IPB products become part of paragraph 1b of the running estimate.

The G-2 (S-2) uses IPB to determine possible threat courses of action and arrange them in probable order of adoption. These courses of action, depicted as situation templates, include threat IRCs. A comprehensive IPB addresses threat offensive and defensive capabilities and vulnerabilities, and it is efficacious to friendly mission analysis to develop situation templates depicting how threats and others may employ these capabilities to achieve advantage.

IPB Support of Targeting
IPB identifies high-value targets (HVTs) and shows where and when they may be anticipated. Some of these HVTs are IO-focused or related, such as a specific population group within an area of operation. The G-2 (S-2) works with the IO officer to develop IO-related HVTs into high-payoff targets (HPTs) for the commander's approval. The IO officer determines which HPTs are related to one or more objectives and develops tasks to engage those targets during COA development and analysis.

See chap. 7, Fires & Targeting.

Other IPB Products
IPB identifies facts and assumptions concerning threats and the operational environment that the IO officer considers during planning. These are incorporated into paragraph 2 of the running estimate. The IO officer submits IRs to update facts and verify assumptions. Working with the G-2 (S-2) and other staff sections, the IO officer ensures IRs are clearly identified and requests for information (RFIs) are submitted to the appropriate agency when necessary. IPB may create priority intelligence requirements (PIRs) pertinent to IO planning. The IO officer may nominate these as commander's critical information requirements (CCIRs) and also identify OPSEC vulnerabilities. The IO officer analyzes these to determine appropriate OPSEC measures.

See pp. 4-17 to 4-34 for an in-depth discussion of information environment analysis (IO & IPB) from ATP 3-13.3.

- Paragraph 3, Execution, especially subparagraphs on IO, tasks to subordinate units, and CCIRs.
- Annexes and appendices that address intelligence, operations, fire support, rules of engagement,IO, IRCs, information collection, assessment, and interagency coordination.

Some IO specified tasks, such as support to the higher headquarters deception plan, become unit objectives. Others, particularly those that address only one IRC, are incorporated under IO objectives as tasks. As the staff identifies specified tasks for the overall operation, the IO officer deduces the steps that are necessary to accomplish these specified tasks. These tasks become IO implied tasks. Once the IO officer identifies specified and implied tasks and understands each task's requirements and purpose, essential tasks are identified. An essential task is a specified or implied task that must be executed to accomplish the mission. If the command must accomplish an IO task to accomplish its mission, that task is an essential task for the command and is included in the recommended mission statement.

D. Review Available Assets & Identify Resource Shortfalls

During this sub-step, the commander and staff determine if they have the assets required to perform the specified, implied, and essential tasks. The IO officer performs this analysis to determine if the requisite capabilities are on hand or available through coordination with higher echelons to achieve the effects in the information environment necessary to support the mission. At echelons below division, units have few organic IRCs other than movement and maneuver; Soldier and leader engagement; and presence, posture, and profile. If additional IRCs are required, the IO officer works with the operations officer to request these capabilities and ensure appropriate authorities exist. (See chapter 9 for further discussion of IO at brigade and below).

The IO officer compares available IRCs with the tasks that need to be accomplished to identify capability shortfalls and additional resources required. The IO officer considers how the following will affect attainment of IO objectives and whether additional capacity is required—

- Changes in task organization.
- Limitations of available units and IRCs.
- Nature of effects that need to be achieved in the information environment and the tasks to accomplish them.
- The need for redundancy or repetition to achieve desired effects.
- The level, quantity, and quality of expertise on hand.

E. Determine Constraints

A constraint is a restriction placed on the command by a higher command. A constraint dictates an action or inaction, thus restricting the freedom of action of a subordinate commander (FM 6-0). IO constraints include legal, moral, social, operational, and political factors. They also include limitations imposed by various authorities, such as the Secretary of Defense or U.S. ambassador. Constraints may be listed in the following paragraphs, annexes or appendices of the higher OPLAN/OPORD—

- Commander's intent and guidance.
- Tasks to subordinate units.
- Rules of engagement (no strike list, restricted target list)
- Civil affairs operations.
- MISO
- Fire support.

Constraints establish limits within which the commander can conduct IO. Constraints may also limit the use of military deception and some OPSEC measures. One output of this sub-step is a list of the constraints that the IO officer believes will affect the scheme of IO.

F. Identify Critical Facts & Develop Assumptions

Sources of facts and assumptions include existing plans, initial guidance, observations, and reports. Some facts concerning friendly forces are determined during the review of the available assets. During IPB, the G-2 (S-2), with assistance from the IO officer and other staff elements, develops facts and assumptions about threats and others, the area of operations, and the information environment. The following categories of information are important to the IO officer—

- Intelligence on threat commanders and other key leaders.
- Threat morale.
- Media and/or press coverage of threat and other relevant audiences in the area of operations.
- The weather.
- Dispositions of adversary, friendly, and other key groups.
- Available troops, unit strengths, and materiel readiness.
- Friendly force IO vulnerabilities.
- Threat and other key group IO vulnerabilities.

The primary output of this sub-step is a list of facts and assumptions that concern IO. These are placed in paragraph 1c of the running estimate. The IO officer prepares and submits to appropriate agencies IO IRs for information that would confirm or disprove facts and assumptions. The IO officer reviews facts and assumptions as information is received and revises facts or converts assumptions into facts.

G. Begin Risk Management

Commanders and staffs assess risk when they identify hazards, regardless of type. The IO officer assesses IO-associated risk throughout the operations process. The G-3 (S-3) incorporates the IO risk assessment into the command's overall risk assessment.

IO-related hazards fall into three categories:

- OPSEC vulnerabilities, including hazards associated with compromise of essential elements of friendly information.
- Mission command vulnerabilities, including those associated with the loss of critical assets or identified during the vulnerability assessment.
- Hazards associated with executing IO tasks.

During mission analysis, the IO officer assesses primarily OPSEC- and mission command-related hazards, as well as hazards associated with IO-related specified and implied tasks identified up to this point in mission analysis. The list of task-associated hazards is refined during COA development, after articulating IRC tasks that support IO objectives. The IO element uses experience in previous operations as a means of identifying known or expected hazards, and IRC representatives often best articulate hazards associated with their tasks.

As with all operations, IO entails risk. Resource constraints, combined with threat reactions and initiatives, reduce the degree and scope of advantage possible in the information environment. Risk assessment is one means commanders use to allocate resources. Staffs identify which hazards pose the greatest threat to mission accomplishment. They then determine the resources required to control them and estimate the benefits gained. This estimate of residual risk gives commanders a tool to help decide how to allocate resources and where to accept risk.

H. Develop Commander's Critical Information Requirements (CCIRs) & Essential Elements Of Friendly Information (EEFIs)

A commander's critical information requirement (CCIR) is an information requirement identified by the commander as being critical to facilitating timely decision making (JP 3-0). CCIRs include priority intelligence requirements (PIRs) and friendly forces information requirements (FFIRs). Staff sections, including the IO officer, recommend CCIRs to the G-3 (S-3). In a time-constrained environment, the staff may collectively compile this information. The G-3 (S-3) presents a consolidated list of CCIRs to the commander for approval. The commander determines the final CCIRs.

Establishing CCIRs is one means commanders use to focus assessment efforts. CCIRs change throughout the operations process because the information that affects decision making changes as an operation progresses.

During planning, staff sections establish IRs to obtain the information they need to develop the plan. Commanders produce CCIRs to support decisions they must make regarding the form the plan takes.

During preparation, the focus of IRs and CCIRs shifts to decisions required to refine the plan. During execution, commanders establish CCIRs that identify the information they need to make execution and adjustment decisions.

During mission analysis, the IO officer derives the information needed by the commander to determine how to employ IO during the upcoming operation. The IO officer recommends the IO IRs to be included in the CCIRs. This sub-step produces no IO-specific product unless the IO officer recommends one or more IO IRs as CCIRs. However, at this point, the IO officer should have assembled a list of IO IRs and submitted friendly-force-related IRs to the G-3 (S-3) and threat-related IRs to the G-2 (S-2).

The following is an example of CCIRs for a stability operation in which an information operation is the decisive operation:

- Who are the municipality's key players in ethnic violence?
- What are the interests of the political parties?
- Who are the formal and informal leaders within the political parties?
- How can friendly forces exploit political party interests to garner support?
- Which party represents the majority of the people, but also actively support progress within the municipality?
- What is the status of IRCs within the area of operations?

In addition to nominating CCIRs to the commander, the staff also identifies and nominates essential elements of friendly information, or EEFIs. EEFIs are elements of information to protect rather than to collect, and identify those elements of friendly force information that, if compromised, would jeopardize mission success. Although EEFIs are not CCIRs, they have the same priority as CCIRs and require approval by the commander. Like CCIRs, EEFIs change as an operation progresses (FM 6-0).

4-60. Submission of IO-focused requirements for potential inclusion as CCIRs, along with other CCIRs, enable the staff to develop the initial information collection plan. Approval of EEFIs enable the staff to plan and implement friendly force information protection measures, such as provided by military deception and OPSEC.

I. Develop the Initial Information Collection Plan

The staff identifies information gaps, especially those needed to answer IRs. The IO officer identifies gaps in information needed to support IO planning, execution and assessment. These are submitted to the G-2 (S-2) as IO IRs. The initial information collection plan sets the priorities for information collection in order to answer CCIRs. The G-3 (S-3) issues the information collection plan as part of a WARNORD, a FRAGORD or an OPORD. Within these orders, the information collection plan is found in Annex L.

J. Update Plan for the Use of Available Time

At this point, the G-3 (S-3) refines the initial time plan developed during receipt of mission. The IO officer provides input specifying the long lead-time items associated with certain IRC tasks (such as military deception and MISO). Upon receiving the revised timeline, the IO officer compares the time available to accomplish IRC tasks with the command's and threat's time lines, and revises the IO time allocation plan accordingly. The IO product for this sub-step is a revised time plan.

K. Develop Initial Themes and Messages

4-63. Gaining and maintaining the trust of relevant audiences and actors is an important aspect of operations. Faced with a diverse array of individuals, organizations, and publics who affect or are affected by their unit's operations, commanders identify and engage entities vital to operational success. The behaviors of these entities can aid or complicate the friendly forces' challenges as commanders strive to accomplish missions.

The IO officer does not develop themes and messages. This is done by the public affairs officer and MISO element. The public affairs officer adjusts and refines themes and messages received from higher headquarters for use by the command. These themes and messages are designed to inform specific domestic and foreign audiences about current or planned military operations. The Office of the Secretary of Defense, Department of State, or geographic combatant commander (depending on the operation) provides applicable themes to MISO forces, which then develop actions and messages. The highest level MISO element in theater adjusts or refines the themes depending on the situation. It employs themes and messages as part of planned activities designed to influence specific foreign targets and audiences for various purposes that support current or planned operations.

The commander and the chief of staff approve all themes and messages used to support operations in their area of responsibility. Although the IO officer does not develop themes and messages, they do assist the G-3 (S-3) and the commander to de-conflict and synchronize IRCs used specifically to execute actions for psychological effect and deliver messages during operations.

L. Develop a Proposed Problem Statement

Problem statements are typically developed during design. If this did not occur prior to mission analysis, it is accomplished during this step of the MDMP. If done during design, the commander and staff revise the problem statement based on their enhanced understanding of the situation. The key is identifying the right problem to solve, because it leads to the formulation of specific solution-sets. In identifying the problem, the commander and staff compare the current situation to the desired end state and list issues that impede the unit from achieving this end state.

Given the increasing impact of the information environment, the prevailing problem or impeding issues are likely to be information-related. Also, information-related problems can be more complex and multi-dimensional than geographical or technological problems or impediments. Therefore, it is essential to spend the time necessary to articulate the problem and impediments as carefully and clearly as possible.

M. Develop a Proposed Mission Statement

The G-3 (S-3) or executive officer develops the proposed restated mission based on the force's essential tasks, which the commander approves or modifies. The IO officer provides input based on the current IO running estimate. The mission statement includes any identified IO essential tasks.

Mission statements should use tactical mission tasks, which are specific activities performed by units while executing a form of tactical operation or form of maneuver (Refer to ATP 3-90.1). IO tasks do not always neatly fit into this framework, as they are rarely terrain- or combined arms-based. However, if they are framed in terms of friendly force actions (for example, influence the population in a certain area) or effects on threat forces (deceive the threat's reserve forces commander), and if they support the commander's intent and planning guidance, then they can be integrated effectively into the restated mission.

The IO officer also develops an IO mission statement that guides IO execution and ensures IO objectives are accomplished. The IO mission statement is explicitly stated in Appendix 15 (Information Operations) to Annex C (Operations) of the base order.

Refer to FM 6-0, Appendix C, for additional details on functional area mission statements.

N. Present the Mission Analysis Briefing

The staff briefs the commander on the results of its mission analysis. The mission analysis briefing is an essential means for the commander, staff, subordinates and other partners to develop a shared understanding of the upcoming operation and the interrelationships among the mission variables and elements of combat power. IO input is based on its running estimate, analysis in the foregoing steps, and how IO impacts or is impacted by other areas and functions. Time permitting, the staff employs the outline provided in figure 4-4.

See facing page, fig. 4-4. Information operations input to mission analysis briefing.

O. Develop and Issue Initial Commander's Intent

The commander's intent is a clear and concise expression of the purpose of the operation and the desired military end state that supports mission command, provides focus to the staff, and helps subordinate and supporting commanders act to achieve the commander's desired results without further orders, even when the operation does not unfold as planned (JP 3-0). The IO officer develops recommended input to the commander's intent and submits it to the G-3 (S-3) for the commander's consideration. When developing recommended input to the commander's intent, the IO officer assists the commander in visualizing and understanding the information environment, ways it will affect operations, and ways that IO can affect the information environment to the commander's advantage.

P. Develop and Issue Initial Planning Guidance

After approving the restated mission and issuing the intent, commanders provide additional guidance to focus staff planning activities. As appropriate, the commander includes their visualization of IO in this guidance. Commanders consider the following when developing their IO planning guidance:

- Aspects of higher headquarters IO policies or guidance that the commander wants to emphasize.
- Aspects of the mission for which IO is most likely to increase the chance of success or which maybe IO-dominant.
- Risks they are willing to take with respect to IO.
- IO decisions for which they want to retain or delegate authority.

IO Input to Mission Analysis Briefing

Ref: *FM 3-13, Information Operations (Dec '16), p. 4-4.

Outline	Information Operations Input
Mission and commander's intent of headquarters two echelons up.	IO specified and implied tasks
Mission commander's intent, concept of operations of headquarters one echelon up.	IO specified and implied tasks
Proposed problem statement	Information-related problems within the IE.
Proposed mission statement	IO essential tasks
Review of commander's initial guidance	• Guidance concerning IO • EEFI and CCIR • Essential narrative elements
Initial IPB products	Information overlays
Specified, implied, and essential tasks	Specified, implied, and essential tasks for IO
Constraints	Any constraints placed on the command affecting IO
Initial risk assessment	• Recommended OPSEC planning guidance • Recommended controls to protect information-related vulnerabilities and critical assets. • Recommended controls for risk associated with IO tasks
Proposed themes and messages	Possible overlaps or conflicts among IRCs used to disseminate approved themes and messages.
Proposed timeline	• Time required to accomplish IO • Analysis of time needed versus time available

CCIR	EEFI	IO	IPB
Commander's Critical Information Requirements	Essential element of friendly information	Information Operations	Intelligence preparation of the battlefield

Figure 4-4. Information operations input to mission analysis briefing.

Mission Analysis
(Summary of IO Inputs, Actions & Outputs)

Ref: *FM 3-13, Information Operations (Dec '16), table 4-1, pp. 4-14 to 4-18.

Table 4-1 provides a summary of the inputs, actions and outputs required of the IO officer. Only those sub-steps within mission analysis with significant IO activity are listed.

MDMP Sub Step	Inputs	IO Officer Actions	IO Officer Outputs
Conduct IPB	• Higher HQ IPB • Higher HQ running estimates • Higher HQ OPLAN or OPORD • Higher HQ combined information overlay	• Develop IPB products • Analyze and describe the information environment in the unit's area of operations and its effect on friendly, neutral, adversary, and enemy information efforts • Identify threat information capabilities and vulnerabilities • Identify gaps in current intelligence on threat information efforts • Identify IO-related high-value targets • Determine probable threat information-related COAs • Assess the potential effects of IO on friendly, neutral, adversary, and enemy operations • Determine threat's ability to collect on friendly critical information • Determine additional EEFIs (OPSEC)	• Input to IPB products • IRs to G-2 (S-2), as well as the foreign disclosure officer • Refined EEFIs (OPSEC)
Determine Specified, Implied, and Essential Tasks	• Specified tasks from higher HQ OPLAN or OPORD • IPB and combined information overlay products	• Identify specified tasks in the higher HQ OPLAN or OPORD • Develop implied tasks • Determine if there are any essential tasks • Develop input to the command targeting guidance • Assemble critical and defended asset lists, especially low density delivery systems • Determine additional EEFIs (OPSEC)	• Specified, implied and essential tasks • List of IRCs to G-3 (S-3) • Input to command targeting guidance • Refined EEFIs (OPSEC)
Review Available Assets	• Current task organization for information related capabilities • Higher HQ task organization for information related capabilities • Status reports • Unit standard operating procedure	• Identify friendly IRCs (include capabilities that are joint, interorganizational, and multinational) • Analyze IRC command and support relationships • Determine if available IRCs can perform tasks necessary to support lines of operation or effort • Identify additional resources (such as air assets) needed to execute or support IO	• List of available IRCs [IO running estimate paragraph 1b(4)] • Request for additional IRCs, if required
Determine Constraints	• Commander's initial guidance • Higher HQ OPLAN or OPORD	• Identify IO-related constraints	• List of constraints (IO appendix to Annex C; scheme of IO or coordinating instructions)

MDMP Sub Step	Inputs	IO Officer Actions	IO Officer Outputs
Identify Critical Facts and Develop Assumptions	• Higher HQ OPLAN or OPORD • Commander's initial guidance • Observations and reports	• Identify facts and assumptions affecting IRCs • Submit IRs that will confirm or disprove assumptions • Identify facts and assumptions regarding OPSEC indicators that identify vulnerabilities	• List of facts and assumptions (IO running estimate paragraph 1c.) • IRs that will confirm or disprove facts and assumptions
Begin Risk Management	• Higher HQ OPLAN or OPORD • IPB • Commander's initial guidance	• Identify and assess hazards associated with IO • Propose controls • Identify OPSEC indicators • Assess risk associated with OPSEC indicators to determine vulnerabilities • Establish OPSEC measures	• List of assessed hazards • Input to risk assessment • Develop risk briefing matrix • List of provisional OPSEC measures
Develop Initial CCIRs and EEFIs	• IO IRs	• Determine information the commander needs in order to make critical decisions concerning IO efforts • Identify IRs to recommend as commander's critical information requirements	• Submit IRs
Determine Initial Information Collection Plan	• Initial IPB • PIRs or IO IRs	• Identify gaps in information needed to support planning, execution, and assessment of early initiation actions • Confirm that the initial information collection plan includes IRs concerning enemy capability to collect EEFIs	
Update Plan for the Use of Available Time	• Revised G-5 (S-5)/G-3 (S-3) plans timeline	• Determine time to accomplish IO planning requirements • Assess viability of planning timeline vis-à-vis higher HQ timeline and threat timeline as determined during IPB • Refine initial time allocation plan	• Timeline (provided to G-5 (S-5), with emphasis on the effect(s) of long-lead time events
Develop Initial Themes and Messages	• Public affairs themes and messages adjusted and refined from higher HQ • MISO actions and messages adjusted and refined from higher HQ	• Assess impact of initial themes and messages on the information environment • Assess whether planned IO effects will reinforce themes and messages • Contribute to development of talking points aimed at influencing perceptions and behaviors	• PA themes/ messages and MISO actions/ messages de-conflicted • Initial list of talking points • IRC actions to disseminate approved messages/ talking points
Issue a Warning Order	• Commander's intent and guidance • Approved restated mission and initial objectives • Mission analysis products	• Prepare input to the warning order. Input may include — - Early tasking to subordinate units - Initial mission statement - OPSEC planning guidance - Reconnaissance and surveillance tasking • Military deception guidance	• Input to mission, commander's intent, commander's critical information requirements, and concept of the operations

COA course of action
EEFI essential element of friendly information
G-2 assistant chief of staff, intelligence
G-3 assistant chief of staff, operations
G-5 assistant chief of staff, plans
HQ headquarters
IO information operations
IPB intelligence preparation of the battlefield
IR information requirements
IRC information related capability
MISO military information support operations
OPLAN operations plan
OPORD operations order
OPSEC operations security
PA public affairs
PIR priority intelligence requirement
S-2 battalion or brigade intelligence officer
S-3 battalion or brigade operations staff officer
S-5 battalion or brigade plans staff officer

(INFO Planning) III. IO & the MDMP   4-49

Planning guidance focuses on the command's essential tasks. Commanders may give guidance for IO separately or as part of their overall guidance. This guidance includes any identified or contemplated IO objectives, stated in finite and measurable terms. It may also include OPSEC planning guidance, military deception guidance, and targeting guidance.

Factors that the IO officer considers when recommending input to initial planning guidance include:
- The extent that the command is vulnerable to hostile information-based warfare.
- Specific IO actions required for the operation.
- The command's capability to execute specific actions or weighted efforts.
- Additional information needed to conduct IO.

Q. Develop Course of Action Evaluation Criteria

Course of action (COA) evaluation criteria are used during course of action analysis and comparison to measure the relative effectiveness and efficiency of COAs to another. They are developed during this sub-step to enhance objectivity and lessen the chances of bias. Typically, the chief of staff will develop the criterion and associated weight. The IO officer will propose possible refinement to ensure consideration of IO factors affecting success or failure and then employ approved criteria to score each COA.

R. Issue Warning Order

As the mission and operation dictate, the WARNORD will include essential IO tasks within the mission statement. It will note changes to task organization involving IRC or IO units and address IO factors in other relevant paragraphs, sections, or annexes, as appropriate.

See previous pages (pp. 4-48 to 4-49) for a summary of the inputs, actions and outputs required of the IO officer during mission analysis.

Step III. Course of Action Development

After the mission analysis briefing, the staff begins developing COAs for analysis and comparison based on the restated mission, commander's intent, and planning guidance. During COA development, the staff prepares feasible COAs that integrate the effects of all combat power elements to accomplish the mission. Based on the unit's approved mission statement, the IO officer develops a distinct scheme of IO, IO objectives, and IRC tasks for each COA.

The IO officer is involved early in COA development. The focus is on determining how to achieve decisive advantage in and through the information environment at the critical times and places of each COA. Depending on the time available, planning products may be written or verbal.

See pp. 4-54 to 4-55 for a summary of the inputs, actions and outputs required of the IO officer/element during COA development.

A. Assess Relative Combat Power

IO synchronization of IRCs enhances the combat power, constructive and destructive, of friendly forces in numerous ways. Some examples include:
- Military deception influences application (or misapplication) of threat forces and capabilities at places and times that favor friendly operations.
- Countering the effects of propaganda degrades threat propaganda efforts by exposing lies and providing accurate information.
- MISO and civil military operations favorably influence foreign audiences by emphasizing the positive actions of U.S. forces.
- Movement and maneuver destroys or disrupts threat communicators, controls territory through which information flows, and influences affected populations.

- Electronic warfare jams threat communications and command and control signals.
- Fires destroys threat communication infrastructure.

The IO officer ensures that the staff considers IO when analyzing relative combat power. IO can be especially valuable in reducing resource expenditures by other combat power elements. For example, commanders can use electronic warfare to jam a communications node instead of using fires to destroy it.

IO contributions are often difficult to factor into numerical force ratios. With IO officer support, staff planners consider the effects of IO on the intangible factors of military operations as they assess relative combat power. Intangible factors include such things as the uncertainty of war and the will of friendly forces and the threat. Varied approaches and methods may be used to achieve IO effects. One method is to increase the relative combat power assigned to forces who effectively employ organic IRCs. For example, strict OPSEC discipline by friendly forces increases the difficulty the threat has in collecting information. Units with a Theater IO Group OPSEC support detachment may further increase their relative combat power as a result of this augmentation.

B. Generate Options

Options are expressed as COAs. Given the increasing impact of the information environment on operations and the threat's use of information-focused warfare to gain advantage, staffs recognize that, in certain COAs, IO may be the main effort.

The IO officer assists the staff in considering the ways that IO can support each COA. This requires the IO officer to determine which IRCs to employ and the tradeoffs associated with each. In brainstorming options, the IO officer thinks first in an unconstrained manner, then refines available options based on the running estimate and knowledge of available assets and those that are anticipated. During this substep, the IO officer also develops input to military deception COAs, if applicable. The main output of this effort is an initial scheme of IO by phase for each COA.

C. Array Forces

The staff arrays forces to determine the forces necessary to accomplish the mission and to develop a knowledge base for making decisions concerning concepts of operations. The IO officer ensures planners consider the impact of available IRCs on force ratios as they determine the initial placements. IRCs may reduce the number of maneuver forces required or may increase the COA options available. Planners consider the deception story during this step because aspects of it may affect unit positioning.

Although the staff considered IRC availability when developing COAs, this step allows them to further validate if the required capabilities are present and, if not, determine if they can be obtained and positioned in time to achieve required effects. It also enables the IO officer to determine if available IRCs are properly positioned and task-organized.

D. Develop a Broad Concept

The broad concept concisely expresses the "how" of the commander's visualization and will eventually provide the framework for the concept of operations and summarizes the contributions of all warfighting functions (FM 6-0). The IO officer develops schemes of IO and IO objectives for each COA that nest with the broad concept. With input from IRC representatives, the IO officer considers how IRCs can achieve the IO objectives.

IO schemes of support are further expressed in terms of the weighted efforts required to support the overall concept of operations. Depending on proportion of offense, defense, and stability tasks, the IO officer determines the best mix of attack, defend, and stability IO efforts needed to ensure achievement of objectives. The IO officer then determines which IRCs to allocate to each effort and possible tasking conflicts.

See following pages (pp. 4-52 to 4-53) for further discussion.

Develop a Broad Concept

Ref: *FM 3-13, Information Operations (Dec '16), pp. 4-19 to 4-24.

The broad concept concisely expresses the "how" of the commander's visualization and will eventually provide the framework for the concept of operations and summarizes the contributions of all warfighting functions (FM 6-0). The IO officer develops schemes of IO and IO objectives for each COA that nest with the broad concept. With input from IRC representatives, the IO officer considers how IRCs can achieve the IO objectives.

IO schemes of support are further expressed in terms of the weighted efforts required to support the overall concept of operations. Depending on proportion of offense, defense, and stability tasks, the IO officer determines the best mix of attack, defend, and stability IO efforts needed to ensure achievement of objectives. The IO officer then determines which IRCs to allocate to each effort and possible tasking conflicts.

During this sub-step, the IO officer develops control measures, critical and defended asset lists, and additional EEFIs for each COA, as well as determines OPSEC vulnerabilities and measures. Most importantly, the IO officer produces five essential, often time-intensive, outputs. These are—

COA Worksheets

The IO officer employs COA worksheets to prepare for COA analysis and focus IRC efforts. These worksheets can be narrative or graphical or a combination of both. The IO officer prepares one worksheet for each IO objective in each scheme of IO. IO worksheets include the following information, as a minimum:

- A description of the COA.
- The scheme of IO in statement form.
- The IO objective in statement form.
- Information concerning IRC tasks that support the objective, listed by IRC.
- Anticipated adversary counteractions for each IRC task.
- Measures of performance and effectiveness for each IRC task.
- Information required to assess each IRC task.

The COA worksheet needs to show how each IRC contributes to the IO objective and the scheme of IO for that COA. When completed, the work sheets help the IO officer tie together the staff products developed to support each COA. IO planners also use the worksheets to focus task development for all IRCs. They retain completed work sheets for use during subsequent steps of the MDMP.

Synchronization Matrix

The IO officer develops an IO synchronization matrix for each COA to determine when to execute IRC tasks. IO synchronization matrices show estimates of the time it takes for friendly forces to execute an IRC task; the adversary to observe, process and analyze the effect(s) of the executed task; and the adversary to act on those effect(s). The IO officer synchronizes IRC tasks with other combined arms tasks. The G-2 (S-2)and G-3 (S-3) time lines are used to reverse-plan and determine when to initiate IRC tasks. Due to the lead time required, some IRC tasks must be executed early in an operation. Regardless of when the IRC tasks start, they are still synchronized with other combined arms tasks. Many IRC tasks are executed throughout an operation; some are both first to begin and last to end. IO synchronization matrices vary in format, depending on commander preference and unit standard operating procedures. At a minimum, the synchronization matrix should include—

- IO objectives.
- IRC tasks.

- The operational timeline to execute the IRC tasks.
- The depiction of how IRC synchronization integrates with lines of operations or lines of effort.

Target Nominations

The IO officer uses information derived during mission analysis, IPB products, and the high-value target list to nominate high-pay-off targets (HPTs) for each friendly COA. HPTs are selected to be added to the high-payoff target list. HPTs are developed in conjunction with the IRC tasks employed to affect them. Targets attacked by nonlethal means, such as jamming or MISO broadcasts, may require assessment by means other than those normally used in battle damage assessment. The IO officer submits IRs for this information to the G-2 (S-2) when nominating them. If these targets are approved, the IRs needed to assess the effects on them become PIRs that the G-2 (S-2) adds to the information collection plan. If the command does not have the assets or resources to answer the IO IRs, the target is not engaged unless the attack guidance specifies otherwise or the commander so directs. The targeting team performs this synchronization.

Risk Assessment

The assessment of IO-associated risk during COA development and COA analysis focuses primarily on hazards related to executing the scheme of IO and its associated IRC tasks. However, the IO officer assesses all hazards as they emerge. The IO officer also monitors identified hazards and evaluates the effectiveness of controls established to counter them.

The IO officer examines each COA and its scheme of IO to determine if they contain hazards not identified during mission analysis. The IO officer then develops controls to manage these hazards, determines residual risk, and prepares to test the controls during COA analysis. The IO officer coordinates controls with other staff sections as necessary. Controls that require IRC tasks to implement are added to the IO COA worksheet for the COA.

The IO officer considers two types of hazards associated with the scheme of IO: those associated with the scheme of IO itself and its supporting IRC tasks; and those from other aspects of the concept of operations that may affect execution of IO. The IO officer identifies as many of these hazards as possible so the commander can consider them in decisions.

Measures of Performance and Effectiveness

Measures of performance and measures of effectiveness drive information requirements necessary to measure the degree to which operations accomplish the unit's mission. As COA development continues, the IO officer considers how to assess IO effectiveness, by determining:

- IRC tasks that require assessment.
- Measures of performance for IRC tasks and measures of effectiveness for IO objectives, as well as baselines to measure the degree of change, and associated IO-related targets.
- The information needed to make the assessment.
- How to collect the information.
- Who or what will collect the information.
- How the commander will use the information to support decisions.

The responses to these considerations are recorded on the IO COA worksheets and added to the IO portion of the operations assessment plan. Information required to assess IO effects becomes IRs. The IO officer submits IRs for the COA that the commander approves to the G-2 (S-2). The IO officer establishes measures of performance and effectiveness based on how IRC tasks contribute to achieving one or more IO objectives. If a task's results are not measurable, the IO officer eliminates the task.

COA Development
(Summary of IO Inputs, Actions & Outputs)
Ref: *FM 3-13, Information Operations (Dec '16), table 4-2, pp. 4-22 to 4-25.

Table 4-2 provides a summary of the inputs, actions and outputs required of the IO officer/element. Only those sub-steps within COA development with significant IO activity are listed.

MDMP Sub-Step	Inputs	IO Officer Actions	IO Officer Outputs
Assess Relative Combat Power	• IPB or combined information overlay • Task organization • IO running estimate • Vulnerability assessment	For each COA — • Analyze IRC effects on friendly and threat capabilities, vulnerabilities, and combat power	For each COA — • Description of the potential effects of relative combat power stated by IRC
Generate Options	• Commander's intent and guidance • IPB or combined information overlay • Friendly, neutral, and enemy information related capabilities, resources, and vulnerabilities	• Determine different ways for IO to support each COA • Determine IRCs to employ. • Determine how to focus IRCs on the overall objective • Determine IO's role in the decisive and shaping operations for each COA • Determine possible tradeoffs among IRCs • Develop input to military deception COAs (deception stories)	• Scheme of IO for each COA • Input to military deception COAs
Array Forces	• Restated mission • Commander's intent and guidance • IPB or combined information overlay • Input to military deception plan or concept	• Allocate IRCs for each scheme • Identify requirements for additional IRCs • Examine effect of possible military deception COAs on force positioning • Identify military deception means	• Initial IRC location and task organization • Additional IRC requirements

INFO Planning

MDMP Sub-Step	Inputs	IO Officer Actions	IO Officer Outputs
Develop a Broad Concept	• COAs • IPB or combined information overlay • High value target list • IO mission statement • Initial scheme of IO for each COA	For each COA — • Develop scheme of IO • Develop objectives • Develop control measures • Identify and prioritize IRC tasks • Nominate selected HPTs • Determine initial IO task execution timeline • Refine input to risk assessment • Develop IO portion of assessment plan • Identify additional EEFIs • Identify and assess OPSEC indicators to determine vulnerabilities • Develop OPSEC measures to shield vulnerabilities • Determine residual risk associated with each vulnerability after OPSEC measures are applied • Determine feedback required for assessment of military deception COAs	For each COA — • Refined scheme, objectives, and control measures; IRC tasks; and tasks to subordinate units • IO COA worksheets • Synchronization matrices • Execution time line • IO-related high-payoff target nominations • Critical and defended asset lists • Input to risk management plan, including residual risk associated with each OPSEC vulnerability • Success criteria to support assessment • Additional EEFIs • OPSEC vulnerabilities • OPSEC measures to shield vulnerabilities
Assign Headquarters	• IPB/combined information overlay • IO running estimate • IO vulnerability assessment • IO tasks by IRC and subordinate unit	For each COA — • Assess mission command strengths and weaknesses to determine vulnerabilities of specific headquarters regarding ability to execute IO • Assess mission command strengths and weaknesses to determine vulnerabilities of subordinate commands • Reevaluate critical and defended asset lists	For each COA — • Recommendations for allocation of G-3 (S-3) IO personnel to headquarters in light of mission command vulnerability assessment • Recommendations of grouping of IRCs to subordinate commands in light of mission command vulnerability assessment • Updated critical and defended asset lists • Initial list of IRCs to tasks assigned
Develop COA Statements and Sketches	• COA statement • A scheme of IO and objectives for each COA	• Submit input for each COA statement and sketch to G-3 (S-3) • Prepare scheme statement and sketch for each COA	• Input for each COA statement and sketch • Scheme of IO and sketches for each COA, stating the most important objectives

COA course of action
EEFI essential element of friendly information
HPT high-payoff target
IO information operations
IPB intelligence preparation of the battlefield
IRC information-related capability
IPB intelligence preparation of the battlefield
OPSEC operations security

E. Assign Headquarters

Headquarters are typically assigned based on their ability to integrate the warfighting functions. Their capacity to plan, prepare, execute, and assess IO varies, depending on such variables as organic capabilities, mission essential tasks, and training. When commanders determines that the decisive operation or a shaping operation is IO-dominant, they turn to the IO officer to assess potential mission command vulnerabilities and ways to mitigate them. Higher headquarters, in particular, conduct this assessment for subordinate headquarters being assigned IO-dominant missions and provides additional assets, as required.

F. Develop Course of Action Statements & Sketches

The G-3 (S-3) prepares a COA statement and supporting sketch for each COA for the overall operation. Together, the statement and sketch cover who, what, when, where, how, and why for each subordinate unit. They also state any significant risks for the force as a whole. The IO officer provides IO input to each COA statement and sketch. At a minimum, each COA statement and sketch should include its associated scheme of IO. COA statements may also identify select IO objectives and IRC tasks when they address specific commander concerns or priorities.

G. Conduct Course of Action Briefing

Given the increasing impact of the information environment on operations, commanders benefit from ensuring the IO officer is present during all MDMP briefings. For this specific briefing, the IO officer is able to provide essential rationale for the scheme of IO and respond to IO-related questions from the commander or G-3 (S-3).

H. Select or Modify Course of Action for Continued Analysis

Whether the commander selects a given COA or COAs, modifies COAs, or creates a new COA altogether, the IO officer prepares for COA analysis and war-gaming. If the commander rejects all COAs, the IO officer develops new schemes of support, mindful of the commander's revised planning guidance.

See previous pages (pp. 4-54 to 4-55) for a summary of the inputs, actions and outputs required of the IO officer/element during COA development.

Step IV. Course of Action Analysis & War-Gaming

COA analysis (war-gaming) enables commanders and staffs to identify difficulties or coordination problems as well as probable consequences of planned actions for each COA being considered. It helps them think through the tentative plan. War-gaming is a disciplined process that staffs use to envision the flow of battle. Its purpose is to stimulate ideas and provide insights that might not otherwise be discovered. Effective war-gaming allows the staff to test each COA, identify its strengths and weaknesses, and alter it if necessary. During war-gaming, new hazards may be identified, the risk associated with them assessed, and controls established. OPSEC measures and other risk control measures are also evaluated.

War-gaming helps the IO officer synchronize IRC operations and helps the staff integrate IO into the overall operation. During the war game, the IO officer addresses how each IRC contributes to the scheme of IO for that COA and its associated time lines, critical events, and decision points. The IO officer revises the schemes of IO as needed during war-gaming.

The IO officer uses the synchronization matrices and worksheets for each COA as scripts for the war game. The IRCs are synchronized with each other and with the concepts of operations for the different COAs. To the extent possible, the IO officer also includes planned counter-actions to anticipated threat reactions.

COA Analysis & War-Gaming
(Summary of IO Inputs, Actions & Outputs)

Ref: *FM 3-13, Information Operations (Dec '16), table 4-2, pp. 4-22 to 4-25.

MDMP Step	Inputs	IO Officer Actions	IO Officer Outputs
Course of Action Analysis	• Updated running estimate. • IPB/combined information overlay • Updated assumptions **For each COA —** • Scheme of IO and objectives for each COA sketch • Execution timeline	• Develop evaluation criteria for each COA • Gather the tools • List all friendly IRCs • List assumptions • Synchronize tasks performed by different IRCs and subordinate commands • Coordinate IO with cyber electromagnetic activities • Integrate scheme of IO into the concept of operations for each COA • Synchronize scheme of IO with higher and adjacent headquarters • Identify enemy information warfare capabilities and likely actions and reactions • War game friendly IRCs against enemy vulnerabilities and display the results • War game friendly IRC impacts on various audiences and populations and display the results • War game enemy information warfare capabilities against friendly vulnerabilities and display the results • Synchronize and de-conflict targets • Determine whether modifications to the COA result in additional EEFIs or OPSEC vulnerabilities; if so recommend OPSEC measures to shield them • Assign attack measures to HPTs. • Test OPSEC measures • Determine decision points for executing tasks • War game each military deception COA • Identify each military deception COA's potential branches; assess risk to the COA • List the most dangerous or beneficial branch on the decision support template or synchronization matrix • Participate in the war game briefing (optional)	• Potential decision points • Initial assessment measures • Updated assumptions • An evaluation of each military deception COA in terms of criteria established before the war game **For each COA —** • An evaluation in terms of criteria established before the war game • Recorded input to war game results • Refined scheme of IO • Refined tasks • Refined input to attack guidance matrix and target support matrix • IRs and requests for information identified during war game • Refined EEFIs and OPSEC vulnerabilities and OPSEC measures • Paragraph 4 of the running estimate • Input to the G-3 (S-3) synchronization matrix • Input to the HPTL

COA course of action
EEFIs essential elements of friendly information
HPT high-payoff target
HPTL high-payoff target list
IO information operations
IPB intelligence preparation of the battlefield
IR information requirement
IRC information-related capability
OPSEC operations security
MDMP military decisionmaking process

Table 4-3 provides a summary of the inputs, actions and outputs required of the IO officer during course of action analysis.

During preparation for war-gaming, the IO officer gives the G-2 (S-2) likely threat information-related actions and reactions to friendly IO, to include possible threat responses in the information environment to friendly operations. The IO officer also continues to provide input to the G-2 (S-2) for HPT development and selection.

Before beginning the war game, staff planners develop criteria to evaluate the effectiveness and efficiency of each COA during COA comparison. These criteria are listed in paragraph 3c of the IO running estimate and become the outline for the COA analysis in paragraph 4. The IO officer develops the criteria for evaluating the schemes of IO. Using IO-specific criteria allows the IO officer to explain the advantages and disadvantages of each COA.

Evaluation criteria that may help discriminate among various COAs could include:
- Lead time required for implementation.
- The number of decision points that require support.
- The cost of achieving an IO objective versus the expected benefits.
- The risk to friendly assets posed by threat information activities.

During war-gaming the IO officer participates in the action-reaction-counteraction process. For example, the action may be patrols designed to enforce curfew; the threat reaction is messaging accusing U.S. forces of causing damage and casualties; the counteraction is assigning combat camera to document U.S. force patrols and interactions with the indigenous population and incorporating the documentation with another IRC in order to provide appropriate content to the target audience. The IO officer uses the synchronization matrices and COA worksheets to insert IRC tasks into the war game at the time planned. A complete COA worksheet allows the IO officer to state the organization performing the task and its location. The IO officer remains flexible throughout the process and is prepared to modify input to the war game as it develops. The IO officer is also prepared to modify the scheme of IO, IO objectives, and IRC tasks to mitigate possible threat actions discovered during the war game. The IO officer notes any branches and sequels identified during the war game. Concepts of support for these branches or sequels are developed as time permits.

The results of COA analysis are a refined scheme of IO and associated products for each COA. During war-gaming, the IO officer refines IRs, EEFIs, and HPTs for each COA, synchronizing them with that COA's concept of operations. Staff planners normally record war-gaming results, including IRC effects, on the G-3 (S-3) synchronization matrix. The IO officer may also record the results on the COA worksheets. These help the IO officer subsequently synchronize IRCs. The worksheets and synchronization matrices provide the basis for IO input to paragraph 3 of the OPLAN/OPORD, paragraph 3 of the IO and IRC appendices.

Step V. Course of Action Comparison

During COA comparison, the staff compares feasible COAs to identify the one with the highest probability of success against the most likely adversary COA and the most dangerous adversary COA. Each staff section evaluates the advantages and disadvantages of each COA from the staff section's perspective, and presents its findings to the staff. The staff outlines each COA in terms of the evaluation criteria established before the war game and identifies the advantages and disadvantages of each with respect to the others. The IO officer records this analysis in paragraph 4 of the IO estimate.

The IO officer determines the COA that IO can best support based on the evaluation criteria established during war-game preparation. The results of this comparison become paragraph 5 of the IO estimate.

Table 4-4 (facing page) provides a summary of the inputs, actions and outputs required of the IO officer during course of action comparison.

COA Comparison
(Summary of IO Inputs, Actions & Outputs)

Ref: *FM 3-13, Information Operations (Dec '16), table 4-2, pp. 4-22 to 4-25.

MDMP Task	Inputs	IO Officer Actions	IO Officer Outputs
Course of Action Comparison	• Updated IO running estimate • Refined COAs • COA evaluation criteria • COA evaluations from COA analysis • Updated assumptions	• Compare the COAs with each other to determine the advantages and disadvantages of each • Determine which COA is most supportable from an IO perspective • Determine if any OPSEC measures require the commander's approval	• Advantages and disadvantages for each COA • Most supportable COA from an IO perspective • Input to COA decision matrix • Updated assumptions • Paragraph 4, IO running estimate

COA course of action   **IO** information operations   **MDMP** military decisionmaking process   **OPSEC** operations security

Table 4-4 provides a summary of the inputs, actions and outputs required of the IO officer during course of action comparison.

Step VI. Course of Action Approval

After completing the COA comparison, the staff identifies its preferred COA and recommends it to the commander in a COA decision briefing, if time permits. The concept of operations for the approved COA becomes the concept of operations for the operation itself. The scheme of IO for the approved COA becomes the scheme of IO for the operation. Once a COA is approved, the commander refines the commander's intent and issues additional planning guidance. The G-3 (S-3) then issues a WARNORD and begins orders production.

The WARNORD issued after COA approval contains information that executing units require to complete planning and preparation. Possible IO input to this WARNORD includes:

- Contributions to the commander's intent/concept of operations.
- Changes to the CCIRs.
- Additional or modified risk guidance.
- Time-sensitive reconnaissance tasks.
- IRC tasks requiring early initiation.
- A summary of the scheme of IO and IO objectives.

During the COA decision briefing, the IO officer is prepared to present the associated scheme of IO for each COA and comment on the COA from an IO perspective. If the IO officer perceives the need for additions or changes to the commander's intent or guidance with respect to IO, they ask for it.

MDMP Step	Inputs	IO Officer Actions	IO Officer Outputs
Course of Action Approval	• Updated IO running estimate • Evaluated COAs • Recommended COAs • Updated assumptions	• Provide input to COA recommendation • Re-evaluate input to the commander's intent and guidance • Refine scheme of IO, objectives, and tasks for approved COA and update synchronization matrix • Prepare input to the WARNORD • Participate in the COA decision briefing • Recommend the COA that IO can best support • Request decision on executing any OPSEC measures that entail significant resource expenditure or high risk	• Finalized scheme of IO for approved COA • Finalized tasks based on approved COA • Input to WARNORD • Updated synchronization matrix

COA course of action **IO** information operations **MDMP** military decisionmaking process **WARNORD** warning order

Step VII. Orders Production, Dissemination, and Transition

Based on the commander's decision and final guidance, the staff refines the approved COA and completes and issues the OPLAN/OPORD. Time permitting, the staff begins planning branches and sequels. The IO officer ensures input is placed in the appropriate paragraphs of the base order and its annexes, especially the IO appendix to the operations annex. When necessary, the IO officer or appropriate special staff officers prepare appendixes for one or more IRCs/

See p. 4-61 for table 4-6 (summary of IO inputs to orders production, dissemination and transition) along with an annotated format of appendix 15 (Information Operations) to Annex C (Operations).

IV. Appendix 15 (IO) to Annex C (Operations)

Ref: *FM 3-13, Information Operations (Dec '16), annex A. (*See note p. 1-2.)

Based on the commander's decision and final guidance, the staff refines the approved COA and completes and issues the OPLAN/OPORD. Time permitting, the staff begins planning branches and sequels. The IO officer ensures input is placed in the appropriate paragraphs of the base order and its annexes, especially the IO appendix to the operations annex. When necessary, the IO officer or appropriate special staff officers prepare appendixes for one or more IRCs.

See p. 4-10 for related discussion of IO input to operation orders and plans.

MDMP Task	Inputs	IO Officer Actions	IO Officer Outputs
Orders Production, Dissemination and Transition	• Approved COA • Refined commander's guidance • Refined commander's intent • IO running estimate • Execution matrix • Finalized mission statement, scheme of IO, objectives, and tasks	• Ensure input is placed in tasks to subordinate units and coordinating instructions • Produce Appendix 14 (MILDEC) to Annex C (Operations) • Produce Appendix 15 (IO) to Annex C (Operations) • Produce Appendix 3 (OPSEC) to Annex E (Protection) • Coordinate tasks with IRC staff officers • Conduct other staff coordination. • Refine execution matrix • Transition from planning to operations	• Synchronization matrix • Approved Paragraph 3.k. (10) • Approved Appendix 14 to Annex C • Approved Appendix 15 to Annex C • Approved Appendix 3 to Annex E • IO input to AGM and TSM • Subordinates understand the IO portion of the plan or order

AGM attack guidance matrix **COA** course of action **IO** information operations **IRC** information-related capability **MDMP** military decisionmaking process **MILDEC** military deception **OPSEC** operations security **TSM** trunk signaling mission

Ref: FM 3-13, table 4-6. Orders production, dissemination and transition.

Appendix 15 (Information Operations) to Annex C (Operations)

Commanders and staffs use Appendix 15 (Information Operations) to Annex C (Operations) to operation plans and orders to describe how information operations (IO) will support operations described in the base plan or order. The IO officer is the staff officer responsible for this appendix. Products or guidance:

- Combined information overlay. *(See pp. 4-32 to 4-33.)*
- Synchronization matrix. *(See p. 4-16.)*
- Instructions for IRCs not covered by other appendices, such as operations security, visual information, and combat camera.

See following pages (pp. 4-62 to 4-64) for an annotated format of appendix 15 (Information Operations) to Annex C (Operations). The figure is a guide and should not limit the information contained in an actual Appendix 15. Appendix 15 should be specific to the operation being conducted; thus, the content of actual Appendix 15s will vary greatly.

Appendix 15 (IO) to Annex C (Operations)

Ref: *FM 3-13, Information Operations (Dec '16), fig. A-1. Appendix 15 (IO) to Annex C (Operations).

[CLASSIFICATION]

Place the classification at the top and bottom of every page of the OPLAN or OPORD. Place the classification marking at the front of each paragraph and subparagraph in parentheses. Refer to AR 380-5 for classification and release marking instructions.

                                         Copy ## of ## copies
                                         Issuing headquarters
                                                   Place of issue
                                  Date-time group of signature
                                    Message reference number

Include heading if attachment is distributed separately from the base order or higher-level attachment.

APPENDIX 15 (INFORMATION OPERATIONS) TO ANNEX C (OPERATIONS) TO OPERATION PLAN/ORDER [number] [(code name)]- [issuing headquarters] [(classification of title)]

(U) **References**: Refer to higher headquarters· OPLAN or OPORD and identify map sheets for operation (optional). Add any other specific references to IO if needed.

1. (U) **Situation**. Include information affecting information operations (IO) that paragraph 1 of the OPLAN or OPORD does not cover or that needs expansion.

    a. (U) <u>Area of Interest</u>. Describe the information environment as it relates to IO. Refer to Tab 1 (Combined Information Overlay) to Appendix 15 (Information Operations) to Annex C (Operations) as required.

    b. (U) <u>Area of Operations</u>. Refer to Appendix 2 (Operation Overlay) to Annex C (Operations).

        (l) (U) <u>Information Environment</u>. Describe the physical. informational. and cognitive dimensions of the information environment that affect IO. Refer to Tab 1 (Combined Information Overlay) to Appendix 15 (Information Operations) to Annex C (Operations) as required.

        (2) (U) <u>Weather</u>. Describe aspects of weather that impact information operations. Refer to Annex B (Intelligence) as required

    c. (U) <u>Enemy Forces</u>. List known and templated locations and activities of enemy information unitsfor one echelon up and two echelons down. List enemy maneuver and information-related capabilities that will impact friendly operations. State probable enemy courses of action and employment of enemy information assets. Describe the informational and cognitive dimensions of the information environment that affect enemy actions. Refer to Tab 1 (Combined Information Overlay) to Appendix 15 (Information Operations) to Annex C (Operations) as required.

    d. (U) <u>Friendly Forces</u>. Outline the higher headquarters· plan as it pertains to IO. List designation. location. and outline of plan of higher. adjacent. and other jUnctional area assets that support or impact the issuing headquarters or require coordination and additional support. Identify friendly IO/IRC assets and resources that affect subordinate commander IO planning. Identify friendly forces IO vulnerabilities. Identify friendly foreign forces with which subordinate commanders may operate. Identify potential conflicts within the information environment especially if conducting joint or multinational operations. Identijj; and deconflict IRC employment and information environment effects.

[page number]
[CLASSIFICATION]

[CLASSIFICATION]

APPENDIX 15 (INFORMATION OPERATIONS) TO ANNEX C (OPERATIONS) TO OPERATION PLAN/ORDER [number] [(code name)]- [issuing headquarters] [(classification of title)]

 e. (U) <u>Interagency Intergovernmental and Nongovernmental Organizations</u>. *Identify and describe other organizations in the area of operations that may impact the conduct of IO or implementation of IO-specific equipment and tactics.*

 f. (U) <u>Civil Considerations</u>. *Describe critical aspects of the civil situation that impact IO. See Tab C (Civil Considerations) to Appendix 1 (Intelligence Estimate) to Annex B (Intelligence) and Annex K (Civil Affairs Operations) as required Also refer to Tab 1 (Combined Information Overlay) to Appendix 15 (Information Operations) to Annex C (Operations) as required.*

 g. (U) <u>Attachments and Detachments</u>. *List IRCs or IO units only as necessary to clarify task organization. Examples include Tactical MISO Teams, Mobile Public Affairs Detachments, and Visual Information Teams. Refer to Annex A (Task Organization) as required.*

 h. (U) <u>Assumptions</u>. *List any IO-specific assumptions.*

2. (U) **Mission**. *State the IO mission.*

3. (U) **Execution**.

 a. (U) <u>Scheme of Support</u>. *Describe how IO supports the commander's intent and concept of operations. Establish the priorities of support to units for each phase of the operation. Establish IO objectives to employ IRCs to achieve the desired endstate. Describe how IO weighted efforts will support offense, defense, and stability tasks. Identify target sets and effects, by priority. Describe the general concept for the integration of IO. List the staff sections, elements, and working groups responsible for aspects of IO. Include IO collection methods for information developed in staff sections, elements, and working groups outside the IO element and working group. Ensure subordinate units and higher headquarters receive the IO synchronization plan. Describe the plan for the integration of unified action and nongovernmental partners and organizations. Refer to Annex C (Operations) as required This section is designed to provide insight and understanding a/how IO is integrated across the operational plan.*

 b. (U) <u>Assessment</u>. *Describe the priorities for assessment and Identify the measures of performance and effectiveness and indicators used to assess information operations objectives against end state conditions. Refer to Annex M (Assessment) as required.*

 c. (U) <u>Tasks to Subordinate Units</u>. *List IO tasks assigned to specific subordinate units not contained in the base order.*

 d. (U) <u>Coordinating Instructions</u>. *List only IO instructions applicable to two or more subordinate units not covered in the base order. Identify and highlight any IO-specific rules of engagement risk reduction control measures, environmental considerations, coordination requirements between units, and CCIRs and EEFIs that pertain to IO.*

4. (U) **Sustainment**. *Identify priorities of sustainment for IO key tasks and speciJj; additional instructions as required Refer to Annex F (Sustainment) as required*

 a. (U) <u>Logistics</u>. *Use subparagraphs to Identify priorities and specific instruction for logistics pertaining to IO. See Appendix 1 (Logistics) to Annex F (Sustainment) and Annex P (Host-Nation Support) as required.*

 b. (U) <u>Personnel</u>. *Use subparagraphs to Identify priorities and specific instruction for human resources support pertaining to IO. See Appendix 2 (personnel Services Support) to Annex F (Sustainment) as required.*

 c. (U) <u>Health System Support</u>. *See Appendix 3 (Army Health System Support) to Annex F (Sustainment) as required*

Continued on next page

[CLASSIFICATION]

[Classification]

APPENDIX 15 (INFORMATION OPERATIONS) TO ANNEX C (OPERATIONS) TO OPERATION PLAN/ORDER [number] [(code name)]- [issuing headquarters] [(classification of title)]

Continued from previous page

5. (U) <u>Command and Signal</u>.

    a. (U) Command.

        (1) (U) <u>Location of Commander</u>. *State the location of key IO leaders.*

        (2) (U) <u>Liaison Requirements</u>. *State the IO liaison requirements not covered in the unit's SOPs.*

    b. (D) <u>Control</u>.

        (I) (U) <u>Command Posts</u>. *Describe IO integration into command posts (CPs), including the location of each CP and its time of opening and closing.*

        (2) (U) <u>Reports</u>. *List IO-specific reports not covered in SOPs. See Annex R (Reports) as required*

    c. (U) <u>Signal</u>. *Address any IO-specific communications requirements. See Annex H (Signal) as required*

ACKNOWLEDGE: *Include instructions for the acknowledgement of the OPLAN/OPORD by addressees. The word "acknowledge" may suffice. Refer to the message reference number if necessary. Acknowledgement of a plan or order means that it has been received and understood.*

                                          [Commander's last name]
                                          [Commander's rank]

The commander or authorized representative signs the original copy. If the representative signs the original, add the phrase "For the Commander." The signed copy is the historical copy and remains in the headquarters' files.

OFFICIAL:
[Authenticator's name]
[Authenticator's position]

Use only if the commander does not sign the original order. If the commander signs the original, no further authentication is required. If the commander does not sign, the signature of the preparing staff officer requires authentication and only the last name and rank of the commander appear in the signature block.

ATTACHMENT: List lower-level attachments (tabs and exhibits).

Tab A-Combined Information Overlay

Tab B- Information-Related Capabilities Synchronization Matnx

Tab C-Presence, Posture, and Profile

Tab D-Combat Camera

Tab E-Soldier and Leader Engagement

DISTRIBUTION: Show only if distributed separately from the base order or higher-level attachments.

                                          [page number]
                                          [CLASSIFICATION]

Chap 5

(Information) PREPARATION

Ref: *FM 3-13, Information Operations (Dec '16), chap. 5. (*See note p. 1-2.)

See p. 1-62 for discussion of preparation as related to information advantage (ADP 3-13).

> **Preparation** consists of those activities performed by units and Soldiers to improve their ability to execute an operation (ADP 5-0). Preparation creates conditions that improve friendly force opportunities for success. Because many IO objectives and IRC tasks require long lead times to create desired effects, preparation for IO often starts earlier than for other types of operations.

IO Preparation Activities

Peacetime preparation by units or capabilities involves building contingency plan databases about the anticipated area of operations. These databases can be used for IO input to IPB and to plan IO to defend friendly intentions, such as network protection and operations security (OPSEC). IO portions of contingency plans are continuously updated.

During peacetime, IO officers prepare for future operations by analyzing anticipated area(s) of operations' information environment and likely threat information capabilities. Examples of factors to consider include, but are not limited to—

- Religious, ethnic, and cultural mores, norms, and values.
- Non-military communications infrastructure and architecture.
- Military communication and command and control infrastructure and architecture.
- Military training and level of proficiency (to determine susceptibility to denial, deception, and IO).
- Literacy rate.
- Formal and informal organizations exerting influence and leaders within these organizations.
- Ethnic factional relationships and languages.

Preparation includes assessing unit readiness to execute IO. Commanders and staffs monitor preparations and evaluate them against criteria established during planning to determine variances. This assessment forecasts the effects these factors have on readiness to execute the overall operation as well as individual IRC tasks.

Preparation for IO takes place at three levels: staff (IO officer), IRC units or elements, and individual. The IO officer helps prepare for IO by performing staff tasks and monitoring preparations by IRC units or elements. These units perform preparation activities as a group for tasks that involve the entire unit, and as individuals for tasks that each soldier and leader must complete.

Refer to BSS7: The Battle Staff SMARTbook, 7th Ed., updated for 2023 to include FM 5-0 w/C1 (2022), FM 6-0 (2022), FMs 1-02.1/.2 (2022), and more. Focusing on planning & conducting multidomain operations (FM 3-0), BSS7 covers the operations process; commander/ staff activities; the five Army planning methodologies; integrating processes (IPB, information collection, targeting, risk management, and knowledge management); plans and orders; mission command, command posts, liaison; rehearsals & after action reviews; operational terms & military symbols.

Chapter 3 of ADP 5-0 provides a comprehensive overview of preparation activities. The activities most relevant to conducting IO include—

A. Improve Situational Understanding

The IO officer/element must understand and share their understanding of the information environment with the commander and staff. During preparation, information collection begins, which helps to validate assumptions and improve situational understanding. Coordination, liaison, and rehearsals further enhance this understanding. Given the information environment's complexity, this task is never-ending and depends on everyone, not just the IO officer, to update and refine understanding of the information environment.

B. Revise and Refine Plans and Orders

Plans are not static; the commander adjusts them based on new information. This information may be the result of analysis of unit preparations, answers to IO IRs, and updates of threat information capacity and capability.

During preparation, the IO officer adjusts the relevant portions of the operation plan (OPLAN) or operation order (OPORD) to reflect the commander's decisions. The IO officer also updates the IO running estimate so that it contains the most current information about adversary information activities, changes in the weather or terrain, and friendly IRCs.

The IO officer ensures that IO input to IPB remains relevant throughout planning and preparation. To do this, they ensure that IO input to the information collection plan is adjusted to support refinements and revisions made to the OPLAN/OPORD.

IO preparation begins during planning. As the IO appendix begins to take shape, IO officer coordination with other staff elements is vital because IO affects every other warfighting function. For example, planning an attack on a command and control (C2) high-payoff target requires coordination with the targeting team. A comprehensive attack offering a high probability of success may involve air interdiction and therefore needs to be placed on the air tasking order. It may involve deep attack: rocket and missile fires have to be scheduled in the fire support plan. Army jammers and collectors have to fly the missions when and where needed. The IO officer ensures the different portions of the OPLAN/OPORD contain the necessary coordinating instructions for these actions to occur at the right time and place.

Effective IO is consistent at all echelons. The IO officer reviews subordinate unit OPLANs/OPORDs to ensure IO has been effectively addressed and detect inconsistencies. The IO officer also looks for possible conflicts between the command's OPLAN/OPORD and those of subordinates. When appropriate, the IO officer reviews adjacent unit OPLANs/OPORDs for possible conflicts. This review allows the IO officer to identify opportunities to mass IO effects across units.

OPLAN/OPORD refinement includes developing branches and sequels. Branches and sequels are normally identified during war-gaming (COA analysis). However, the staff may determine the need for them at any time. The G-3 (S-3) prioritizes branches and sequels. The staff develops them as time permits. The IO officer participates in their development as with any other aspect of planning.

A key focus during preparation is on assessment of the current state of the information environment. This assessment is performed to establish baselines, which are subsequently used when assessing whether IO objectives and IRC tasks were effective in creating desired effects.

C. Conduct Coordination and Liaison

IO requires all units and elements to coordinate with each other continuously, as well as liaise. Coordination begins during planning; however, input to a plan alone does not constitute coordination. Coordination involves exchanging the information needed to synchronize operations. The majority of coordination takes place during preparation. It is then that the IO officer follows through on the coordination initiated during planning. Exchanging information is critical to successful coordination and execution. Coordination may be internal or external and is enhanced through liaison.

Internal Coordination

Internal coordination occurs within the unit headquarters. The IO officer initiates the explicit and implicit coordinating activities with other staff sections, as well as within the IO element, if one exists. Much of this coordination occurs during IO working group meetings; however, IO working group members do not wait for a meeting to coordinate. They remain aware of actions that may affect, or be affected by, their functional responsibilities. They initiate coordination as soon as they become aware of a situation that requires it. The IO officer remains fully informed of IO-related coordination. The IO officer corrects or resolves problems of external coordination revealed by command and staff visits and information gathering. During internal coordination, the IO officer resolves problems and conflicts and ensures that resources allocated to support IO arrive and are distributed.

See following pages (pp. pp. 5-4 to 5-5) for examples of internal coordination.

External Coordination

External coordination includes coordinating with or among subordinate units and higher headquarters, as well as IO support units, IRCs, and resources that may not be under the unit's control during planning but are necessary to execute the plan. External coordination also includes coordinating with adjacent units or agencies. (Adjacent refers to any organization that can affect a unit's operations in and through the information environment.) This coordination is necessary to integrate IO throughout the force.

See following pages (pp. 5-4 to 5-5) for examples of external coordination.

Liaison

Establishing and maintaining liaison is one of the most important means of external coordination. The IO officer may perform direct liaison but units may select another staff member to be part of the liaison team. Establishing liaison during planning enhances subsequent coordination during preparation and execution.

Practical liaison can be achieved through personal contact between IO officers or between the IO officer and agencies/organizations involved in affecting the information environment. This coordination is accomplished through exchanging personnel, through agreement on mutual support between adjacent units or organizations, or by a combination of these means. Liaison should, when possible, be reciprocal between higher, lower, and adjacent units/organizations. Liaison must be reciprocal between IO sections when U.S. forces are operating with or adjacent to multinational partners.

D. Initiate Information Collection

Execution requires accurate, up-to-date situational awareness. During preparation, the IO officer updates IRs to ensure the most current information possible. The IO officer also works with the G-2 (S-2) to update collection asset taskings necessary to assess IO.

E. Initiate Security Operations

Security operations serve to protect the force from surprise and threat attacks during preparation. While often considered in terms of specific missions that physically screen, guard, cover, or provide area or local security, security operations should also include IRC tasks that provide these same protections in the informational and cognitive dimensions of the information environment. Military deception, OPSEC,

Examples of Internal/External Coordination

Ref: *FM 3-13, Information Operations (Dec '16), pp. 5-3 to 5-4.

IO requires all units and elements to coordinate with each other continuously, as well as liaise. Coordination begins during planning; however, input to a plan alone does not constitute coordination. Coordination involves exchanging the information needed to synchronize operations. The majority of coordination takes place during preparation. It is then that the IO officer follows through on the coordination initiated during planning. Exchanging information is critical to successful coordination and execution. Coordination may be internal or external and is enhanced through liaison.

Internal Coordination

Internal coordination occurs within the unit headquarters. The IO officer initiates the explicit and implicit coordinating activities with other staff sections, as well as within the IO element, if one exists. Much of this coordination occurs during IO working group meetings; however, IO working group members do not wait for a meeting to coordinate. They remain aware of actions that may affect, or be affected by, their functional responsibilities. They initiate coordination as soon as they become aware of a situation that requires it. The IO officer remains fully informed of IO-related coordination. The IO officer corrects or resolves problems of external coordination revealed by command and staff visits and information gathering. During internal coordination, the IO officer resolves problems and conflicts and ensures that resources allocated to support IO arrive and are distributed. Examples of internal coordination include, but are not limited to:

- Deconflicting military information support operations (MISO) with public affairs activities and products.
- Monitoring the progress of answers to IO RFIs.
- Monitoring RFIs to higher headquarters by the G-3 (S-3) current operations.
- Checking the air tasking order for missions requested by the IO officer/element.
- Monitoring the movements and readiness of IRCs.
- Determining space asset status and space weather implications.
- Participating in the integration of IO-related targets into the targeting process.
- Continuous monitoring and validation of OPSEC procedures, particularly in preparation for military deception. This could include a short statement on physical security, particularly during movement.

The IO officer remains mindful that training is conducted during planning and preparation. This training occurs as new soldiers and IRCs are integrated into the command and its battle rhythm. Additionally, the IO officer provides training to subordinate elements, as requested, to fill gaps in their IO capacity.

Internal coordination is especially important to ensure requisite staff support to various IRCs in order to enhance their readiness and effectiveness. Examples include but are not limited to—

- Electronic warfare (EW).
- G-2 (S-2)—Coordinates intelligence gathering in support of the EW mission. Recommend the use of EW against adversary systems that use the electromagnetic spectrum.
- G-3 (S-3)—Coordinates and prioritizes EW targets.
- G-4 (S-4)—Coordinates distribution of EW equipment and supplies, less cryptographic support.
- IO officer—Coordinates EW tasks with those of other IRCs and assists with preparation of the cyberspace electromagnetic activities appendix.

- EW officer—Monitors the preparation of military intelligence units to support EW missions; prepare cyber effects request forms and electronic attack request forms; monitors other staff functions that support or affect EW.
- MISO.
- G-2 (S-2)—Prepares intelligence estimate and analysis of the area of operation.
- G-3 (S-3)—Requests additional MISO units as required.
- IO officer—Identifies requirements for additional MISO units to the G-3 (S-3).
- G-4 (S-4)—Prepares logistic support of MISO.
- Psychological Operations (PSYOP) officer—Prepares the MISO appendix to Annex C. Prepares the MISO estimate.
- OPSEC.
- G-2(S-2)—Provides data on threat intelligence collection capabilities.
- IO officer—Determines the EEFIs.
- G-4 (S-4)—Advises on the vulnerabilities of supply, transport, and maintenance facilities, and lines of communications.
- G-5 (S-5)—Determines availability of civilian resources for use as guard forces.
- OPSEC officer—Prepares the OPSEC estimate and appendix.
- Provost marshal—Advises on physical security measures.
- Military deception.
- G-2 (S-2)—Determines adversary surveillance capabilities.
- G-3 (S-3)—Coordinates movement of units participating in military deception.
- G-4 (S-4)—Coordinates logistic support to carry out assigned deception tasks.
- G-9 (S-9)—Coordinates host-nation support to implement the military deception plan.
- Military deception officer—Prepares to monitor execution of military deception operation.

External Coordination

External coordination includes coordinating with or among subordinate units and higher headquarters, as well as IO support units, IRCs, and resources that may not be under the unit's control during planning but are necessary to execute the plan. External coordination also includes coordinating with adjacent units or agencies. (Adjacent refers to any organization that can affect a unit's operations in and through the information environment.) This coordination is necessary to integrate IO throughout the force. Examples of external coordination include:

- Assessing unit OPSEC posture.
- Making sure the military deception operation is tracking with preparation for the overall operation.
- Periodically validating assumptions.
- Ensuring military deception operations are synchronized with those of higher, lower, and adjacent units.

The IO officer remains aware of the effectiveness of cybersecurity tasks taken by the G-6 (S-6). Proper protection of plans and orders, and refinements to them, are essential during operations.

Coordination with joint, interorganizational, and multinational partners is essential to the conduct of IO, as these entities and organizations affect the information environment and are affected by it. The IO working group is the primary means for this coordination but direct, face-to-face coordination is frequently necessary to ensure unity of effort.

space operations, and cyberspace operations all support security operations. Not including these IRC effects into plans potentially puts the force at risk.

F. Initiate Troop Movements

During preparation, IRCs are positioned or repositioned, as necessary, to ensure they can fulfill their assigned tasks. IO unit augmentation and integration also occurs during preparation.

G. Initiate Network Preparation

IO supports the commander's ability to optimize the information element of combat power. In terms of establishing and readying the network, units must think in terms of both technical and human networks. Technical networks have to be set up, engineered, tailored, and tested to meet the specific needs of each operation. Similarly, human networks have to be initiated, cultivated, and refined during preparation. The IO officer coordinates the establishment of networks that help shape the information environment favorable to friendly objectives.

H. Manage and Prepare Terrain

Terrain management is the process of allocating terrain by establishing areas of operation, designating assembly areas, and specifying locations for units and activities to deconflict activities that might interfere with each other (ADRP 5-0). While terrain is physical and geographic, it is a subset of the operational and information environments. When commanders designate areas of operation, they are simultaneously assigning responsibility to specific portions of the information environment. One of the most important reasons for managing physical terrain is to avoid fratricide. The same rationale exists for the information environment: to avoid information fratricide. For example, the IO officer can ensure control measures are established to deconflict EW activities with MISO efforts to inform the local populace through radio broadcasts.

Analysis of the information environment during IPB leads to an understanding of aspects of the information environment in which friendly forces have an advantage and in which they are disadvantaged. During preparation, the IO officer, in concert with the IO working group and its members, undertake actions to exploit the advantages and overcome the disadvantages. For example, if cellular phone communication is essential to strengthen coordination between U.S. forces and an indigenous ally and cell towers are non-existent or degraded, mobile towers could be deployed.

I. Conduct Confirmation Briefings

A confirmation brief is a briefing subordinate leaders give to the higher commander immediately after the operation order is given. It is the leaders' understanding of the commander's intent, their specific tasks, and the relationship between their mission and the other units in the operation. The IO officer assists subordinate commanders and their IO representatives with these briefings when the commander's intent and specific tasks are IO-focused or have aspects related to IO. They also assist subordinate commanders to deduce IO implied tasks and to understand the information environment in their area of operations.

J. Conduct Rehearsals

The IO officer participates in unit rehearsals to ensure IO is integrated with overall operation and to identify potential problems during execution. The IO officer may conduct further rehearsals of tasks and actions to ensure coordination and effective synchronization of IRCs. Before participating in a rehearsal, the IO officer reviews the plans or orders of subordinate and supporting commands.

Chap 6

(Information) EXECUTION

Ref: *FM 3-13, Information Operations (Dec '16), chap. 6. (*See note p. 1-2.)

See p. 1-63 for discussion of execution as related to information advantage (ADP 3-13) and pp. 2-52 to 2-54 as related to operations in the information environment (JP 3-04).

> **Execution** is the act of putting a plan into action by applying combat power to accomplish the mission and adjusting operations based on changes in the situation (ADP 5-0). In execution, commanders, staffs, and subordinate commanders focus their efforts on translating decisions into actions. They direct action to apply combat power at decisive points and times to achieve objectives and accomplish missions. Inherent in execution is deciding whether to execute planned actions (such as phases, branches, and sequels) or to modify the plan based on unforeseen opportunities or threats.

Execution of IO includes IRCs executing the synchronization plan and the commander and staff monitoring and assessing their activities relative to the plan and adjusting these efforts, as necessary. The primary mechanism for monitoring and assessing IRC activities is the IO working group. There are two variations of the IO working group. The first monitors and assesses ongoing planned operations and convenes on a routine, recurring basis. The second monitors and assesses unplanned or crisis situations and convenes on an as-needed basis.

I. Information Operations Working Group

The IO working group is the primary means by which the commander, staff and other relevant participants ensure the execution of IO. The IO working group is a collaborative staff meeting led by the IO officer, and periodically chaired by the G-3 (S-3), executive officer, chief of staff or the commander. It is a critical planning event integrated into the unit's battle rhythm.

Purpose

The IO working group is the primary mechanism for ensuring effects in and through the information environment are planned and synchronized to support the commander's intent and concept of operations. This means that the staff must assess the current status of operations relative to the end state and determine where efforts are working well and where they are not. More specifically, they must ensure targets are identified and nominated at the right place and time to achieve decisive results. The IO working group occurs regularly in the unit's battle rhythm and always before the next targeting working group. The only exception is a crisis IO working group (also referred to as consequence management or crisis action working group), which occurs as soon as feasible before or after an event or incident.

Inputs/Outputs

The example in figure 6-1 *(following page)* is not exhaustive. In terms of inputs, it identifies those documents, products, and tools that historically and practically have provided the IO working group the information necessary to achieve consensus and make informed recommendations to the G-3 (S-3) and commander.

One tool that the IO working group uses to affirm and adjust the synchronized employment of IRCs is the IO synchronization matrix. An updated synchronization matrix is the working group's key output and essential input to the next targeting meeting. (See p. 4-16.)

IO Working Group
Roles & Responsibilities
Ref: *FM 3-13, Information Operations (Dec '16), table 6-1, pp. 6-3 to 6-4.

Representative	Responsibility
Information Operations	• Distribute read-ahead packets • Lead working group • Establish and enforce agenda • Lead information environment update • Recommend commander's critical information requirements • Keep records, track tasks, and disseminate meeting notes
Cyber Electromagnetic Activities	• Provide cyber electromagnetic activities-related information and capabilities to support information operations analysis and objectives • Coordinate, synchronize and deconflict information operations efforts with cyberspace electromagnetic activities efforts or cyberspace electromagnetic activities efforts with information operations efforts
Military Information Support Operations	• Advise on both psychological effects (planned) and psychological impacts (unplanned) • Advise on use of lethal and nonlethal means to influence selected audiences to accomplish objectives • Develop key leader engagement plans • Monitor and coordinate assigned, attached, or supporting military information support unit actions • Identify status of influence efforts in the unit, laterally, and at higher and lower echelons • Provide target audience analysis
G-2 (S-2)	• Provide an intelligence update • Brief information requirements and priority information requirements • Develop the initial information collection plan • Provide foreign disclosure-related guidance and updates
G-3 (S-3)	• Provide operations update and significant activity update • Task units or sections based on due outs • Update fragmentary orders • Maintain a task tracker
Subordinate unit information operations	• Identify opportunities for information operations support to lines of effort • Provide input to assessments • Provide input to information environment update
Public Affairs	• Develop media analysis products • Develop media engagement plan • Provide higher headquarters strategic communication plan • Provide changes to themes and messages from higher headquarters • Develop command information plan
G-9 (S-9)	• Provides specific country information • Ensures the timely update of the civil component of the common operational picture through the civil information management process • Advise on civil considerations within the operational environment • Identify concerns of population groups within the projected joint operational area/area of operations and potential flash points that can result in civil instability • Provide cultural awareness briefings • Advise on displaced civilians movement routes, critical infrastructure, and significant social, religious, and cultural shrines, monuments, and facilities • Advise on information impacts on the civil component • Identify key civilian nodes
Information-related capabilities (IRCs) representatives	• Serve as subject-matter expert for their staff function or capability • Identify opportunities for information-related capability support to lines of effort or operations

Agenda
Ref: *ATP 3-13.1, The Conduct of Information Operations (Oct '18), pp. 4-3 to 4-4.

The IO working group has a purpose, agenda and proposed timing, inputs and outputs, and structure and participants. Figure 4-1 below illustrates these components. To enhance the IO working group's effectiveness, the IO officer and element (if one exists) consider a number of best practices before, during, and after the meeting. Because it relies on information from the commander's daily update briefing and feeds the targeting process, the IO working group occurs between the two events in the unit's battle rhythm.

IO Working Group (Agenda/Components)

Purpose	Agenda and Proposed Timing	
Prioritize, request, and synchronize IRCs and IO augmentation to optimize effects in and through the information environment. **Battle rhythm:** Before targeting working group	Part 1: Operations and intelligence update	30 min
	• Intelligence update	5 min
	• Information environment update	3 min
	• Operations update or significant activities	7 min
	• Review plans, future operations, and current operations	5 min
	• Assessment update (information requirements, indicators)	5 min
	• Calendar update, due outs, and responsibilities from previous meeting	5 min
	Part 2: Stabilize efforts, if any	
	Part 3: Defend efforts • Review and update synchronization matrix	6 min / 12 min
	Part 4: Attack efforts • Guidance and comments	12 min

Inputs and Outputs		Structure and Participants
Inputs: • Higher headquarters orders and guidance • Commander's intent, concept of operations, and narrative • IRC status (running estimates) • Intelligence collections assets • CIO and IPB • Media monitoring analysis • Cultural calendar • Engagements schedule • Audience analysis • Scheme of IO and synchronization matrix • Commander's objectives for IO • Measures of effectiveness and performance	**Outputs:** • Updated scheme of IO • Updated IO synchronization matrix • Key leader engagement recommendations • Refined themes and messages • Refined operational products • Target nominations • Updated CIO • Plans and orders update • Information requirements	**Lead:** IO officer or representative [Chair: G-3 (S-3), executive officer, deputy commanding officer, or commander] **Core participants:** MISO, G-2 (S-2), subordinate unit representatives, G-3 (S-3), fires, G-9 (S-9), operations security, public affairs, CEMA (CO and EW) **Other participants (mission and situation dependent):** G-1 (S-1), G-4 (S-4), G-5 (S-5), G-6 (S-6), space operations, MILDEC, combat camera, FAO, FDO, special forces liaison, KM officer, engineer, STO chief, chaplain, staff judge advocate, unified action partner representatives

CEMA	cyberspace electromagnetic activities	IPB	intelligence preparation of the battlefield
CIO	combined information overlay	IRC	information-related capability
CO	cyberspace operations	KM	knowledge management
EW	electronic warfare	MILDEC	military deception
FAO	foreign area officer	min	minute
FDO	foreign disclosure officer	MISO	military information support operations
G-1	assistant chief of staff, personnel	S-1	personnel staff officer
G-2	assistant chief of staff, intelligence	S-2	intelligence staff officer
G-3	assistant chief of staff, operations	S-3	operations staff officer
G-4	assistant chief of staff, logistics	S-4	logistics staff officer
G-5	assistant chief of staff, plans	S-5	plans staff officer
G-6	assistant chief of staff, signal	S-6	signal staff officer
G-9	assistant chief of staff, civil affairs operations	S-9	civil affairs operations staff officer
IO	information operations	STO	special technical operations

Ref: ATP 3-13.1, fig. 4-1. Components of an information operations working group.

II. IO Responsibilities Within the Various Command Posts

*Ref: *FM 3-13, Information Operations (Dec '16), pp. 6-3 to 6-5.*

IO execution involves monitoring and assessing IO as the operation unfolds and requires coordination among the tactical command post (CP) and main CP, which can be challenging. Each monitors different parts of the operation and not all have an assigned functional area 30 or IO officer. Continuous exchange of information among those assigned responsibility for IO at these CPs is essential to ensuring the effective execution of IO.

The tactical CP directs IO execution and adjusts missions as required. The IO representative or responsible agent—

- Maintains the IO portion of the common operational picture to support current operations.
- Maintains information requirement status.
- Coordinates preparation and execution of IO with maneuver and fires.
- Recommends adjustments to current IO.
- Tracks IRCs and recommends repositioning, as required.
- Tracks applicable targets in conjunction with the G-2 (S-2).
- Nominates targets for attack.
- Provides initial assessment of effectiveness.

The main CP plans, coordinates, and integrates IO. It—

- Creates and maintains IO aspects of the common operational picture.
- Maintains the IO estimate.
- Incorporates answers to IRs and requests for information into the IO estimate.
- Maintains a current IO order of battle.
- Deconflicts IO internally and externally.
- Requests/coordinates IO support with other warfighting function representatives, outside agencies, higher headquarters, and augmenting forces.
- Identifies future objectives based on successes or failures of current operations.

The IO officer monitors IRCs and keeps the G-3 (S-3) informed on overall IO status. The IO officer also recommends to the G-3 (S-3) changes to IRC taskings for inclusion in fragmentary orders, as warranted.

IO Working Group in Anticipation of/Response to Crisis or Significant Incident

The IO working group convenes as soon as feasible before or after an event. Anyone can request the convening of the IO working group to deal with crisis or incident through the IO officer who, in consultation with the G-3 (S-3) and commander, determines the merits of the request and those personnel who should comprise the working group's initial membership. The working group's purpose is to determine the additional measures, activities, and effects that must be undertaken or generated in order to sustain operational advantage in the information environment. The group also seeks to mitigate possible negative consequences resulting from crisis events or incidents, particularly those that would adversely affect U.S. and coalition credibility. Its membership is more ad-hoc than the routine IO working group but also situation dependent.

III. Assessing During Execution

Assessment precedes and guides the other activities of the operations process. It involves continuous monitoring of the current situation and evaluation of the current situation against the desired end state to determine progress and make decisions and adjustments.

The IO officer compiles information from all CPs, the G-2 (S-2), and higher headquarters to maintain a continuous IO assessment in the IO estimate. The primary objective of assessment is to determine whether IO is achieving planned effects. As the situation changes, the IO officer and G-3 (S-3) make sure IO remains fully synchronized with the overall operation.

Assessment is continuous; it precedes and guides every operations process activity and concludes each operation or phase of an operation. During planning, the commander and staff determine those IO objectives to be assessed, measures of performance and effectiveness, and the means of obtaining the information necessary to determine effectiveness. During orders production, the IO officer uses this information to prepare the IO portion of the overall assessment plan. During execution, the IO officer uses established measures of performance and effectiveness, as well as baselines and indicators, to assess IO objectives.

A. Monitoring IO

The IO officer monitors IRCs to determine progress towards achieving the IO objectives. Once execution begins, the IO officer monitors the threat and friendly situations to track IRC task accomplishment, determine the effects of IO during each phase of the operation, and detect and track any unintended consequences.

Monitoring the execution of defend-weighted tasks is done at the main CP because it is the focal point for intelligence analysis and production, and because the headquarters mission command nodes are monitored there. The IO officer works closely with the intelligence cell, G-2 (S-2), and IO working group representatives to provide a running assessment of the effectiveness of threat information efforts and keeps the G-3 (S-3) and various integrating cells informed.

With G-2 (S-2), G-3 (S-3), and fire support representatives, the IO officer monitors attack-weighted IO execution in the tactical CP and the main CP. For example, during combined arms maneuver, the IO officer is concerned with attacking threat command and control nodes with airborne and ground-based jammers, fire support, attack helicopters, and tactical air. After preplanned IO-related HPTs have been struck, the strike's effectiveness is assessed. Effective IO support of current operations depends on how rapidly the tactical CP can perform the targeting cycle to strike targets of opportunity. The G-3 (S-3) representative in the tactical CP keeps the main CP informed of current operations, including IO.

IV. Intelligence Support

Ref: Adapted from *ATP 3-13.1 (Oct '18), pp. 5-1 to 5-3 and ADP 3-13 (Nov '23).

The intelligence warfighting function contributes to the integration of all the information activities by providing relevant information and intelligence to decision makers. It directly contributes to developing situational understanding and informs decision making. The intelligence warfighting function contributes to understanding the human, information, and physical dimensions of an OE, to include identifying relevant actors, their relationships, and patterns of thinking.

> ### Key Terms *(See p. 4-9.)*
> - **Information Requirement.** Any information elements the commander and staff require to successfully conduct operations (ADRP 6-0).
> - **Intelligence Requirement.** A requirement for intelligence to fill a gap in the command's knowledge or understanding of the operational environment or threat forces (JP 2-0).
> - **Priority Intelligence Requirement.** An intelligence requirement that the commander and staff need to understand the threat and other aspects of the operational environment (JP 2-01). The commander designates PIRs. Information requirements not designated by the commander as PIRs become intelligence requirements.
> - **Intelligence Estimate.** The appraisal, expressed in writing or orally, of available intelligence relating to a specific situation or condition with a view to determining the courses of action open to the enemy or adversary and the order of probability of their adoption (JP 2-0).

An important synergy exists between IO and the intelligence and fires warfighting functions. Among the doctrinal tasks of the intelligence warfighting function is providing support to IO, IRCs, and targeting. The integration of IO into the targeting process—a task managed within the fires warfighting function—is important to mission accomplishment across the range of military operations.

Intelligence support to IO is continuous and typically requires long-lead times. The intelligence necessary to affect the perceptions and decision making of enemies, adversaries, or other audiences often requires that units position and employ specific sources and methods to collect the information and conduct the analyses needed for the information operation. The challenge is to get the right information and intelligence at the right time.

Intelligence "Push" and "Pull"

Intelligence is disseminated by the "push" or "pull" principle. For "push," IO planners coordinate with the intelligence staff to get access to the dissemination means that have IO-pertinent products. This is accomplished by working with the intelligence analysts to get IO-specific information requirements injected into the collection cycle, nominating PIRs for either the information environment or adversary actions in the information environment, and coordinating with higher headquarters' IO staffs to routinely receive distribution of intelligence products. To "pull" intelligence from the intelligence staff, IO planners coordinate for access to those assets and systems that have IO-relevant information and intelligence, attend intelligence staff updates and fusion meetings, and coordinate with units.

Information Operations and the Intelligence Process

All intelligence for the commander and staff, including that needed for IO, is produced as part of the intelligence process. By working closely with the intelligence staff officer during the intelligence process, IO planners can minimize intelligence gaps and maximize available intelligence and collection assets to develop a reasonably accurate understanding of the information environment and a representative and reliable model of adversary operations in the information environment. To integrate into the intelligence process, IO planners—

- Identify IO-specific intelligence gaps concerning the information environment and adversary operations in the information environment, recommend intelligence requirements as PIRs, and submit requests for information to fill the gaps.
- Become familiar with available collection assets, capabilities, and support relationships (direct support or general support). Planners determine time requirements for each collection asset and consider the capabilities and limitations of the assets that will perform the mission.
- Coordinate with the collection manager to ensure information requirements for IO are considered for inclusion as collection tasks. Ensure that IO-specific information requirements are matched to the correct information collection asset.
- Establish relationships with key intelligence personnel.
- Vet all intelligence products developed from reachback support and other external sources through the intelligence staff officer to avoid disconnected analysis.
- Provide feedback on the quality of intelligence provided and its usefulness to facilitate refinement.
- Assess the intelligence support provided to improve the working relationship with the intelligence staff while providing feedback to the intelligence analyst for improvements.

Intelligence Preparation of the Operational Environment (IPOE) *(See p. 1-24.)*

The basis of intelligence support to IO is the IPB process, a prerequisite to planning any operation. IPOE is the systematic process of analyzing the mission variables of enemy, terrain, weather, and civil considerations in an area of interest to determine their effect on operations. This includes informational considerations pertaining to the enemy, terrain, weather, and civil considerations. Continuous holistic IPOE contributes to creating and exploiting information advantages by improving understanding of an OE. A holistic approach—

- Describes the totality of relevant aspects of an OE that may impact friendly, threat, and neutral forces.
- Accounts for all relevant domains that may impact friendly and threat operations.
- Identifies windows of opportunity to leverage friendly capabilities against threat forces.
- Allows commanders to leverage positions of relative advantage at a time and place most advantageous for mission success with the most accurate information available.

Refer to ATP 2-01.3 for a detailed discussion of IPOE . *Doctrinally, this was formerly described/defined as intelligence preparation of the battlefield (IPB).*

Information Collection *(See p. 1- 24.)*

Information collection is an activity that synchronizes and integrates the planning and employment of sensors and assets as well as the processing, exploitation, and dissemination systems in direct support of current and future operations (FM 3-55). It integrates the functions of the intelligence and operations staffs that focus on answering CCIRs. Information collection includes acquiring information and providing it to processing elements. It has three steps:

- Plan requirements and assess collection
- Task and direct collection
- Execute collection

V. Decision Making During Execution
Ref: *FM 3-13, Information Operations (Dec '16), pp. 6-6 to 6-7.

Decision making during execution includes:
- Executing IO as planned.
- Adjusting IO to a changing friendly situation.
- Adjusting IO to an unexpected enemy reaction.

A. Executing IO as Planned
Essential to execution is a continuous information flow among the various functional and integrating cells. The IO officer tracks execution with intelligence and current operations cells, as well as with the targeting staff. The IO officer, in concert with the IO working group, maintains a synchronization matrix. This matrix is periodically updated and provided to the headquarters' functional and integrating cells. Using the matrix, the IO officer and working group keep record of completed IRC tasks. As tasks are completed, the IO officer passes the information to the intelligence cell. The IO officer and working group use this information to keep IO synchronized with the overall operation.

The IO officer determines whether the threat commander and other identified leaders are reacting to IO as anticipated during course of action analysis. The IO officer, in concert with the IO working group, looks for new threat vulnerabilities and for new IO-related targets. The IO officer proposes changes to the operation order (OPORD) to deal with variances throughout execution. The G-3 (S-3) issues FRAGORDs pertaining to IO, as requested by the IO officer. These FRAGORDs may implement changes to the scheme of IO, IO objectives, and IRC tasks. The IO officer updates the IO synchronization matrix and IO assessment plan to reflect these changes.

Given the flexibility of advanced information systems, the time available to exploit new threat command and control vulnerabilities may be limited and requires an immediate response from designated IRCs. Actions to defeat threat information efforts need to be undertaken before exploitation advantage disappears. The G-3 (S-3) may issue a verbal FRAGORD when immediate action is required.

B. Adjusting IO to a Changing Friendly Situation
As IO is executed, it often varies from the plan. Possible reasons for a variance include:
- An IO task is aborted or assets redirected.
- An IO-related target did not respond as anticipated.
- The threat effectively countered an IO attack.
- The threat successfully disrupted friendly mission command.
- The initial plan did not identify an emergent IO-related target or target of opportunity.

The IO officer's challenge is to rapidly assess how changes in IO execution affect the overall operation and to determine necessary follow-on actions. Based on the commander's input, the IO officer, in coordination with the rest of the headquarters' functional and integrating cells, considers COAs, conducts a quick COA analysis, and determines the most feasible COA.

If the selected COA falls within the decision-making authority of the G-3 (S-3), IO execution can be adjusted without notifying the commander. When changes exceed previously designated limits, the IO officer obtains approval from the commander. At this point, a more formal decision-making process may be required before issuing a FRAGORD, especially if a major adjustment to the operation order (OPORD) is needed. In such a case, the IO officer, working with the G-3 (S-3), participates in a time-constrained military decisionmaking process (MDMP) to develop a new COA.

C. Adjusting IO to an Unexpected Threat Reaction

The threat may react in an unexpected manner to IO or to the overall operation. If threat actions diverge significantly from those anticipated when the OPORD was written, the commander and staff look first at branch and sequel plans. If branch or sequel plans fail to adequately address the new situation, a new planning effort may be required.

The IO officer prepares branches that modify defend weighted efforts when threat actions cause new friendly vulnerabilities, or when friendly attack or stabilize efforts prove ineffective. The intelligence and current operations integration cells work with the IO officer to maintain a running assessment of threat capability to disrupt friendly mission command, and look for ways to lessen friendly vulnerabilities. Concurrently, they look for opportunities to reestablish IO effectiveness. Under these conditions, the IO officer determines the adequacy of existing branches and sequels. If none fit the situation, they create a new branch or sequel and disseminate it by FRAGORD.

If a new plan is needed, time available dictates the length of the decision-making process and the amount of detail contained in an order. The IO officer may only be able to recommend the use of IRCs that can immediately affect the overall operation: for example, electronic warfare, and MISO. Other IRCs proceed as originally planned and are adjusted later, unless they conflict with the new plan.

Other Execution Considerations
Other considerations include, but are not limited to—

IO Execution Begins Early
Potential adversary and enemy commanders begin forming perceptions of a situation well before they encounter friendly forces. Recognizing this fact, commanders establish a baseline of IO that is practiced routinely in garrison and training. Selected IRCs (for example, MISO, OPSEC, combat camera, and military deception) begin contributing to an IO objective well before a deployment occurs. To support early execution of the overall operation, IO planning, preparation, and execution frequently begin well before the staff formally starts planning for an operation.

IO Execution Requires Flexibility
Actions by threat decision makers sometimes take surprising turns, uncovering unanticipated weaknesses or strengths. Similarly, friendly commanders may react unexpectedly in response to threat activities. Flexibility is key to success in IO execution. Effective commanders and well-trained staffs are flexible enough to expect the unexpected and exploit threat vulnerabilities/friendly strengths and protect against threat strengths/friendly vulnerabilities.

Refer to BSS7: The Battle Staff SMARTbook, 7th Ed., updated for 2023 to include FM 5-0 w/C1 (2022), FM 6-0 (2022), FMs 1-02.1/.2 (2022), and more. Focusing on planning & conducting multidomain operations (FM 3-0), BSS7 covers the operations process; commander/ staff activities; the five Army planning methodologies; integrating processes (IPB, information collection, targeting, risk management, and knowledge management); plans and orders; mission command, command posts, liaison; rehearsals & after action reviews; operational terms & military symbols.

To organize and portray IO execution, the IO officer and working group use several tools, to include:
- IO synchronization matrix.
- Decision support template.
- High-payoff target list.
- Critical asset list and defended asset list.

IO officer and working group use either the synchronization matrix from the IO appendix or an extract containing current and near-term IO objectives and IRC tasks, depending on the complexity of the operation. The synchronization matrix is used to monitor progress and results of IO objectives and IRC tasks and keep IO execution focused on contributing to the overall operation. The decision support template produced by the G-3 (S-3) is used by the IO officer to monitor progress of IO in relation to decision points and any branches or sequels. The IO officer maintains a list or graphic (for example, a link and node diagram) that tracks the status of IO-related HPTs identified during planning. The IO officer uses the critical asset list and defended asset list to monitor the status of critical friendly information nodes and the status of critical systems supporting IO, for example: electronic warfare systems, military information support operations (MISO) assets, and deep attack assets.

B. Evaluating IO

During execution, the IO officer works with the intelligence cell and integrating cells to obtain the information needed to determine individual and collective IO effects. Evaluation not only estimates the effectiveness of task execution, but also evaluates the effect of the entire IO effort on the threat, other relevant audiences in the area of operations, and friendly operations. Task execution is evaluated using measures of performance. Task effectiveness is evaluated using measures of effectiveness, which compare achieved results against a baseline.

See chap. 8, IO Assessment, for additional discussion.

Based on the IO effects evaluation, the IO officer adjusts IO to further exploit enemy vulnerabilities, redirects actions yielding insufficient effects, or terminates actions after they have achieved the desired result. The IO officer keeps the G-3 (S-3) and commander informed of IO effects and how these impact friendly and adversary operations. Some of the possible changes to IO include:
- Strike a target or continue to protect a critical asset to ensure the desired effect.
- Execute a branch or sequel.

I. Fires (INFO Considerations)

Chap 7

Ref: ADP 3-19, Fires (Jul '19), chap. 1 and ADP 3-0, Operations (Jul '19), p. 5-5.

Success in large-scale combat operations is dependent on the Army's ability to employ fires. Fires enable maneuver. Over the past two decades, potential peer threats have invested heavily in long-range fires and integrated air defense systems, making it even more critical that the U.S. Army possess the ability to maneuver and deliver fires in depth and across domains.

I. The Fires Warfighting Function

The fires warfighting function is the related tasks and systems that create and converge effects in all domains against the threat to enable actions across the range of military operations (ADP 3-0). **These tasks and systems create lethal and nonlethal effects delivered from both Army and Joint forces, as well as other unified action partners.** The fires warfighting function does not wholly encompass, nor is it wholly encompassed by, any particular branch or function. Many of the capabilities that contribute to fires also contribute to other warfighting functions, often simultaneously. For example, an aviation unit may simultaneously execute missions that contribute to the movement and maneuver, fires, intelligence, sustainment, protection, and command and control warfighting functions. Additionally, air defense artillery (ADA) units conduct air and missile defense (AMD) operations in support of both fires and protection warfighting functions.

Commanders must execute and integrate fires, in combination with the other elements of combat power, to create and converge effects and achieve the desired end state. Fires tasks are those necessary actions that must be conducted to create and converge effects in all domains to meet the commander's objectives. The tasks of the fires warfighting function are:

Integrate Army, multinational, and joint fires through:
- Targeting.
- Operations process.
- Fire support.
- Airspace planning and management.
- **Electromagnetic spectrum management.**
- Multinational integration.
- Rehearsals.
- Air and missile defense planning and integration.

Execute fires across all domains and in the information environment, employing:
- Surface-to-surface fires.
- Air-to-surface fires.
- Surface-to-air fires.
- **Cyberspace operations and EW.**
- **Space operations.**
- Multinational fires.
- Special operations.
- **Information operations.**

See pp. 7-4 to 7-5 for an overview and further discussion.

II. Fires Overview
Ref: ADP 3-19, Fires (Jul '19).

Success in large-scale combat operations is dependent on the Army's ability to employ fires. Fires enable maneuver. Over the past two decades, potential peer threats have invested heavily in long-range fires and integrated air defense systems, making it even more critical that the U.S. Army possess the ability to maneuver and deliver fires in depth and across domains.

Fires in Support of Unified Land Operations

The Army operational concept for conducting operations as part of a joint team is unified land operations. Unified land operations is the simultaneous execution of offense, defense, stability, and defense support of civil authorities across multiple domains to shape operational environments, prevent conflict, prevail in large-scale ground combat, and consolidate gains as part of unified action (ADP 3-0). The goal of unified land operations is to achieve the JFC's end state by applying landpower as part of unified action. Commanders employ fires to set conditions for the successful employment of other elements of combat power to conduct unified land operations. The targeting process can help commanders and staffs to prioritize and integrate assets to create effects that allow for achievement of the commander's objectives within unified land operations.

The Army's primary mission is to organize, train, and equip its forces to conduct prompt and sustained land combat to defeat enemy ground forces and seize, occupy, and defend land areas. During the conduct of unified land operations, Army forces support the joint force through four strategic roles:

- Shape OEs.
- Prevent conflict.
- Prevail during large-scale ground combat.
- Consolidate gains.

Fires in Support of Large-Scale Combat Operations

The Army, as part of the joint force, conducts large-scale combat operations. The preponderance of large-scale combat operations will consist of offensive and defensive operations initially, although some. Stability operations will occur simultaneously as part of consolidating gains. Commanders employ fires as part of large-scale combat operations by creating effects to enable joint force freedom of action.

Commanders use Army and joint targeting to select and prioritize targets, integrating lethal and nonlethal effects from different capabilities in support of large-scale combat operations. Commanders may converge effects from multiple systems, either simultaneously or in close succession, to create an even greater effect than would have been achieved if each effect was created individually. Convergence is the massing of capabilities from multiple domains to create effects in a single domain. Convergence overwhelms the enemy, giving them too many dilemmas to address simultaneously, which creates gaps for exploitation by the joint force. The convergence of multiple effects within an area requires careful synchronization prior to execution to ensure effects don't interfere with one another or pose a risk to the force.

To effectively enable joint force freedom of action during large-scale combat operations, commanders must synchronize the effects created with fires with the actions of the rest of the joint force. This synchronization initially takes place during planning, where commanders and their staffs determine the timing of the creation of the effect and link that timing to a clearly defined, conditions-based trigger. Commanders must also plan for assessment of the effects and determine alternate courses of action if the effects are not created as planned.

Fires Logic Diagram

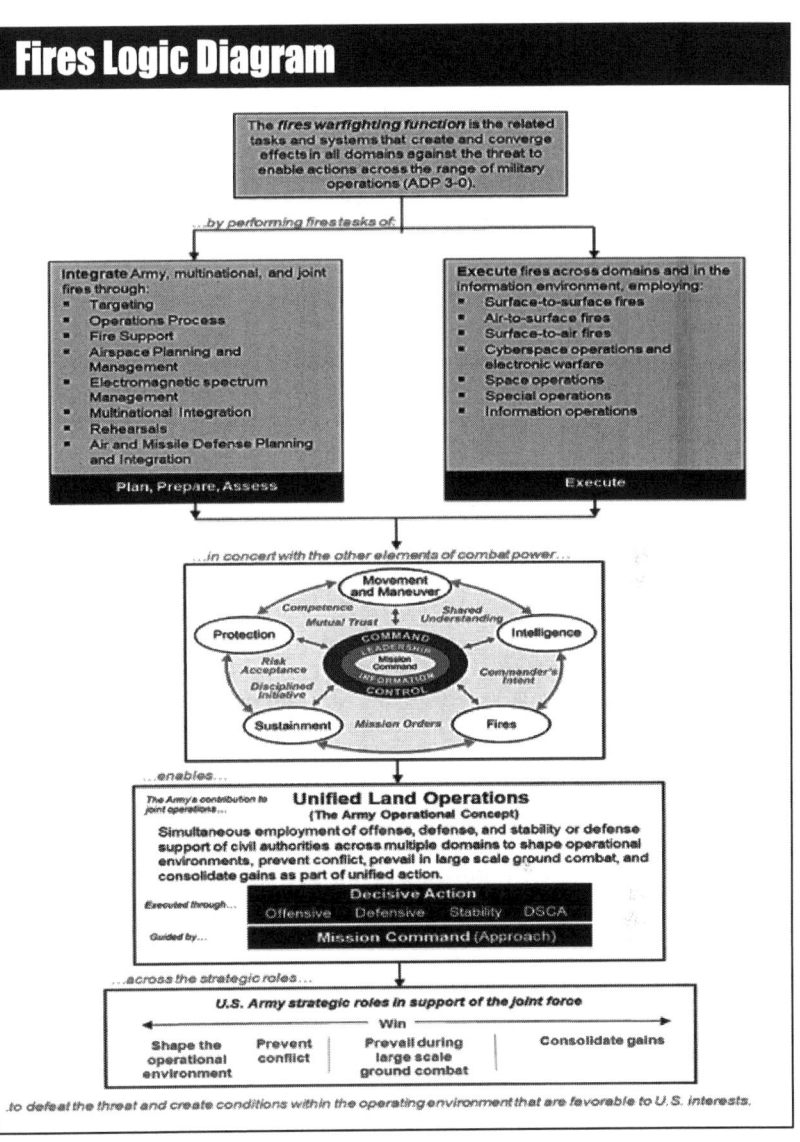

The **fires warfighting function** is the related tasks and systems that create and converge effects in all domains against the threat to enable actions across the range of military operations (ADP 3-0).

...by performing fires tasks of:

Integrate Army, multinational, and joint fires through:
- Targeting
- Operations Process
- Fire Support
- Airspace Planning and Management
- Electromagnetic spectrum Management
- Multinational Integration
- Rehearsals
- Air and Missile Defense Planning and Integration

Plan, Prepare, Assess

Execute fires across domains and in the information environment, employing:
- Surface-to-surface fires
- Air-to-surface fires
- Surface-to-air fires
- Cyberspace operations and electronic warfare
- Space operations
- Special operations
- Information operations

Execute

...in concert with the other elements of combat power...

(Diagram: Movement and Maneuver, Intelligence, Fires, Sustainment, Protection, surrounding Command/Leadership/Information/Mission Command/Control core; with Competence, Mutual Trust, Shared Understanding, Commander's Intent, Mission Orders, Disciplined Initiative, Risk Acceptance)

...enables...

Unified Land Operations
(The Army Operational Concept)

The Army's contribution to joint operations... Simultaneous employment of offense, defense, and stability or defense support of civil authorities across multiple domains to shape operational environments, prevent conflict, prevail in large scale ground combat, and consolidate gains as part of unified action.

Executed through... **Decisive Action**: Offensive | Defensive | Stability | DSCA

Guided by... **Mission Command (Approach)**

...across the strategic roles...

U.S. Army strategic roles in support of the joint force

| Shape the operational environment | Prevent conflict | Prevail during large scale ground combat | Consolidate gains |

— Win —

...to defeat the threat and create conditions within the operating environment that are favorable to U.S. interests.

Ref: ADP 3-19, Fires (Jul '19), introductory figure, ADP 3-19 Logic chart.

Refer to AODS7: The Army Operations & Doctrine SMARTbook (Multidomain Operations). Completely updated with the 2022 edition of FM 3-0, AODS7 focuses on Multidomain Operations and features rescoped chapters on generating and applying combat power: command & control (ADP 6-0), movement and maneuver (ADPs 3-90, 3-07, 3-28, 3-05), intelligence (ADP 2-0), fires (ADP 3-19), sustainment (ADP 4-0), & protection (ADP 3-37).

III. Execute Fires Across the Domains
Ref: ADP 3-19, Fires (Jul '19), chap. 2.

The commander is responsible for the integration of fires within the AO. The commander consults the fire support coordinator, chief of fires, air liaison officer, fire support officer, and experts on AMD, **cyberspace, EW, space**, special operations, and **information operations** for advice on the allocation, integration, and use of available fires resources. Fires in all domains require detailed coordination and planning to support the commander's objectives. Employment of these systems requires the use of common terminology and coordination measures across the joint force. It includes **surface-to-surface fires, air-to-surface fires, and nonlethal means** that the commander uses to support the concept of the operation.

Surface-to-Surface Fires
Army surface-to-surface indirect fires includes cannon, rocket, and missile systems as well as mortars organic to maneuver elements. Field Artillery is the equipment, supplies, ammunition, and personnel involved in the use of cannon, rocket, or surface-to-surface missile launchers (JP 3-09). The role of the field artillery (FA) is to destroy, neutralize, or suppress the enemy by cannon, rocket, and missile fire and to integrate and synchronize all fire support assets into operations. Fire support is fires that directly support land, maritime, amphibious, and special operations forces to engage enemy forces, combat formations, and facilities in pursuit of tactical and operational objectives (JP 3-09).

Air-to-Surface Fires
Army and joint forces employ various types of air-to-surface capabilities, to include fixed-wing aircraft, rotary-wing aircraft, and unmanned aircraft systems (UASs). These systems provide lethal and nonlethal effects, standoff weapons, and target acquisition capabilities that can be employed to detect and create integrated effects against adversary targets.

Surface-to-Air Fires
Surface-to-air fires capabilities include active defense weapons that are employed in both area and point defenses. ADA delivers precision surface-to-air missiles to defend friendly forces, fixed and semi-fixed assets, population centers and key infrastructure against air and missile threats. ADA executes Army AMD operations in support of joint counterair efforts. The role of ADA is to deter and defeat the range of aerial threats in order to assure allies, ensure operational access, and defend critical assets and deployed forces in support of unified land operations (FM 3-01).

Space Operations *(See pp. 3-61 to 3-70.)*
Many lethal and nonlethal fires capabilities depend on space capabilities to support, integrate, and deliver fires. Army space capabilities are integrated throughout the fires warfighting function, providing robust and reliable planning, contributing to target development, and providing positioning, navigation, and timing (PNT), satellite communications, imagery, geolocation, weather, and terrain capabilities.

- GPS enables precision guided munitions, command and control systems, and near real-time situational awareness for lethal and nonlethal fires.
- Satellite communications enables real time communications between commanders and forces to enable immediate redirection of fires over extended distances to shape the operations.
- Weather satellites provide a variety of data points necessary for predicting effects of meteorological conditions on fires.
- Combined, PNT and satellite communications supports fires through the systems interfaces on the Advanced Field Artillery Tactical Data System.

In space operations, the fires warfighting function includes space control operations that create a desired effect on enemy space systems and across multiple domains. Space control plans and capabilities use a broad range of response options to provide continued, sustainable use of space. Space control contributes to space deterrence by employing a variety of measures to assure the use of space and attribute enemy attacks. These include terrestrial fires to defend space operations and assets. A capability for, or employment of, fires may deter threats and/or contain and de-escalate a crisis.

Offensive space control are offensive operations conducted for space negation (JP 3-14). Negation in space operations, are measures to deceive, disrupt, degrade, deny, or destroy space systems. (JP 3-14). Offensive space control actions targeting an enemy's space-related capabilities and forces could employ reversible or nonreversible means, and are considered a form of fires.

Cyberspace Operations & Electronic Warfare *(See pp. 3-45 to 3-60.)*

Friendly, enemy, adversary, and host nation networks, communications systems, computers, cellular phone systems, social media websites, and technical infrastructures are all part of cyberspace. Cyberspace operations are the employment of cyberspace capabilities where the primary purpose is to achieve objectives in or through cyberspace (JP 3-0). The interrelated cyberspace missions are Department of Defense information network operations, defensive cyberspace operations, and offensive cyberspace operations.

Electronic attack involves the use of electromagnetic energy, directed energy, or anti-radiation weapons to attack personnel, facilities, or equipment with the intent of degrading, neutralizing, or destroying enemy combat capability and is considered a form of fires. Electronic attack includes:

- Actions taken to prevent or reduce an enemy's effective use of the electromagnetic spectrum.
- Employment of weapons that use either electromagnetic or directed energy as their primary destructive mechanism.
- Offensive and defensive activities, including countermeasures.

* Information Operations *(See pp. 3-1 to 3-4.)*

Information operations is the integrated employment, during military operations, of information-related capabilities in concert with other lines of operation to influence, disrupt, corrupt, or usurp the decision-making of adversaries and potential adversaries while protecting our own (JP 3-13). Information operations, as an integration and synchronization staff function, plans and oversees the coordinated delivery of information-related capabilities to achieve cognitive effects against adversary and enemy decision-makers across the conflict continuum while simultaneously establishing the conditions that allow for more timely and better-informed friendly decision-making.

Commanders can also designate other enabling information related capabilities (both lethal and nonlethal) to control the flow of information to adversary/enemy decision-makers and protect friendly command and control means. These activities and capabilities include:

- Physical attack (to include lethal fires and maneuver).
- Presence, posture, and profile.
- Communication synchronization.
- Cybersecurity.
- Foreign disclosure.
- Physical security.
- Special access programs.
- Civil military operations.
- Intelligence.

See chap. 3, Information Capabilities.

IV. Joint Fires (OIE Considerations) (See p. 2-43.)

Ref: JP 3-04, Information in Joint Operations (Sept '22), pp. III-4 to III-6.

Fires is the use of weapon systems or other actions to create specific lethal or nonlethal effects on a target. The nature of the target or threat, the conditions of the mission variables (i.e., mission, enemy, terrain and weather, troops and support available, time available, and civil considerations), and desired outcomes determine how lethal and nonlethal capabilities are employed. Operations in the information environment (OIE) may **leverage the inherent informational aspects of joint fires.** Fires in and through the IE encompass a number of tasks, actions, and processes, including targeting, coordination, deconfliction, and assessment (e.g., BDA).

OIE tasks and capabilities **leverage information** through fires to create specific effects. To integrate effectively, information planners participate in the joint targeting process by selecting and prioritizing targets for fires or TAs for other actions. OIE units create fires that typically result in nonlethal effects. OIE can also indirectly create effects that result in physical destruction (e.g., manipulating computers that control physical processes). Additionally, OIE can leverage the inherent informational aspects of fires to reinforce the psychological effect of those fires. OIE may rely on joint fires support to transmit information to relevant actors and to deliver **nonlethal payloads** to affect information, information systems, and information networks (e.g., leveraging CO to deliver computer code designed to deny network access to an adversary, PA releases to inform friendly audiences, or MISO products to influence foreign audiences).

Joint Force Capabilities, Operations, and Activities for <u>Leveraging Information</u> (See p. 2-19.)

Ref: JP 3-04, Information in Joint Operations (Sept '22), pp. II-6 to II-15.

When commanders leverage information, they expand their range of options for the employment of military capabilities beyond the use of or threatened use of physical force. JFCs leverage information in two ways. First, by planning and conducting all operations, activities, and investments to deliberately leverage the inherent informational aspects of such actions. Second, by conducting OIE.

INFORM Domestic, International, and Internal Audiences

Inform activities are the release of accurate and timely information to the public and internal audiences, to foster understanding and support for operational and strategic objectives by putting joint operations in context; facilitating informed perceptions about military operations; and countering misinformation, disinformation, and propaganda. Inform activities help to ensure the trust and confidence of the US population, allies, and partners in US and MNF efforts; and to deter and dissuade adversaries and enemies from action. PA is the primary means the joint force uses to inform; however, civil-military operations (CMO), key leader engagement (KLE), and military information support operations (MISO) also support inform efforts.

INFLUENCE Relevant Actors

The purpose of the influence task is to affect the perceptions, attitudes, and other drivers of relevant actor behavior. Regardless of its mission, the joint force considers the likely psychological impact of all operations on relevant actor perceptions, attitudes, and other drivers of behavior. The JFC then plans and conducts every operation to create desired effects that include maintaining or preventing behaviors or inducing changes in behaviors. This may include the deliberate selection and use of specific capabilities for their inherent informational aspects (e.g., strategic bombers); adjustment of the location, timing, duration, scope, scale, and even visibility of an operation (e.g., presence, profile, or posture of the joint force); the use of signature management and

MILDEC operations; the employment of a designated force to conduct OIE; and the employment of individual information forces (e.g., CA, psychological operations forces, cyberspace forces, PA, combat camera [COMCAM]) to reinforce the JFC's efforts. US audiences are not targets for military activities intended to influence.

ATTACK AND EXPLOIT Information, Information Networks, and Information Systems

The joint force targets information, information networks, and information systems to affect the ability of adversaries and enemies to use information in support of their own objectives. This activity includes manipulating, modifying, or destroying data and information; accessing or collecting adversary or enemy information to support joint force activities or operations; and disrupting the flow of information to gain military advantage. Attacking and exploiting information, information networks, and information systems supports the influence task when it undermines opponents' confidence in the sources of information or the integrity of the information that they rely on for decision making. Activities used to attack and exploit information include offensive cyberspace operations (OCO), electromagnetic warfare (EW), MISO, and CA operations. PA also contributes to this task by publicly exposing malign activities.

Nonlethal Effects
Ref: JP 3-0, Joint Campaigns and Operations (Jun '22), p. III-36.

Joint force capabilities can create nonlethal effects. Some capabilities can produce non-lethal effects that limit collateral damage, reduce risk to civilians, and reduce exploitation opportunities for enemy or adversary propaganda. They may also reduce the number of casualties associated with excessive use of force, limit reconstruction costs, and maintain the goodwill of the local populace. Some capabilities are nonlethal by design and include blunt impact and warning munitions, acoustic and optical warning devices, and vehicle and vessel stopping systems.

Cyberspace Attack *(See p. 3-54.)*
Cyberspace attack actions create various direct denial effects in cyberspace (i.e., degradation, disruption, or destruction) or manipulation that leads to denial that appears in the physical domains.

Electromagnetic Attack (EA) *(See p. 3-56.)*
EA involves the use of EM energy, DE, or antiradiation weapons to attack personnel, facilities, or equipment to degrade, neutralize, or destroy enemy combat capability. EA can be against a computer when the attack occurs through the EMS. Integration and synchronization of EA with maneuver, C2, and other joint fires are essential. EW is a component of JEMSO used to exploit, attack, protect, and manage the EME to achieve the commander's objectives. EW can be a primary capability or used to facilitate OIE through the targeting process.

Military Information in Suport of Operations (MISO) *(See p. 3-33.)*
MISO actions and messages can generate effects that gain support for JFC objectives; reduce the will of the enemy, adversary, and sympathizer; and decrease the combat effectiveness of enemy forces. MISO are effective throughout the competition continuum. JFCs and their component commanders are the key players in fully integrating MISO into their plans and operations. MISO require unique budget, attribution, and authorities that are coordinated and approved prior to employment. Commanders carefully review and approve MISO programs that comply with mission-tailored, product approval guidelines from national-level authorities. An approved program does not necessarily constitute authority to execute a mission. Commanders obtain required authorities through a MISO-specific execute order (EXORD) or as a task specified in an EXORD for an operation.

V. Foundations of Fire Support (FS) (INFO Considerations)

Ref: FM 3-09, Fire Support & Field Artillery Operations (Apr '20), pp. 1-1 and 2-16 to 2-21.

Fire support is a rapid and continuous integration of surface to surface indirect fires, target acquisition, armed aircraft, and other **lethal and nonlethal attack/delivery systems that converge against targets across all domains** in support of the maneuver commander's concept of operations.

Fire support (FS) is inherently joint, conducted in all domains, and simultaneously executed at all echelons of command. Lethal FS attack and delivery systems consist of indirect fire weapons and armed aircraft to include FA, mortars, naval surface fire support, and air-delivered munitions from fixed wing and rotary wing aircraft. Field artillery is equipment, supplies, ammunition, and personnel involved in the use of cannon, rocket, or surface-to-surface missile launchers. **Nonlethal capabilities include cyberspace electromagnetic activities (CEMA), information related activities, space, and munitions such as illumination and smoke.** Fires are the use of weapons systems to create a specific lethal or nonlethal effect on a target (JP 3-0). A nonlethal weapon is a weapon, device, or munition that is explicitly designed and primarily employed to incapacitate personnel or materiel immediately, while minimizing fatalities, permanent injury to personnel, and undesired damage to property in the target area or environment (JP 3-09).

The commander employs these capabilities to support the scheme of maneuver, to mass firepower, and to destroy, neutralize, and suppress enemy forces. Enemy a party identified as hostile against which the use of force is authorized (ADP 3-0). FS is a critical component of the fires warfighting function tasks of integrate and execute that allow the commander to **converge effects across all domains to achieve positions of relative advantage** in the context of large-scale ground combat operations (see ADP 3-19). Large-scale ground combat operations are sustained combat operations involving multiple corps and divisions (ADP 3-0). In large-scale ground combat operations, FS could be the principal means of destroying enemy forces. In this event, the scheme of maneuver would be designed specifically to capitalize on the effects of FS. The commander will utilize organic and joint attack/delivery assets and capabilities to provide joint FS. Joint fire support is joint fires that assist air, land, maritime, and special operations forces to move, maneuver, and control territory, populations, airspace, and key waters (JP 3-0).

See related discussion of positions of relative advantage on p. 1-5.

Attack & Delivery Capabilities

This section discusses lethal and nonlethal weapon systems capabilities.

- **Joint fire support (FS) surface to surface and air to surface capabilities.** *For more technical information refer to ATP 3-09.32/MCRP 3-31.6/NTTP 3-09.2/AFTTP 3-2.6, JFIRE, multi-service TTPs for joint application of firepower.*
- **CEMA**.
- **Space operations**.
- **Information related activities**.

Cyberspace Electromagnetic Activities (CEMA) (See pp. 3-45 to 3-46.)

Cyberspace electromagnetic activities is the process of planning, integrating, and synchronizing cyberspace and electronic warfare operations in support of unified land operations (ADP 3-0). Cyberspace operations are the employment of cyberspace capabilities where the primary purpose is to achieve objectives in or through cyberspace (JP 3-0). Both the offensive cyberspace operations (OCO) and defensive cyberspace operations response action (DCO-RA) missions may rise to the level of use of force, where physical damage or destruction of enemy systems require use of fires in cyberspace. OCO are intended to project power by the application of force in and through cyberspace and DCO-RA uses defensive measures, including fires, outside the defended network to protect it.

Cyberspace attack actions are a form of fires, are taken as part of an OCO or DCO-RA mission, are coordinated with other USG departments and agencies, and are carefully synchronized with planned fires in the physical domains. *See pp. 3-47 to 3-54.*

Space Operations *(See pp. 3-61 to 3-68.)*

Many lethal and nonlethal fires capabilities depend on space capabilities to support, integrate, and deliver fires. Army space capabilities are integrated throughout the fires warfighting function, providing robust and reliable planning, contributing to target development, and providing positioning, navigation, and timing, satellite communications, imagery, geolocation, weather, and terrain capabilities.

As a FS attack/delivery capability, space control operations that create a desired effect on enemy space systems across all domains by employing a variety of measures to assure the use of space and attribute enemy attacks. These include terrestrial fires to defend space operations and assets. A capability for, or employment of, fires may deter threats and/or contain and de-escalate a crisis.

* Information Related Activities (IRCs) *(See chap. 3.)*

The integration and synchronization of FS with information-related activities through the targeting process is fundamental to creating the necessary synergy between information-related activities and more traditional maneuver and strike operations. **Some information-related activities supporting joint fires include:**

Military deception is actions executed to deliberately mislead adversary military, paramilitary, or violent extremist organization decision makers, thereby causing the adversary to take specific actions (or inactions) that will contribute to the accomplishment of the friendly mission (JP 3-13.4). Deception applies to all levels of warfare, across the range of military operations, and is conducted during all phases of military operations. Physical attack/destruction can support military deception by shaping an enemy's intelligence collection capability through destroying or nullifying selected ISR capabilities or sites. Attacks can mask the main effort from the enemy. When properly integrated with operations security (OPSEC) and other information-related capabilities, deception can be a decisive tool in altering how the enemy views, analyzes, decides, and acts in response to friendly military operations.

OPSEC is a capability that identifies and controls critical information and indicators of friendly force actions attendant to military operations and incorporates countermeasures to reduce the risk of an adversary exploiting vulnerabilities. OPSEC identifies critical information and actions attendant to friendly military operations to deny observables to the threat intelligence systems. For example, camouflage and concealment are OPSEC measures and survivability operations tasks used to protect friendly forces and activities from enemy detection and attribution.

Military information support operations which are planned operations to convey selected information and indicators to foreign audiences to influence their behavior and ultimately the behavior of their governments. Selected audiences may include enemies, adversaries, unified action partners, and neutral groups or populations. Psychological operations support forces devise actions and craft messages using visual, audio, and audiovisual formats, which can then be delivered by air, land, and maritime means, and through cyberspace, to selected individuals and groups. Many actions of the FS system, such as strikes, have psychological impact, but they are not military information support operations unless their primary purpose is to influence the attitudes, rules, norms, beliefs, and subsequent behavior of a target audience.

Fire support requirements should be deconflicted and synchronized with **special technical operations (STO)**. Detailed information related to special technical operation and their contribution to fire support can be obtained from the special technical operations planners at combatant command or service component HQ.

VI. Scheme of Information Operations

Ref: *ADP 3-19, Fires (Jul '19), pp. 1-4 to 1-5. (*See note p. 1-2.)

IO brings together IRCs at a specific time and in a coherent fashion to **create effects in and through the information environment** that advance the ability to deliver operational advantage to the commander. While IRCs create individual effects, IO stresses aggregate and synchronized effects as essential to achieving operational objectives.

Information-Related Capabilities (IRCs) *(See chap. 3.)*

An information-related capability (IRC) is a tool, technique, or activity employed within a dimension of the information environment that can be used to **create effects and operationally desirable conditions** (JP 1-02). The formal definition of IRCs encourages commanders and staffs to employ all available resources when seeking to affect the information environment to operational advantage. For example, if artillery fires are employed to destroy communications infrastructure that enables enemy decision making, then artillery is an IRC in this instance. In daily practice, however, the term IRC tends to refer to those tools, techniques, or activities that are inherently information-based or primarily focused on affecting the information environment.

All unit operations, activities, and actions affect the information environment. Even if they primarily affect the physical dimension, they nonetheless also affect the informational and cognitive dimensions. For this reason, whether or not they are routinely considered an IRC, a wide variety of unit functions and activities can be adapted for the purposes of conducting information operations or serve as enablers to its planning, execution, and assessment.

Scheme of Information Operations *(See pp. 4-12 to 4-13.)*

The scheme of IO is a clear, concise statement of where, when, and how the commander intends to employ and synchronize IRCs to create effects in and through the information environment to achieve the mission and support decisive operations.

Based on the commander's planning guidance, the IO officer develops a separate **scheme of IO** for each COA the staff develops. **Schemes of IO are written in terms of IO objectives—and their associated weighted efforts (attack, defend, stabilize)—and IRC tasks required to achieve these objectives.** For example, the overall scheme may be weighted heavily on defending friendly information but also include attack and stabilize objectives.

IO Weighted Efforts & Enabling Activities

Decisive action is the continuous, simultaneous combination of offensive, defensive, and stability or defense support of civil authorities tasks (ADRP 3-0). IO contributes to decisive action through the **continuous and simultaneous combination and synchronization of IRCs** in support of offense, defense, and stability tasks. IO itself is not offensive, defensive, or stabilizing, but contributes to all of these simultaneously by **weighting its efforts** in such a way that it achieves requisite effects in and through the information environment in support of the commander's intent.

IO weighted efforts are broad orientations used to focus the integration and synchronization of IRCs to create effects that seize, retain, and exploit the initiative in the information environment. Commanders, supported by their staffs, visualize and describe how IO will support the concept of operations by aligning and balancing the **efforts of defend, attack, and stabilize.**

To support decisive action effectively, the commander and staff undertake three enabling activities—analyze and depict the information environment, determine IRCs and IO organizations available, and optimize IRC effects.

II. Targeting (IO Integration)

Ref: *FM 3-13, Information Operations (Dec '16), chap. 7 and *ATP 3-13.1, The Conduct of Information Operations (Oct '18), chap. 5. (*See note p. 1-2.)

Targeting is the process of selecting and prioritizing targets and matching the appropriate response to them, considering operational requirements and capabilities (JP 3-0). IO is integrated into the targeting cycle to produce effects in and through the information environment that support objectives. The targeting cycle facilitates the engagement of the right target with the right asset at the right time. The IO officer or representative is a part of the targeting team, responsible to the commander and staff for all aspects of IO.

Targeting Methodology

Army targeting methodology is based on four functions: decide, detect, deliver, and assess (D3A) *(see figure 7-1)*. The decide function occurs concurrently with planning. The detect function occurs during preparation and execution. The deliver function occurs primarily during execution, although some IO-related targets may be engaged while the command is preparing for the overall operation. The assess function occurs throughout.

Operations Process Activity	Targeting Process Function	Targeting Task
ASSESSMENT / PLANNING	DECIDE	**Mission Analysis** Develop IO-related HVTs Provide IO input to targeting guidance and targeting objectives **COA Development** Designate potential IO-related HPTs Contribute to the threat and vulnerability assessment Deconflict and coordinate potential HPTs **COA Analysis** Develop high priority target list Establish target selection standards Develop AGM Determine criteria of • Successful BDA • Requirements **Orders Production** Finalize high-payoff target list Finalize target selection standards Finalize AGM Submit IO information requirements/requests for information to G-2 (S-2)
PREPARATION / EXECUTION	DETECT	• Execute collection plan • Update PIRs/IO IRs as they are answered • Update high-payoff target list and AGM
	DELIVER	• Execute attacks in accordance with the AGM
	ASSESS	• Evaluate effects of attacks • Monitor targets attacked with nonlethal IO

AGM attack guidance matrix **BDA** battle damage assessment **COA** course of action **HPT** high-payoff target **HVT** high-value target **IO** Information operations **PIR** priority intelligence requirements

Ref: FM 3-13, fig. 7-1. The operations process, targeting cycle and IO-related tasks.

The targeting process is cyclical. The command's battle rhythm determines the frequency of targeting working group meetings. IO-related target nominations are developed by the IO officer and by the IO working group, which validates all IO-related targets before they are nominated to the targeting working group. Therefore, the IO working group is always scheduled in advance of the targeting working group.

I. Decide, Detect, Deliver, Assess (D3A)

Army targeting methodology is based on four functions: decide, detect, deliver, and assess (D3A). The decide function occurs concurrently with planning. The detect function occurs during preparation and execution. The deliver function occurs primarily during execution, although some IO-related targets may be engaged while the command is preparing for the overall operation. The assess function occurs throughout.

D - Decide

The decide function is part of the planning activity of the operations process. It occurs concurrently with the military decisionmaking process (MDMP). During the decide function, the targeting team focuses and sets priorities for intelligence collection and attack planning. Based on the commander's intent and concept of operations, the targeting team establishes targeting priorities for each phase or critical event of an operation. The following products reflect these priorities—

- High-payoff target list.
- Information collection plan.
- Target selection standards.
- Attack guidance matrix.
- Target synchronization matrix.

The high-payoff target list is a prioritized list of targets whose loss to the enemy will significantly contribute to the success of the friendly course of action. High-payoff targets (HPTs) are those high-value targets (HVTs) identified during COA development and validated in subsequent steps that must be acquired and successfully attacked for the success of the friendly commander's mission. Examples of IO-related HPTs are threat command and control nodes and intelligence collection assets/capabilities.

The information collection plan, prepared by the G-3 (S-3) and coordinated with the entire staff, synchronizes the four primary means information collection to provide intelligence to the commander. The G-2 (S-2) ensures all available collection assets provide the required information. Information requirements submitted by the IO officer can require longer lead times to detect targets and dwell times to assess the effects of IRCs directed against these targets.

Target selection standards establish criteria for deciding when targets are located accurately enough to attack. These criteria are often more complicated for IO, especially when attempting to identify actors and audiences with precision.

The attack guidance matrix addresses how and when targets are to be engaged and desired effects of the engagement. For IO-related targets, effects are diverse, running the gamut from destruction of assets to changed behaviors.

The target synchronization matrix is a list of HPTs by category and the agencies responsible for detecting them, attacking them, and assessing the effects of the attacks. It combines data from the high-payoff target list, information collection plan and attack guidance matrix.

The targeting team develops or contributes to these products throughout the MDMP. The commander approves them during COA approval. The IO officer ensures they include information necessary to engage IO-related targets. IO-related vulnerability analyses done by the G-2 (S-2) and IO officer provide a basis for deciding which IO-related targets to attack.

See following pages (pp. 7-16 to 7-21) for further discussion of "Decide" targeting tasks during the MDMP.

D - Detect

This function involves locating HPTs accurately enough to engage them. It primarily entails execution of the information collection plan. All staff agencies, including the IO officer, are responsible for passing to the G-2 (S-2) information collected by their assets that answer IRs. Conversely, the G-2 (S-2) is responsible for passing combat information and intelligence to the agencies that identified the IRs. Sharing information allows timely evaluation of attacks, assessment of IO, and development of new targets. Effective information and knowledge management are, therefore, essential.

The information collection plan focuses on identifying HPTs and answering PIRs. These are prioritized based on the importance of the target or information to the commander's concept of operation and intent. When designated by the commander, PIRs can include requirements concerning IO; obtaining answers to these requirements will assist the IO officer in assessing IO. Thus, there is some overlap between detect and assess functions. Detecting targets for nonlethal attacks may require information collection support from higher headquarters. The targeting team adjusts the high-payoff target list and attack guidance matrix to meet changes as the situation develops. The IO officer submits new IO IRs/RFIs as needed.

During the detect function, the IO officer updates the high-payoff target list and target synchronization matrix. In addition to the information collection plan, the IO officer will use other information sources, particularly culturally-attuned ones that have unique access to or knowledge of the information environment and its various audiences. Examples include atmospheric teams; cultural attaches or advisors; joint, interorganizational or multinational partner cultural experts; interpreters, or indigenous leaders.

D - Deliver

This function occurs primarily during execution, although some IO-related targets may be engaged while the command is preparing for the overall operation. The key to understanding the deliver function is to know which assets are available to perform a specific function or deliver a specific effect and to ensure these assets are ready and capable. Examples of delivery methods include but are not limited to: corps/division/brigade commander, provincial reconstruction team member or other unified action partner, host nation government leader, loudspeaker, media broadcast., social media posts and videos, and patrols.

During this step, the IO officer executes relevant portions of the target synchronization matrix. As IO-related delivery means and methods are multi-faceted and often involve human interaction, this step includes recording the delivery act and keeping detailed accounts or notes of actions taken or the proceedings, discussions, and commitments involved. The IO officer will ensure that required reporting procedures are explained and disseminated in the operations order or as part of the unit's standard operating procedures.

A - Assess

There are multiple types and levels of assessment. Assessment within D3A specifically focuses on whether the commander's targeting guidance was met for a specific target. From an IO perspective, such guidance may speak in terms of influence or degraded decision making, which are difficult to quantify. In the case of engagements, for example, assessment will help determine whether messages were retained by the target, whether these messages resulted in changed behavior, and whether reengagement may be necessary. An ongoing consideration in the information environment is that there may be a significant lag between the time of delivery, the effect taking place, and determination of an effect.

During this step, the IO officer and IRCs evaluate measures of effectiveness and performance to determine if desired effects were achieved. If not, it recommends re-engagement or other actions.

II. Targeting Tasks during the MDMP
Ref: *FM 3-13, Information Operations (Dec '16), pp. 7-2 to 7-6.

A. Mission Analysis
The two targeting-related IO products of mission analysis are a list of IO-related HVTs and recommendations for the commander's targeting guidance. The IO officer works with the G-2 (S-2) during IPB to develop IO-related HVTs, and with other members of the targeting team to develop IO targeting guidance recommendations.

Intelligence Preparation of the Battlefield (IPB)
IPB includes preparing templates that portray threat forces and assets unconstrained by the environment. The intelligence cell adjusts threat templates based on terrain and weather to create situational templates that portray possible threat COAs. These situational templates allow the intelligence to identify HVTs. The IO officer works with the intelligence cell throughout IPB to identify threat information-related capabilities and vulnerabilities and other key groups in the area of operations. These capabilities and vulnerabilities become IO-related HVTs.

See pp. 4-17 to 4-34 for related discussion of IO & IPB (information environment analysis).

Targeting Guidance
Issued within the commander's guidance is targeting guidance. This guidance describes the desired effects the commander wants to achieve. IO targeting focuses on HVTs that support critical, information-related threat capabilities that underpin their objectives and are vulnerable to friendly IO exploitation.

The IO officer develops input to targeting guidance based on the initial mission and available and anticipated IRCs. The IO officer identifies the functions, capabilities, or units to be attacked; the effects desired; and the purpose for the attack. The IO officer uses the targeting guidance to select IO-related HPTs from among identified HVTs. These HPTs are confirmed during COA analysis.

Targeting guidance is developed separately from IO objectives. IO objectives are generally broad in scope. They encompass all IO weighted efforts (attack, defend, stabilize). The IO officer develops recommendations for targeting guidance that supports achieving objectives.

When developing IO input to the targeting guidance, the IO officer considers the time required to achieve effects and the time required to determine results. Some IRCs require targeting guidance that allows for the acquisition, engagement, and assessment of targets while the unit is preparing for the overall operation. For example, the commander may want to psychologically and electronically isolate the enemy's reserve before engaging it with fires. Doing this requires electronic attack of threat command and control systems and military information support operations (MISO) directed at the threat 24 to 48 hours before lethal fires are initiated. Successfully achieving IO objectives for this phase of the operation requires targeting guidance that gives IO-related targets the appropriate priority.

B. COA Development
Feasible COAs, that integrate the effects of all elements of combat power, are developed by the staff. The IO officer prepares a scheme of IO that identifies objectives and IRC tasks for each COA. The IRC tasks are correlated with targets on the HVT list. A single IRC or multiple IRCs can be planned against a single HVT.

For each COA, the IO officer identifies HVTs that will support attainment of an IO objective. IO-related HVTs that subsequently support friendly IO objectives, and that can be engaged by IRCs, become HPTs. The targeting team also performs target value analysis, coordinates and deconflicts targets, and establishes assessment criteria. The IO officer participates in each of these tasks.

Target Value Analysis (TVA)

The targeting team performs target value analysis for each COA the staff develops. The initial sources for target value analysis are target spreadsheets and target sheets. Target spreadsheets (target folders) identify target sets associated with adversary functions that could interfere with each friendly COA or that are key to adversary success. IO-related targets can be analyzed as a separate target set or incorporated into other target sets. The IO officer establishes any IO-specific target sets. Each target set is assigned a priority based on its contribution to the success of a friendly objective, its impact on an enemy or adversary COA, and friendly capability to service the target.

The targeting team uses target spreadsheets during the war game to determine which HVTs to attack. The IO officer ensures that target spreadsheets include information on threat capabilities and IO-related HVTs and that the IO target set, if designated, is assigned a value appropriate to IO's relative importance to each friendly COA. If an IO target set is not designated, the IO officer ensures that IO-related targets are assigned an appropriate priority within the target sets used.

Target sheets contain the information required to engage a specific target. Target sheets state how attacking the target affects the threat's operation. The IO officer prepares target sheets for HVTs to analyze them from an IO planning perspective. These HVTs are expressed as target subsets, such as decision makers. Information requirements include:

- What influences these decision makers.
- How they communicate.
- With whom they communicate.
- Weaknesses, susceptibilities, accessibility, feasibility, and pressure points.

Deconflicting and Coordinating Targets

The IO officer and working group consider the possible consequences of attacking any target or target set. Their purpose is to identify possible duplication or attenuation of effects. The attack of physical targets always has second- and third-order effects (informational and cognitive) that could diminish or enhance their value to the overall operation. For example, fires that result in the collateral deaths of civilian non-combatants can have a negative cognitive effect, while using fires to destroy the enemy's fiber network so that it relies on radio communications vulnerable to jamming can have a positive informational effect. Also, the effects achieved by one IRC might compete with or diminish the effects of another IRC. Thus, IRC synchronization and the integration of IO into other lines of effort requires methodical coordination and deconfliction efforts.

IO working group members consider all targets from their various perspectives. Deconfliction in this context means ensuring that engaging a target does not produce effects that interfere with the effects of other IRC tasks or IO-related targets, or otherwise inhibit mission accomplishment. Coordination ensures that the effects of engaging different targets complement each other and further the commander's intent.

IO officers at different echelons may seek to engage the same targets and, possibly, desire different effects. Therefore, IO-focused targeting includes coordinating and deconflicting targets with higher and subordinate units before the targeting working group meets. Some IO-related targets may also be nominated by other staff elements. The IO officer presents the effects required to accomplish the IO objective associated with those targets when the targeting team determines how to engage them. IO officers must also coordinate and deconflict targets with unified action partners whose doctrinal use of IRCs and policies governing their employment differ. Such coordination extends the planning horizon and may limit how IRCs are integrated.

One way to achieve this coordination and deconfliction is by beginning parallel planning as early as possible in the MDMP. This means that the IO officer and the targeting team should share all pertinent information with subordinate units and adjacent and higher headquarters.

(Fires & Targeting) II. Targeting 7-15

Targeting Tasks during the MDMP (Cont.)

Ref: *FM 3-13, Information Operations (Dec '16), pp. 7-2 to 7-6.

Assessment Criteria

Generally, the effects of lethal attacks can be evaluated quickly using readily observable and quantifiable criteria, such as the percentage of the target destroyed. Assessing non-lethal attacks often requires monitoring the target over time, using a mix of quantitative and qualitative criteria. Establishing meaningful measures of performance and effectiveness for IO-related targets requires formulating a theory or logic of change in relation to IO objectives and the desired end state. The IO officer and working group essentially ask: will successful attack of a specific target or target set contribute to the attainment of the objective and what will the observable actions or activities leading to the desired outcome look like? The logic of change is expressed in terms of the anticipated causal chain that begins when the target is engaged.

IO-related targets attacked by means such as jamming or MISO broadcasts require assessment by means other than those used in battle damage assessment. The IO officer develops post-attack or post-engagement assessment criteria for these targets and determines the information needed to determine how well they have been met. The IO officer prepares IO IRs or RFIs for this information. If these targets are approved, the IO IRs for the approved targets may be recommended to the commander as priority intelligence requirements. If the command does not have the assets to answer these IO IRs, the target is not engaged unless the attack guidance specifies otherwise or the commander so directs.

C. COA Analysis

COA analysis (war-gaming) is a disciplined process that staffs use to visualize the flow of a battle. During the war game, the staff decides or determines—

- Which HVTs are HPTs.
- When to engage each HPT.
- Which system or capability to use against each HPT.
- The desired effects of each attack, expressed in terms of the targeting objectives.
- Which HPTs require battle damage assessment or post-attack/engagement assessment. The IO officer submits IRs for IO-related targets to the G-2 (S-2) for inclusion in the collection plan.
- Which HPTs require special instructions or require coordination.

Based on the war game, the targeting team produces the following draft targeting products for each COA:

- High-payoff target list.
- Target selection standards.
- Attack guidance matrix.
- Target synchronization matrix.

High-Payoff Target List (HPTL)

During mission analysis, the IO officer identifies potential targets, which are vetted by the IO working group. The IO officer takes nominated targets to the next targeting working group and works within that body to get these targets onto the high-payoff target list and approved by the targeting board.

High-payoff targets (HPTs) are managed in the high-payoff target list. HPTs are a subset of high-value targets—targets the enemy requires for successful completion of its mission. An HPT is a target whose loss to the enemy will significantly contribute to friendly mission success. Table 5-2 provides examples of HPTs that the IO working group would submit to the targeting working group as nominations.

Phase 1 – Isolate the enemy unit		
Priority	Category	High-Payoff Targets
1	Fire Support	Data link between target acquisition radars and fire direction center
2	Command and Control	Enemy leader's social media sites
3	Maneuver	Militia company-level leaders

Ref: ATP 3-13.1, table 5-2. Sample information operations input to high-payoff target list.

Target Selection Standards (TSS)

Target selection standards are applied to enemy activities to decide whether the activity can be engaged as a target. Target selection standards are usually disseminated as a matrix. Military intelligence analysts use target selection standards to determine targets from combat information and pass them to fire support assets for attack. Attack systems' managers, such as fire control elements and fire direction centers, use target selection standards to determine whether to attack a potential target. The intelligence and fires cells determine target selection standards. The IO officer ensures that they consider IO-related targets and establish appropriate standards for engaging them.

For nonlethal attacks or engagements, the IO officer may have to develop descriptive criteria to supplement or replace criteria developed by the fires cell. For example, target selection standards during a security cooperation operation may describe what constitutes a hostile crowd, such as: a group larger than 25 people, armed with sticks or other weapons, and with leaders using radios or cellular telephones to direct it.

Target selection standards address accuracy or other specific criteria that units must meet before they can engage targets. Standards usually consist of several elements—including HPTs, timeliness, and accuracy—although units can develop their own target selection standards worksheets. The HPT refers to the designated HPT that the collection manager is tasked to acquire. Timeliness refers to the time window within which units report valid targets to weapon systems. Accuracy concerns the allowable target location error for the target. The criteria are the least restrictive target location error given the capabilities of available weapon systems. Table 5-3 provides examples of IO-related target selection standards.

Target Selection Standards Worksheet		
High-payoff target	Timeliness	Accuracy
Observation posts	60 minutes	500 meters
Broadcast tower	480 minutes	100 meters

Ref: ATP 3-13.1, table 5-3. Sample information operations input to target selection standards.

Attack Guidance Matrix (AGM)

The targeting team recommends attack guidance based on the results of the war game. Attack guidance is normally disseminated as a matrix. Only one attack guidance matrix is produced for execution at any point in the operation; however, each phase of the operation may have its own matrix. To synchronize effects, all lethal and nonlethal attack systems, including MISO and electronic attack, for example, are placed on the attack guidance matrix. The attack guidance matrix is a synchronization and integration tool. It is normally included as part of the fire support annex. However, it is not a tasking document. Attack tasks for unit assets, including IRCs, are identified as taskings to subordinate units and agencies in the body or appropriate annexes or appendixes of the OPLAN/ OPORD.

(Fires & Targeting) II. Targeting 7-17

Targeting Tasks during the MDMP (Cont.)

Ref: *FM 3-13, Information Operations (Dec '16), pp. 7-2 to 7-6.

The attack guidance matrix (AGM) provides guidance on what HPTs units should engage and when and how units should engage them (see table 5-4). Although units may develop their own AGM format, the matrix typically includes the following elements:

- **High-payoff target**. The high-payoff target column is a prioritized list of HPTs by phase of the operation.
- **When**. This column indicates the time the target should be engaged.
- **How**. This column indicates the weapon system that will engage the target.
- **Effect**. The desired effects on the target or target system are stated in this column.
- **Remarks**. Remarks concerning whether or not assessment is required, whether coordination must occur, and any restrictions are indicated in this column.

High-payoff target	When	How	Effect	Remarks
Ops	I	Field artillery	Destroy	Use search and attack teams in restricted areas
Militia	I	CO	Neutralize	Destroy command and control
Cell phone	A	EA	Disrupt	Disrupt service starting H-2
Violent IDP crowds	A	MISO or MP	Disperse	25 or more constitute MSR blockage

A	as acquired		IDP	internally displaced person
CO	cyberspace operations		MISO	military information support operations
EA	electronic attack		MP	military police
H-2	hour minus two (representing 2 hours before)		MSR	main supply route
I	immediate		OP	observation post

Ref: ATP 3-13.1, table 5-4. Sample information operations input to attack guidance matrix.

Target Synchronization Matrix

The target synchronization matrix lists HPTs by category and the agencies responsible for detecting them, attacking them, and assessing the effects of the attacks. It combines data from the high-payoff target list, information collection plan, and attack guidance matrix. A completed target synchronization matrix allows the targeting team to verify that assets have been assigned to each targeting process task for each target. The targeting team may prepare a target synchronization matrix for each COA, or may use the high-payoff target list, target selection standards, and attack guidance matrix for the war game and prepare a target synchronization matrix for only the approved COA.

See facing page for a sample targeting synchronization matrix (fig. 5-3).

D. COA Comparison, Approval, & Orders Production

After war-gaming all the COAs, the staff compares them and recommends one to the commander for approval. When the commander approves a COA, the targeting products for that COA become the basis for targeting for the operation. The targeting team meets to finalize the high-payoff target list, target selection standards, attack guidance matrix, and input to the information collection plan. The team also performs any additional coordination required. After accomplishing these tasks, targeting team members ensure that targeting factors that fall within their functional areas are placed in the appropriate part of the OPLAN/OPORD.

Sample Targeting Synchronization Matrix

An important output of the IO working group that feeds into the targeting working group is target nominations. The format for these nominations is a targeting synchronization matrix such as the one shown in figure 5-3.

			PHASE III							
	DECIDE				DETECT	DELIVER			ASSESS	
HPT/PRI	TGT set	TGT #	TGT description	Desired effect	Phase	Asset	Asset	How	When	Asset	Measure of effectiveness	Status
1	Arianan forces	IRC032	Arianan resolve; surrender messaging (OBJ COLORADO)	Degrade	IIIB	HUMINT, IMINT, and SIGINT	1/1, 2/1, 3/1 CAV & 110 MEB (loudspeaker and handbills)	Planned	D+9 - D+10	1 ID	15% of Arianan forces desert or surrender to coalition forces	DL - DM
1	Arianan forces	IRC033	Arianan forces (upon retrograde)	Influence	IIIB	HUMINT, IMINT, and SIGINT	Press release or radio and TV broadcast; COMCAM	Planned	D+9 - D+10	CJTF; 1 ID	75% of local media reports Arianan forces retrograde and treatment of EPWs; 5% increase in surrenders and desertions	DL - DM
2	Atropian 348th BDE, commission on refugees or IDPs & RCC directors	IRC034	Assess IDPs mitigation (OBJ OVERLORD)	Assess	IIIB	HUMINT	110 MEB SLE	Planned	D+10	1 ID	75% of acute essential service needs identified and development or coordination of responses	DM
2	Ministry of Internal Affairs & USAID	IRC035	Assess IDPs mitigation (OBJ CEDAR FALLS)	Inform	IIIB	OSINT and HUMINT	300 SB SLE	Planned	D+10	1 ID	1 ID informed on PH IV engagement requirements; 75% coordination between IA, 1 ID, and host nation	DM
1	Local security forces	IRC036	Coordinate and synchronize for PH IV operations (OBJ CEDAR FALLS)	Coordinate	IIIC	OSINT and HUMINT	Victory 6 SLE	Planned	D+11	CMOC, G-9	75% of local security forces engaged and leading local security efforts with minimal coalition support	DN

Figure 5-3. Example targeting synchronization matrix reflecting IO target nominations.

III. Dynamic Targeting (F2T2EA)

Ref: *FM 3-13, Information Operations (Dec '16), p. 7-7.

Dynamic targeting uses the find, fix, track, target, engage, and assess (known as F2T2EA) process (figure 5-2). Table 5-5 summarizes the IO-related inputs or activities that support each phase of the process.

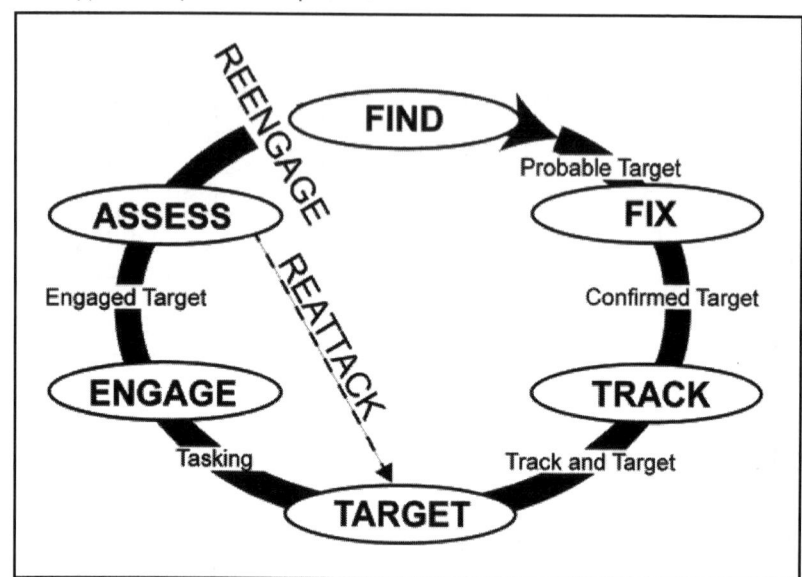

Ref: ATP 3-13.1, fig. 5-2. Dynamic targeting.

Function	IO Input or Activity
Find	• Updated and focused CIO. • IO input to collection plan. • IRCs reporting of potential targeting signatures.
Fix	• IO updates to targeting. • IRCs tasked to report information during mission performance to develop target. • Targets' information-related vulnerabilities.
Track	• Requests for information for target location refinements. • Targets' information-related vulnerabilities updated. • IO input to risk assessment and collateral damage estimate (2nd and 3rd order effects). • IRCs deconflicted.
Target	• IRC tasks developed to achieve desired effect. • MOEs and MOPs also developed.
Engage	• Approved IO tasks in mission order. • IRCs employed to conduct, support, and reinforce engagement. • Initial reports of results from subordinate units as means to monitor MOPs.
Assess	• MOEs assessed against baseline. • CIO updated. • Re-engagement recommendations submitted.

CIO — combined information overlay
IO — information operations
IRC — information-related capability
MOE — measure of effectiveness
MOP — measure of performance

Ref: ATP 3-13.1, table 5-5. Information operations inputs and activities to support F2T2EA.

Chap 8
(Information) ASSESSMENT

Ref: *ATP 3-13.1, The Conduct of Information Operations (Oct '18), chap. 6 and *FM 3-13, Information Operations (Dec '16), chap. 8. (*See note p. 1-2.)

See p. 1-63 for discussion of assessment as related to information advantage (ADP 3-13) and p. 2-54 as related to operations in the information environment (JP 3-04).

I. Assessment Framework

All plans and orders have a general logic. This logic links tasks given to subordinate units with achieving objectives and achieving objectives with attaining the operation's end state. An assessment framework incorporates the logic of the plan and uses measures—MOEs and MOPs—as tools to determine progress toward attaining desired end state conditions, as shown on figure 6-1.

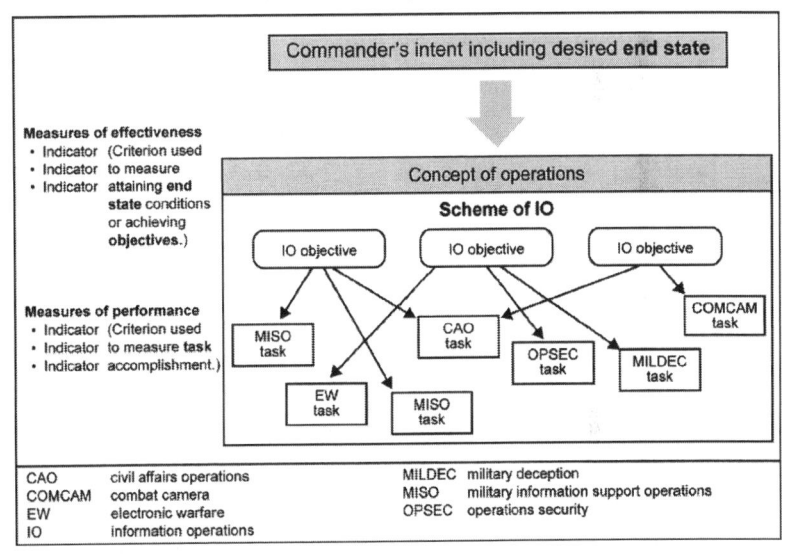

```
                  Commander's intent including desired end state
                                    ↓
Measures of effectiveness       ┌─────────────────────────────────────────┐
 • Indicator (Criterion used    │        Concept of operations            │
 • Indicator  to measure        │          Scheme of IO                   │
 • Indicator  attaining end     │                                         │
              state conditions  │  IO objective  IO objective  IO objective│
              or achieving      │                                         │
              objectives.)      │                               COMCAM    │
                                │        MISO    CAO            task     │
Measures of performance         │        task    task  OPSEC             │
 • Indicator (Criterion used    │                      task   MILDEC     │
 • Indicator  to measure task   │                             task       │
 • Indicator  accomplishment.)  │        EW      MISO                    │
                                │        task    task                    │
                                └─────────────────────────────────────────┘

CAO      civil affairs operations       MILDEC  military deception
COMCAM   combat camera                  MISO    military information support operations
EW       electronic warfare             OPSEC   operations security
IO       information operations
```

Ref: ATP 3-13.1, fig. 6-1. Framework for assessment.

The **purpose of assessment** is to support the commander's decision making. Commanders continuously assess the situation to better understand current conditions and determine how the operation is progressing. Continuous assessment helps commanders anticipate and adapt the force to changing circumstances. Commanders incorporate assessments by the staff, subordinate commanders, and unified action partners into their personal assessments of the situation. Based on their own assessments, commanders modify plans and orders to adapt the force to changing circumstances. Assessment is a staff-wide effort, not simply the product of a working group or a particular staff section or command post cell. Assessment of IO objectives and effects is an integral part of the staff-wide assessment process.

II. IO Assessment Considerations

Ref: *FM 3-13, Information Operations (Dec '16), pp. 8-3 to 8-5. See also pp. 8-7 to 8-9.

Assessment of IO in general and of specific effects in the information environment require careful development of measures of effectiveness and performance, as well as identification of indicators that will best signal achievement of these measures and desired outcomes. Assessment in the information environment is not easy and adherence to the following considerations will aid in making IO assessment more effective.

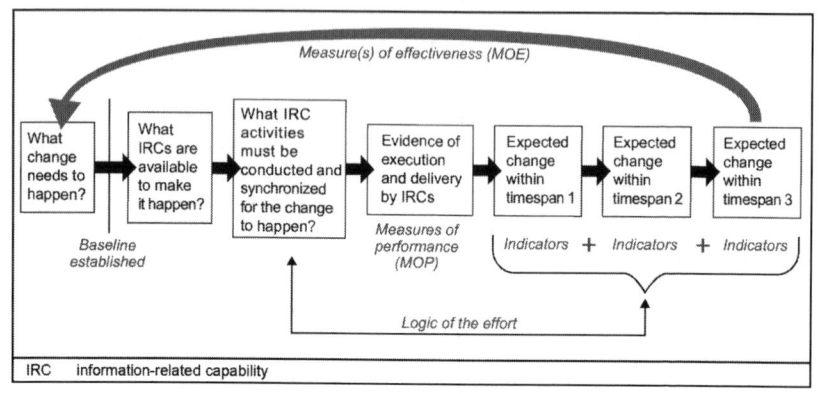

```
                                    Measure(s) of effectiveness (MOE)
```

| What change needs to happen? | What IRCs are available to make it happen? | What IRC activities must be conducted and synchronized for the change to happen? | Evidence of execution and delivery by IRCs | Expected change within timespan 1 | Expected change within timespan 2 | Expected change within timespan 3 |

Baseline established | | | Measures of performance (MOP) | Indicators + Indicators + Indicators

Logic of the effort

IRC — information-related capability

Ref: FM 3-13, fig. 8-2. Logic flow and components of an IO objective. Figure 8-2 portrays the relationship between objectives (the change that needs to happen) and measures of performance, indicators, and measures of effectiveness. The logic of the effort is shown as a relationship between available, selected, and synchronized IRCs and the effects expected over time. While the figure suggests that this logic is generic, it is not. It is unique to every objective and combination of IRCs.

Measures of Effectiveness (MOEs)

A measure of effectiveness is a criterion used to assess changes in system behavior, capability, or operational environment that is tied to measuring the attainment of an end state, achievement of an objective, or creation of an effect (JP 3-0). Measures of effectiveness help measure changes in conditions, both positive and negative. They are commonly found and tracked in formal assessment plans.

Time is a factor when assessing IO and developing measures of effectiveness. The attainment of IO objectives leading to the commander's desired end state often requires days or months to realize. It is essential, therefore, to have a baseline from which to measure change and also to time-bound the change. Time-bounding makes clear how long it will take before the change is observed. It helps to set necessary expectations, foster patience, and avoid a rush to judgment. If a behavioral objective is anticipated to take considerable time, assessment planning may choose to break the objective into smaller increments, each with more immediate observable outcomes. Finally, it is also important to analyze and understand the cultural relevance of time in the area of operations and account for and adapt to it.

Developing informational, behavioral and sentiment baselines often requires significant time and resource investments. Sentiment baselines, such as those determined through surveys or interviews, may require contracted labor to accomplish. The IO officer must factor in the lead time necessary to contract a third-party, provide it time to develop the survey instrument, administer the survey, and tabulate and report on the results.

Commanders and staffs, particularly the IO officer, must account for the order of effects when assessing IO or, more broadly, any effect. For example, an effect in the physical dimension (1st order) can resonate in unexpected ways in the informational and cognitive dimensions (2nd and 3rd orders). Units must account for directness of effect and understand the difference between causational linkages and correlational ones. Certain effects, even desired ones, may not be directly tied to friendly efforts in the information environment; however, friendly forces may still be held accountable for these effects and must react appropriately. This fact underlines the importance of developing a logic of the effort for each IO objective.

Effectiveness in the cognitive dimension typically requires variety and repetition. Rarely does a single tactic, task, method, action, or message change behavior. Assessment plans must therefore build in varied actions and repeated messages and measure their cumulative effect.

Measures of Performance (MOPs)

A measure of performance is a criterion used to assess friendly actions that is tied to measuring task accomplishment (JP 3-0). Measures of performance help answer questions such as "Was the action taken?" or "Were the tasks completed to standard?" A measure of performance confirms or denies that a task has been properly performed.

There is no definitive number of tasks to support a given objective; therefore, there is no definitive number of measures of performance to support any given measure of effectiveness. Again, variety and repetition necessitate that multiple tasks typically support each objective and the corresponding measure of performance is the means to confirm or deny that each task is executed in the first place and properly performed.

Delivery, especially means of delivery, is a critical consideration when developing IRC tasks and their associated measures of performance, particularly when it comes to message delivery. No matter how well-crafted the message, if delivery assets are unavailable or only available in insufficient number, the objective will likely not be achieved. Means of delivery should also be considered in terms of accessibility and acceptability to the target audience. For example, if only a small percentage of the population listens to radio or watches television then these means should not be the only means of delivery considered.

Indicators

An indicator is an item of information that provides insight into a measure of effectiveness or measure of performance (ADRP 5-0). Indicators take the form of reports from subordinates, surveys and polls, and information requirements. Indicators help to answer the question "What is the current status of this measure of effectiveness?" A single indicator can inform multiple measures of effectiveness.

Not everything observed is an indicator and not every indicator is a sign of progress. Indicators of psychological effects or changes in sentiment are not always easy to detect or may not be markers of the desired behavior change. The upshot of these facts is that establishing indicators requires rigorous effort in order to select those observable and measurable signs or signals that are reflective of changed behavior. Often behavior change is incremental and being able to detect the intervening steps to large-scale behavior change is essential to measuring progress. Again, in-depth knowledge is required of those targets or audiences for whom behavior change is required to achieve the commander's desired end state.

Measuring progress requires the ability to detect both micro and macro indicators simultaneously. The IO officer must, therefore, coordinate with the G-2 (S-2) in order to know what collection assets are available and the types of information that each provides and how this information helps create actionable knowledge. Soldiers are a vital collection asset. The IO officer should invest time to train all Soldiers on observation techniques that enable them to spot and discriminate meaningful indicators and ways to report what they see.

The IO officer should employ a variety of means to identify indicators, validate or corroborate conclusions about them, and measure progress. Some of the more commonly used sources are: information collection assets, Military Information Support Operations (MISO) teams, Soldier and leader engagements, civil-military operations, polling and surveys (which primarily measure attitudes, not motivations), and media monitoring and analysis.

III. Assessment Rationale

Assessment or evaluation is a judgment of merit of an action or operation as to whether it achieved its intended outcome(s). It supports planning, improves effectiveness and efficiency of operations, and enforces accountability. These three purposes correspond to three types of evaluation: formative, process, and summative.

Formative evaluation supports planning by examining whether an operation or program is being designed to meet its intended purpose. In terms of IO, it involves testing messages, determining baselines, analyzing audiences, and developing the logic by which the operation will create influence.

Process evaluation occurs primarily during execution and serves to enhance effectiveness and efficiency, as well as facilitate in-process decision making. In terms of IO, it assesses whether the scheme of IO is being executed as planned. If the scheme is not going as planned, process evaluation facilitates decisions that lead to corrective action.

Summative evaluation occurs post-execution and supports decision making and accountability. While process evaluation supports decisions that adjust activities or efforts as the operation unfolds, summative evaluation supports decisions about the overall operation and whether it achieved the commander's intent. It leads to the determination of those aspects of the operation to sustain and those to eliminate or curtail should a similar operation be undertaken in the future.

In addition to supporting users such as the IO officer, the IO working group, IRC managers, other staff sections, and the commander, operations assessment feeds higher headquarters assessment and, oftentimes, external entities, such as governmental leadership. IO efforts, in particular, often elicit congressional scrutiny and commander-led assessment ensures units are ready to demonstrate the effectiveness of their influence efforts.

Assessment is most valuable when operations or operational efforts are not working as planned because it helps the commander and staff figure out why and take corrective action. Units should avoid using assessment to justify decisions already made or merely to check the box. Assessment without the intent to employ its results is a waste of time and resources.

IV. Principles that Enhance the Effectiveness of IO Assessment

Assessment effectiveness is enhanced when it adheres to the following principles or best practices:

- Uses clear, realistic and measurable objectives.
- Begins with planning.
- Employs an explicit logic of the effort.
- Is continual and consistent over time.
- Is iterative.
- Is prioritized and resourced.

Assessment is more effective when IO objectives are specific, measurable, achievable, relevant and time-bound. Creating clear, realistic, and measurable objectives can be challenging early on, as initial guidance from higher might lack clarity. The IO officer asks clarifying questions but also proactively establishes the most specific, measurable, achievable, relevant, and time-bound objectives possible and provides them to higher headquarters for review and refinement. The IO officer also tests its objective statements with relevant stakeholders, most especially the IRCs that contribute to the attainment of these objectives.

V. Assessment Focus
Ref: *ATP 3-13.1, The Conduct of Information Operations (Oct '18), pp. 6-2 to 6-3.

Different levels of headquarters likely have different assessment focuses. Tactical-level units focus primarily on task assessment. Operational-level units focus on environmental (operational and information) assessment.

Assessment aspect	Assessment Focus		
	Task	Environment	Campaign
Source (basis) for criteria	Directed tasks in OPORD or OPLAN.	Desired conditions in OPLAN or OPORD.	End state objectives (success criteria).
Criteria	Primarily MOP.	Primarily MOE.	MOE.
Time horizon	Near (daily).	Mid (weekly or monthly).	Long (monthly, quarterly, or annually).
Indicators	Largely quantitative; may have qualitative commander input.	Mixed-method.	Mixed-method.
Collection means	Reports, SIGACTs, subordinate commanders, circulation.	Reports, polls, media analysis, subordinate commanders, stakeholders, circulation.	Reports, polls, media analysis, subordinate commanders, stakeholders, circulation.
Analysis and evaluation	Current operations centric, after action review, and commander qualitative.	Staff analysis and evaluation through staff-wide efforts, with focused ad hoc assessment cell or working group. Commander parallel evaluation based on qualitative (opinion-based) indicators through commander crosstalk and circulation. Informed by staff efforts.	Combination of the quantitative staff efforts and commander qualitative analysis and evaluation. Trend analysis.
Commander – Staff interface venues	Daily updates, after action review.	Periodic OE or information environment staff assessment updates; commander's circulation reports.	Formal assessment briefings and conferences.
Actions for improvement	Task refinement, changes to quantities or methods of delivery, additional IRC support, and reengagement or repetition of IRC tasks.	Better understanding of local culture, improved information environment analysis, message refinement, and reengagement by alternate means.	Reassessment of campaign strategy, refinement of commander's end state, and expansion of IO planning and execution, including unified action partners.
IO information operations		OE operational environment	
IRC information-related capability		OPLAN operation plan	
MOE measure of effectiveness		OPORD operation order	
MOP measure of performance		SIGACT significant activity	

ATP 3-13.1, table 6-1. Aspects of assessment by level of focus.

A **task assessment** asks whether units or IRCs are performing assigned or implied tasks to standard using MOPs. Task assessment answers the question, "Are we doing things right?" as well as follow-on questions such as, "Was the task completed?" and "Was it completed to standard?"

Environment assessment asks whether units are achieving the necessary objectives and conditions— MOEs-oriented—in the information and operational environments necessary to accomplish the mission. This type of assessment answers the question, "Are we doing the right things?"

Campaign assessment is undertaken at the theater level (such as the geographic combatant command) in the area of responsibility to assess whether units achieve theater strategic or campaign objectives (objective-oriented).

VI. Assessment Methods

Assessments can be quantitative or qualitative or both (mixed method). One method is not necessarily better than another. The various components of the assessment framework—end state, objectives, and tasks, as well as the type of intelligence or information gathered—all govern which method is best suited to yield the feedback necessary to support decision making and operational adjustments.

A. Quantitative

Objectives that are specific, measurable, achievable, realistic, and time-bounded (known as SMART) lend themselves to quantitative assessment because they employ MOEs that are similarly specific and measurable. A quantitative methodology is well-suited for almost all task-level assessments and selected environment and campaign assessments. By its nature, quantitative methodology is data-centric and requires the requisite automated systems and personnel expertise to employ it effectively. Quantitative assessment tends to be staff-centric and is often a check on commanders' more subjective, qualitative assessment.

B. Qualitative

Effective qualitative assessments require the same rigor, if not more, as quantitative assessments and benefit from expertise in their design and conduct. The IO officer, IRCs, functional staff leads, and members of a working group or ad hoc assessment cell assist with the design and conduct of qualitative assessments (just as they do with quantitative assessments). For MOPs, the IRCs are the best resource for whether an activity has started, or is completed, and the standards to which it was performed. For MOEs, the IO officer— working with the intelligence staff and other members of the staff—assesses for the commander whether, and to what extent, units achieve effects in the information environment. To the degree possible, qualitative assessments are enhanced by turning qualitative information into quantitative values. This process helps remove subjectively and facilitates the compilation and reporting of findings.

C. Mixed-Method

Mixed-method or blended assessments combine quantitative and qualitative assessment methodologies to gain the best of both. IO, in particular, benefits from mixed-method assessments due to the diverse range of effects it can create in the information environment, from directly observable physical effects to long-term cognitive effects.

VII. Assessment Process

Assessment involves three inter-related phases: monitoring, evaluation, and adjustment (directing action for improvement). FM 6-0 discusses each of these phases in detail. In the evaluation phase, three sets of criteria are employed to evaluate progress: MOEs, MOPs, and indicators.

A. Monitoring Information Operations

Monitoring is continuous observation of those conditions relevant to the current operation (ADRP 5-0). Monitoring within the assessment process allows staffs to collect relevant information, specifically that information about the current situation that staffs can compare to the forecasted situation described in the commander's intent and concept of operations. Progress cannot be judged, nor effective decisions made, without an accurate understanding of the current situation.

The IO officer monitors IRCs to determine progress towards achieving the IO objectives. Once execution begins, the IO officer monitors the threat and friendly situations to track IRC task accomplishment, determine the effects of IO during each phase of the operation, and detect and track any unintended consequences. The IO officer works closely with the intelligence cell, intelligence staff officer, and IO working group representatives to provide a running assessment of the effectiveness of threat information efforts and keeps the operations staff officer and various integrating cells informed.

B. Evaluating Information Operations

Evaluating is using criteria to judge progress toward desired conditions and determining why the current degree of progress exists (ADRP 5-0). Evaluation is at the heart of the assessment process where most of the analysis occurs. Evaluation helps commanders determine what is working and what is not working, and it helps them gain insights into how to better accomplish the mission.

During execution, the IO officer works with the intelligence cell and integrating cells to obtain the information needed to determine individual and collective IO effects. Evaluation not only estimates the effectiveness of task execution, but also evaluates the effect of the entire IO effort on the threat, other relevant audiences in the AO, and friendly operations. Evaluation assesses whether IO achieved its scheme of IO and subordinate objectives to support the overall mission.

MOEs, MOPs and Indicators

In the evaluation phase, two sets of criteria are employed to evaluate progress: MOEs and MOPs. Task execution is evaluated using measures of performance. Task effectiveness, objective attainment, and mission accomplishment are evaluated using measures of effectiveness, which compare achieved results against a baseline. Progress of both MOEs and MOPs is signaled by indicators.

See following pages (pp. 8-8 to 8-9) for discussion of MOEs, MOPs and indicators (criteria development).

	MOE	MOP	Indicator
Purpose:	Measure attaining an end state condition, achieving an objective, or creating an effect.	Measure task accomplishment.	Provide insight into an MOE or MOP.
Answers:	Are we doing the right things?	Are we doing things right?	What is the status of this MOE or MOP?
Why and What:	Measures *why* (purpose) in the mission statement.	Measures *what* (task completion) in the mission statement.	Information used to make measuring *why* or *what* possible.
Relationship:	No direct hierarchal relationship to MOPs.	No hierarchal relationship to MOEs.	Subordinate to MOEs and MOPs.
Tracking:	Often formally tracked in formal assessment plans.	Often formally tracked in execution matrixes.	Often formally tracked in formal assessment plans.
Level of Challenge:	Typically challenging to choose the appropriate ones.	Typically simple to choose the appropriate ones.	Typically as challenging to select appropriately as the supported MOE or MOP.
MOE measure of effectiveness		MOP measure of performance	

Ref: ATP 3-13.1, table 6-2. Assessment measures and indicators.

No direct hierarchical relationship exists between MOPs and MOEs. MOPs do not feed MOEs or combine in any way to produce MOEs. MOPs simply measure the performance of a task; however, these tasks are essential to fulfilling each objective. For IO, IRC units or staff representatives are responsible for task execution and, in coordination with the IO officer through the IO working group, contribute to MOP development and task assessment.

In the context of assessment, an indicator is an item of information that provides insight into a measure of effectiveness or measure of performance (ADRP 5-0). Indicators take the form of reports from subordinates, surveys, polls, and information requirements. Indicators help to answer the question, "What is the current status of this MOE or MOP?" A single indicator can inform multiple MOPs and MOEs

See following pages (pp. 8-8 to 8-9) for discussion of criteria development.

VIII. Assessment Products

Staff assessment products should directly support the commander's requirements, such as deepening understanding of the operational and information environments, measuring progress toward achieving objectives and accomplishing the mission, and informing the commander's intent and guidance. Efficient staffs also develop, tailor, and optimize products to meet the commander's expectations and ways of receiving information. Campaign assessments are substantially fuller or richer in terms of the scope of information presented than is a task assessment.

As figure 6-5 depicts below, achieving IO objectives depends on producing specific effects in the information environment that ultimately cause the enemy or adversary—as well as many intervening variables, actors, or audiences—to change behavior. Figure 6-6 illustrates several common methods for depicting trends or the status of a given condition in an information environment. Figure 6-7 provides a counterinsurgency example that depicts indicator trends supporting an MOE.

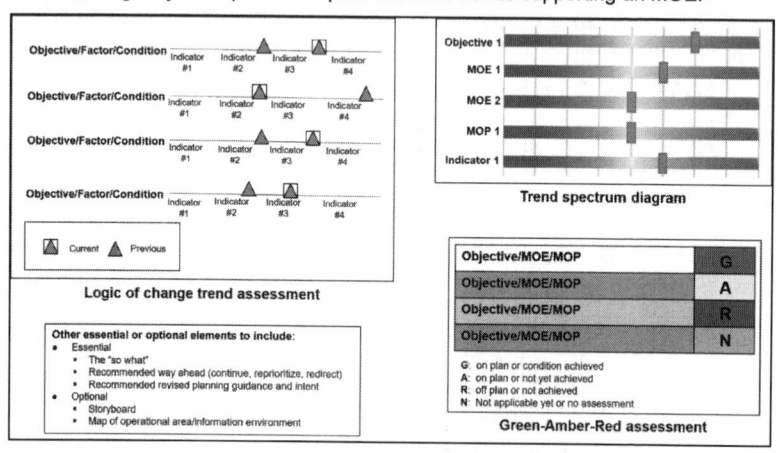

Ref: ATP 3-13.1, fig. 6-6. Sample assessment product templates

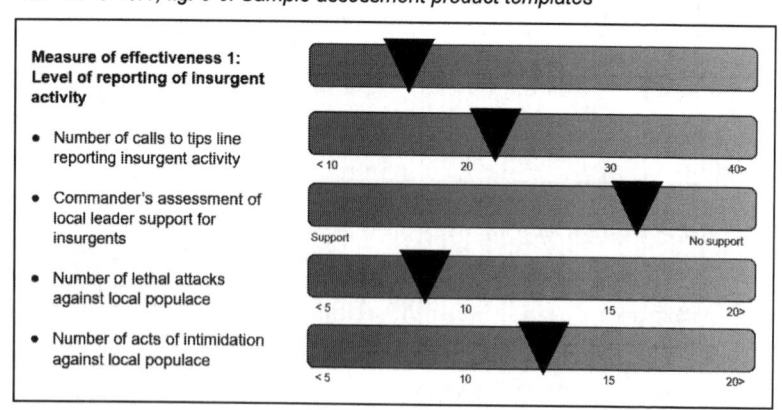

Ref: ATP 3-13.1, fig. 6-7. Example counterinsurgency MOE assessment.

Note. Staffs can use each of these methods to measure progress among any of the various elements of an IO objective, either singly or in combination: the objective itself or the MOE, MOP, and indicators that support it. Also, effective staffs pair a diagram with additional essential or optional information that facilitates decision making, most importantly the bottom line or "so what."

[INFO2] Index

A

Additional Capabilities, 3-69
ADP 3-13, Information (Nov '23), 1-2
Affect Threat Information Warfare Capabilities, 1-53
Appendix 15 (IO) to Annex C (Operations), 4-61
Armed Conflict, 1-20
Army Design Methodology (ADM), 4-2
Army Doctrine, 3-1
Army Forces and the Information Joint Function, 1-59
Army Information Activities During Operations, 1-60
Assessing, 1-63
Assessing During Execution, 6-5
Assessment, 8-1, 2-54
Assignment of Meaning, 0-1
Attack & Delivery Capabilities, 7-8
Attack, 1-49
Attack & Exploit, 2-19, 2-41
Audiences, Stakeholders, and Publics, 3-14
Automated Systems, 0-4

B

Behavior of Relevant Actors, 1-27, 2-15
Build, Protect, and Sustain Joint Force Morale and Will, 2-16

C

C2 of OIE Units, 2-53
Change in Terminology, 1-2
Civil Affairs (CA), 2-32, 3-17
Civil Affairs Operations, 1-43, 1-47, 2-10, 3-17

Civil-Military Operations Center (CMOC), 2-35, 3-24
COA Evaluation Criteria, 2-60
Collaboration, 2-17
Combat Camera (COMCAM), 2-11, 2-33
Combat Power, 1-7
Combatant Commanders (CCDRs), 2-26
Combined Space Tasking Order (CSTO), 3-68
Command and Control,
 Joint Function, 2-42
 Systems, 1-21
 Warfighting Function, 1-14
Commander's Communication Synchronization (CCS), 1-39, 3-11
Commander's Critical Information Requirements (CCIRs), 2-52, 4-9
Commander's Guidance, 4-5
Commander's Initial Planning Guidance, 2-58
Commander's Intent, 4-5
Commander's Narrative, 4-4
Commanders' Responsibilities, 4-3
Common Training and Education, 1-64
Communications Security, 1-34
Community Engagement, 1-40, 1-42
Competition Below Armed Conflict, 1-17
Competition Continuum, 2-8, 2-37
Concept of Operations, 4-5
Coordination and Liaison, 5-3
Correct Misinformation and Counter Disinformation, 1-41, 1-43

Counter Threat Finance (CTF) Cell, 2-31
Counterintelligence (CI), 1-31
Crisis, 1-18
Critical Information, 3-42
Cyberspace, 3-45
Cyberspace Actions, 3-53
Cyberspace Attack, 1-51, 3-54
Cyberspace Defense, 3-53
Cyberspace Domain, 3-48
Cyberspace Electromagnetic Activities (CEMA), 3-46
Cyberspace Exploitation, 3-53
Cyberspace Forces, 2-32
Cyberspace Missions & Actions, 3-51
Cyberspace Operations (CO), 2-11, 3-47
Cyberspace Security, 1-34, 3-53

D

Data and Information, 0-1
Deception, 3-27
Deception Activities, 1-46
Deception in Support of Operations Security (DISO), 1-46, 3-28
Decide, Detect, Deliver, Assess (D3A), 7-12
Decision Making During Execution, 6-8
Defend the Network, Data, and Systems, 1-34
Defense Media Activity (DMA), 2-29
Defensive Cyberspace Operations (DCO), 1-34, 3-50
Degrade Threat Command and Control, 1-52

Index-1

Index

Dept. of State (DOS) Organizations, 2-24
Dept. of Defense Information Network Operations (DODIN), 3-50
Digital Literacy, 1-28
Digital Readiness, 1-28
Dimensions of an OE, 0-6
Disinformation, 0-9
Domains of an OE, 0-6
Domestic Audiences, 1-40
Drivers of Behavior, 0-3
Dynamic of Combat Power, 1-7
Dynamic Targeting (F2T2EA), 7-20

E

Editor's Note, 1-2
Educate Soldiers, 1-38
Electromagnetic Attack (EA), 1-50, 3-56
Electromagnetic Protection (EP),1-34, 3-58
Electromagnetic Spectrum (EMS), 3-45
Electromagnetic Spectrum Operations (EMSO) Forces, 2-33
Electromagnetic Support (ES), 3-60
Electromagnetic Warfare (EW), 2-11, 3-55
Electronic Warfare Reprogramming, 3-60
Enable, 1-21
Enhance Understanding of an Operational Environment, 1-26
Enhancing C2, 1-28
Essential Elements of Friendly Information (EEFIs), 4-9
Executing, 1-63
Execution, 2-52, 6-1

F

Facilitate Shared Understanding, 2-16
Fire Support (FS), 7-8
Fires, 7-1
Joint Function, 2-43

Warfighting Function, 1-15
Foreign Assistance, 3-26
Foreign Audiences, 1-47
Foreign Humanitarian Assistance (FHA), 3-26
Friendly Force Information Requirement (FFIR), 2-62, 4-9
Friendly Information, 1-29

H

Historians, 2-11
Human Advantage, 1-4
Human Dimension, 0-8
Identify and Describe Relevant Actors, 1-27, 2-15

I

Ideas, 3-15
IE Toolbox, 2-50
Indicators, 3-44, 8-3, 8-7
Influence, 1-45, 2-19, 2-41
Inform, 1-35, 2-41
Inform Domestic Audiences, 1-40, 2-19
Inform Internal Audiences, 1-38
Inform International Audiences, 1-42
Information, 0-1
Joint Function, 2-42, 2-7
Information Activities, 1-8, 1-10
Information Activities and the Operations Process, 1-60
Information Advantage, 1-1, 1-5, 2-1, 3-46
Information Advantage (Examples), 1-19
Information Advantage Framework, 1-3
Information Advantages (Across Strategic Contexts), 1-16
Information and Intelligence Sharing, 2-17
Information and KM, 2-54
Information Attack Methods, 1-49
Information Capabilities, 3-1, 3-4

Information Collection, 1-24, 5-3
Information Cross-Functional Team (CFT), 2-30, 2-50
Information Dimension, 0-8
Information Environment (IE), 2-1, 4-17, 4-20
Information Environment Analysis, 4-17
Information Explained, 0-1
Information for Effect, 0-9
Information Forces, 2-21, 2-32, 2-38
Information in Multidomain Operations, 1-4
Information in the Security Environment, 0-5
Information Joint Function, 1-57, 2-12
Information Management (IM), 2-17
Information Operations (IO), 1-2, 3-2
Information Operations Working Group, 6-1
Information Planners, 2-50
Information Planning Cell, 2-30
Information Staff Estimate, 2-54
Information Training and Education, 1-64
Information Within an Operational Environment (OE), 0-6
Informational Considerations, 0-7
Informational Power, 1-6, 2-4
Informational, Physical, and Human Aspects of the Environment, 2-15
Information-Related Capabilities (IRCs), 1-2, 3-2
Inherent Informational Aspects, 0-2
Initial CCIRs, 2-61
Instruments of National Power, 1-6
Integrated Joint Special Technical Operations (IJSTO), 3-70
Integrating Processes, 1-24

Integration, 1-57
Integration of the Information Activities, 1-61
Intelligence,
  Joint Function, 2-42
  Warfighting Function, 1-14
Intelligence "Push" and "Pull", 6-6
Intelligence Preparation of the Battlefield (IPB), 4-17
Intelligence Preparation of the Operating Environment (IPOE), 1-24, 6-7
Intelligence Requirements, 2-53
Intelligence Support to OIE, 2-53
Intelligence Support, 6-6
International Audiences, 1-42
Interorganizational Collaboration, 2-34

J
JEMSOC, 2-31
JFC's Staff, 2-28
Joint and Multinational Information Advantage, 1-57
Joint Fires (OIE Considerations), 7-6
Joint Force, 2-28
Joint Force Capabilities, Operations, & Activities for Leveraging Information, 3-1, 7-6
Joint Force Use of Information, 2-1
Joint Functions and OIE, 2-40
Joint IM Cell, 2-31
Joint Information Advantage, 1-57
Joint Information Operations Warfare Center (JIOWC), 2-29
Joint Informational Power, 1-6
Joint Intelligence Support Element (JISE)/Joint Intelligence Operations Center (JIOC), 2-29
Joint Interagency Coordination Group (JIACG), 2-35

Joint Interagency Task Force (JIATF), 2-35
Joint Organizations, 2-29
Joint Planning Process (JPP), 2-50, 2-55
Joint Planning Support Element-Public Affairs (JPSE-PA), 2-29
JP 3-04, Information in Joint Operations (Sept '22), 2-2

K
Key Leader Engagement (KLE), 2-10, 2-31
Knowledge Management (KM), 1-25, 2-17
Known Facts, 2-58

L
Language, Regional, And Cultural Expertise, 1-48
Large-Scale Combat, 1-20
Leverage Information, 2-10, 2-19, 3-1
Leveraging the Inherent Informational Aspects of Operations, 1-57

M
Malign and Benign Information, 0-9
Measures of Effectiveness (MOE), 8-2, 8-7
Measures of Performance (MOP), 8-3, 8-7
Media Operations Center, 2-31
Messages, 3-15
Military Deception (MILDEC), 1-46, 2-10, 3-27, 3-28
Military Decision-making Process (MDMP), 4-2, 4-35
Military Government, 3-18
Military Information Support Operations (MISO), 1-47, 2-10, 3-33
Military Objectives, 2-59
Misinformation, 0-9
Mission Analysis Brief, 2-62
Mission Statement, 4-11

Mission Variables - METT-TC (I), 1-26
Movement and Maneuver,
  Joint Function, 2-44
  Warfighting Function, 1-14
Multinational Considerations, 1-58, 2-36

N
Narrative, 2-6, 2-54, 3-15
Narrative, Themes, and Messages, 3-15
Nature of Information, 1-1
Nonlethal Effects, 7-7

O
Objectives & IRC Tasks, 4-14
Offensive Cyberspace Operations (OCO), 3-52
OIE Unit Core Activities, 2-40
Operation Orders and Plans, 4-10
Operational Design, 2-50
Operational Environment (OE), 1-26, 2-1, 2-12
Operational Limitations, 2-58
Operational Mission Narrative, 2-56
Operational Variables (PMESII-PT) 1-26
Operations in the Information Environment (OIE), 1-58, 2-37, 2-41
Operations Process, 1-23
Operations Security (OPSEC), 2-10, 3-39, 3-41
Overseas Humanitarian, Disaster, and Civic Aid (OHDACA), 3-26

P
Personnel Recovery (PR), 3-70
Personnel Security Program, 1-31
Physical Advantage, 1-5
Physical Attack, 1-50, 3-70
Physical Destruction, 1-50
Physical Dimension, 0-10
Physical Security, 1-30, 3-71
Planning, 1-60, 2-45
Planning Assumptions, 2-58

Planning Tools and Outputs, 4-3
Planning, 4-1
Police Engagement, 3-72
Populace and Resources Control (PRC), 3-26
Preparation, 5-1
Preparing, 1-62
Presence, Profile, and Posture (PPP), 3-71
Principles of Information, 3-10
Principles of Information Advantage 1-12
Principles of Mission Command, 1-28
Priority Intelligence Requirements (PIRs), 2-62, 4-9
Propaganda, 0-9
Protect, 1-29
Protect Considerations, 1-32
Protect Friendly Information, Information Networks & Systems, 2-16
Protection,
Joint Function, 2-44
Warfighting Function, 1-15
Psychological Operations Forces, 2-32
Public Affairs (PA), 2-10, 2-32, 3-5
Public Communication, 1-40

R
Relative Advantages, 1-4
Relevant Actors, 1-27, 2-6, 2-15
Relevant Automated Systems, 1-27
Relevant Human Actors, 1-27
Requests for Information (RFIs), 2-53
Risk Assessment, 2-60, 3-43, 4-5
Risk Management, 1-25
Running Estimate, 4-6

S
Scheme of Information Operations, 4-12
Secure and Obscure Friendly Information, 1-29

Security Activities, 1-30
Security Operations, 1-30, 5-6
Sequencing, 2-65
Service Organizations, 2-23
Signature Management, 2-11
Social Media 3-72
Soldier and Leader Engagement (SLE), 1-43, 1-47, 3-72
Space Capabilities, 3-62
Space Control, 3-64
Space Forces, 2-33
Space Operations, 1-51, 2-11, 3-61
Space Superiority, 3-64
Special Access Programs (SAP), 3-70
Special Technical Operations (STO), 2-11
Specified, Implied, and Essential Tasks, 2-59
Spectrum Management, 3-57
Staff Estimate, 2-54, 2-62
Staff Responsibilities, 4-3
Strategic Contexts, 1-16
Support Human and Automated Decision Making 2-16
Support to Civil Administration (SCA), 3-26
Sustainment,
Joint Function, 2-44
Warfighting Function, 1-15
Synchronization Matrix, 2-52, 4-16
Synchronization of Information-Related Capabilities, 4-3
Synchronization of OIE Activities, 2-53

T
Tactical Deception (TAC-D), 1-46, 3-28
Targeting, 1-25, 2-53, 7-11
Targeting Methodology, 7-11
Targeting Tasks during the MDMP, 7-16

Technical Training and Education, 1-64
Tenets of Operations, 1-10
Themes, 3-15
Threat Analysis, 3-42
Threat Command and Control, 1-52
Threat Information Warfare, 1-54
Threat Perception and Behaviors, 1-45

U
Unity of Effort (Information Forces), 2-21
USG Organizations, 2-34

V
Vulnerability Analysis, 3-43

W
Warfighting Function Contributions, 1-14

SMARTbooks
INTELLECTUAL FUEL FOR THE MILITARY

Recognized as a "**whole of government**" doctrinal reference standard by military, national security and government professionals around the world, SMARTbooks comprise a **comprehensive professional library** designed with all levels of Soldiers, Sailors, Airmen, Marines and Civilians in mind.

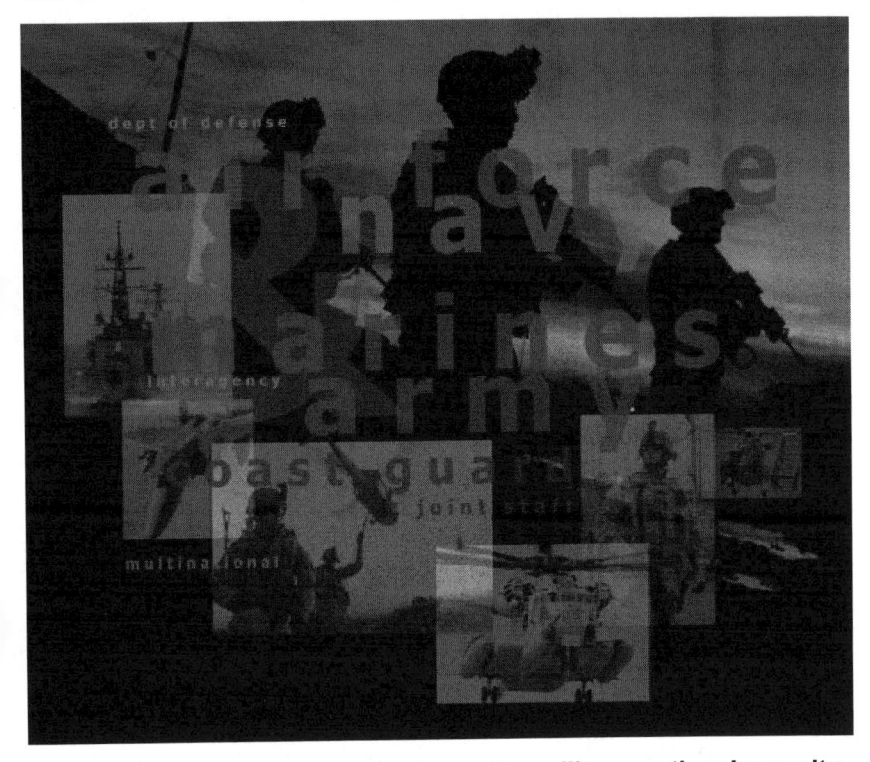

The SMARTbook reference series is used by **military, national security, and government professionals** around the world at the organizational/institutional level; operational units and agencies across the full range of operations and activities; military/government education and professional development courses; combatant command and joint force headquarters; and allied, coalition and multinational partner support and training.

Download FREE samples and SAVE 15% everyday at:
www.TheLightningPress.com

The Lightning Press is a **service-disabled, veteran-owned small business**, DOD-approved vendor and federally registered — to include the SAM, WAWF, FBO, and FEDPAY.

SMARTbooks
INTELLECTUAL FUEL FOR THE MILITARY

MILITARY REFERENCE: SERVICE-SPECIFIC

Recognized as a "whole of government" doctrinal reference standard by military professionals around the world, SMARTbooks comprise a comprehensive professional library.

MILITARY REFERENCE: MULTI-SERVICE & SPECIALTY

SMARTbooks can be used as quick reference guides during operations, as study guides at professional development courses, and as checklists in support of training.

JOINT STRATEGIC, INTERAGENCY, & NATIONAL SECURITY

The 21st century presents a global environment characterized by regional instability, failed states, weapons proliferation, global terrorism and unconventional threats.

The Lightning Press is a **service-disabled, veteran-owned small business,** DOD-approved vendor and federally registered — to include the SAM, WAWF, FBO, and FEDPAY.

RECOGNIZED AS THE DOCTRINAL REFERENCE STANDARD BY MILITARY PROFESSIONALS AROUND THE WORLD.

THREAT, OPFOR, REGIONAL & CULTURAL

In today's complicated and uncertain world, the military must be ready to meet the challenges of any type of conflict, in all kinds of places, and against all kinds of threats.

- OPFOR SMARTBOOK 1 — Chinese Military
- OPFOR SMARTBOOK 2 — North Korean Military
- OPFOR SMARTBOOK 3 — RED TEAM ARMY
- OPFOR SMARTBOOK 4 — Iran the Middle East
- OPFOR SMARTBOOK 5 — Irregular Hybrid Threat

HOMELAND DEFENSE, DSCA, & DISASTER RESPONSE

Disaster can strike anytime, anywhere. It takes many forms—a hurricane, an earthquake, a tornado, a flood, a fire, a hazardous spill, or an act of terrorism.

- HDS1 SMARTBOOK — Homeland Defense & DSCA
- CTS1 SMARTBOOK — Counterterrorism, WMD & HYBRID THREAT
- Disaster Response SMARTBOOK 1 — Federal/National Disaster Response
- Disaster Response SMARTBOOK 2 — Incident Command System (ICS)
- Disaster Response SMARTBOOK 3 — Disaster Preparedness

DIGITAL SMARTBOOKS (eBooks)

In addition to paperback, SMARTbooks are also available in digital (eBook) format. Our digital SMARTbooks are for use with Adobe Digital Editions and can be used on up to **six computers and six devices**, with free software available for **85+ devices and platforms—including PC/MAC, iPad and iPhone, Android tablets and smartphones, Nook, and more!** Digital SMARTbooks are also available for the **Kindle Fire** (using Bluefire Reader for Android).

Download FREE samples and SAVE 15% everyday at:
www.TheLightningPress.com

Purchase/Order

SMARTsavings on SMARTbooks! Save big when you order our titles together in a SMARTset bundle. It's the most popular & least expensive way to buy, and a great way to build your professional library. If you need a quote or have special requests, please contact us by one of the methods below!

View, download FREE samples and purchase online:
www.TheLightningPress.com

Order SECURE Online
Web: www.TheLightningPress.com
Email: SMARTbooks@TheLightningPress.com

24-hour Order & Customer Service Line
Place your order (or leave a voicemail) at 1-800-997-8827

Phone Orders, Customer Service & Quotes
Live customer service and phone orders available
Mon - Fri 0900-1800 EST at (863) 409-8084

Mail, Check & Money Order
2227 Arrowhead Blvd., Lakeland, FL 33813

Government/Unit/Bulk Sales

The Lightning Press is a **service-disabled, veteran-owned small business**, DOD-approved vendor and federally registered—to include the SAM, WAWF, FBO, and FEDPAY.

We accept and process both **Government Purchase Cards** (GCPC/GPC) and **Purchase Orders** (PO/PR&Cs).

Keep your SMARTbook up-to-date with the latest doctrine! In addition to revisions, we publish incremental **"SMARTupdates"** when feasible to update changes in doctrine or new publications. These SMARTupdates are printed/produced in a format that allow the reader to insert the change pages into the original GBC-bound book by simply opening the comb-binding and replacing affected pages. Learn more and sign-up at: **www.thelightningpress.com/smartupdates/**